LONDON MATHEMATICAL SOCIETY LECTURE NOTE SERIES

Managing Editor: Professor J.W.S. Cassels, Department of Pure Mathematics and Mathematical Statistics,
University of Cambridge, 16 Mill Lane, Cambridge CB2 1SB, England

The titles below are available from booksellers, or, in case of difficulty, from Cambridge University Press.

London Mathematical Society Lecture Note Series. 245

Geometry, Combinatorial Designs and Related Structures

Proceedings of the first Pythagorean conference

Edited by

J. W. P. Hirschfeld
University of Sussex at Brighton

S. S. Magliveras
University of Nebraska at Lincoln

M. J. de Resmini
Università di Roma "La Sapienza"

CAMBRIDGE
UNIVERSITY PRESS

CAMBRIDGE UNIVERSITY PRESS
Cambridge, New York, Melbourne, Madrid, Cape Town, Singapore, São Paulo

Cambridge University Press
The Edinburgh Building, Cambridge CB2 2RU, UK

Published in the United States of America by Cambridge University Press, New York

www.cambridge.org
Information on this title: www.cambridge.org/9780521595384

First published 1997

A catalogue record for this publication is available from the British Library

ISBN-13 978-0-521-59538-4 paperback
ISBN-10 0-521-59538-X paperback

Transferred to digital printing 2006

Contents

Preface

This volume contains articles based on talks given at the First Pythagorean Conference on Geometry, Combinatorial Designs and Related Structures, held on the island of Spetses in Greece from 1 to 7 June 1996.

There were 80 invited participants and 48 talks, including hour-long expositions by the keynote speakers: Peter Cameron, John Conway, Jean Doyen, Dieter Jungnickel, Curt Lindner, Rudi Mathon, Ernie Shult, Jef Thas.

It was a conference which will live long in the memory. Apart from the weather and the seaside setting, several other events contributed to an outstanding week. Jean Doyen opened the conference by giving a historical talk on Pythagoras. At a meeting of the Institute of Combinatorics and its Applications, Jef Thas was awarded the 1994 Euler Medal for a distinguished lifetime career. The excursion around the Peloponnesian peninsula included the theatre at Epidaurus and the hilltop ruins of Mycenae.

The editors, who were also the conference organisers, would like to thank the Institute of Combinatorics and its Applications, the Greek Ministry of Civilization, and the Greek Ministry of Education for their financial support. We would also like to thank the Greek Tourist Organization for providing a beautiful illustrated book of Greece for all the participants. Special thanks are due to Simos Magliveras for working out many organizational and tactical problems, and also to Sakis Simopoulos for considerable help in further organizational matters.

James Hirschfeld

Spyros Magliveras

Marialuisa de Resmini

April, 1997

Introduction

All the papers presented here come into the category of incidence structures. Overwhelmingly, they have a geometrical point of view, which both motivates and illustrates.

Although all the topics of all the papers are on interconnected themes, they are here subdivided into categories.

A $t - (v, k, \lambda)$ *design* is an incidence structure $\mathcal{S} = (\mathcal{P}, \mathcal{B}, I)$ such that

(a) $\mathcal{P}$ is a set of v points;

(b) each block B in $\mathcal{B}$ is incident with a fixed number k of points;

(c) any t points are incident with precisely λ blocks.

The *order* of a design is $n = k - \lambda$. A *Steiner system* is a t–design with $\lambda = 1$, and a *Steiner triple system* also has $t = 2, k = 3$.

Now let $\mathcal{S}$ be a connected incidence structure of points and lines such that the lines cover the points and such that each line contains at least two points. Suppose also that

(a) there are $s + 1$ points on a line, with $s > 0$;

(b) there are $t + 1$ points through a point, with $t > 0$;

(c) two points are incident with at most one line;

(d) two lines are incident with at most one point;

(e) given a non-incident point-line pair, there are α lines through the point meeting the line, with $\alpha > 0$.

Then $\mathcal{S}$ is a *partial geometry*; in particular, when $\alpha = 1$, it is a *generalized quadrangle*.

Geometries and groups

Cameron contributes a magisterial survey of the current state of finite geometry from a group-theoretical viewpoint; this essay alone is worth the price of the volume!

1

Designs

A continuing problem is to find t–designs for a given t with small k and λ.

Betten, Laue and *Wassermann* find new 7–designs on the projective lines $PG(1, 23)$, $PG(1, 25)$, $PG(1, 32)$ by computation. They have parameters as follows:

(a) $7 - (24, 8, \lambda)$, for $\lambda = 4$, 5, 6, 7, 8;

(b) $7 - (24, 9, \lambda)$, for $\lambda = 40$, 48, 64;

(c) $7 - (26, 8, 6)$;

(d) $7 - (26, 9, \lambda)$, for $\lambda = 54$, 63, 81;

(e) $7 - (33, 8, 10)$.

Brouwer, Haemers and *Tonchev* consider whether, given a partial geometry $\mathcal{P}$ with v points and k points on a line, one can add to the line set a set of k–subsets of points such that the extended family of k–subsets is a $2-(v, k, 1)$ design (or a Steiner system $S(2, k, v)$). Various conditions are given and it is shown in particular that the partial geometry $PQ^+(4m - 1, 2)$ is embeddable in a Steiner 2–design if and only if $m \leq 2$.

A biplane is a symmetric 2–design with $\lambda = 2$, and are rare objects indeed. *Key* and *Tonchev* consider the five known biplanes of order 9, that is $2 - (56, 11, 2)$ designs, and investigate them via the associated ternary codes. The computations, carried out using the computer language Magma, show, among other results, that none of these biplanes can be extended to a $3 - (57, 12, 2)$ design. This had been suspected but never proved before; all the necessary arithmetic conditions for extendability are satisfied. Also, each of these biplanes is the only one to be found among the weight-11 vectors of its ternary code. All the coding-theoretic parameters of the five codes are computed as well as those of the sixteen non-isomorphic residual $2 - (45, 9, 2)$ designs provided by the five biplanes. Sixteen non-isomorphic ternary codes are obtained.

A *spread* of a design, where the blocks are regarded as subsets of points, is a partition of the point set $\mathcal{P}$ into blocks. A *packing* or *parallelism* is a partition of the block set $\mathcal{B}$ into spreads. *Mathon* presents new algorithms for finding spreads and packings of sets with applications to combinatorial designs and finite geometries. An efficient deterministic method for spread enumeration is used to settle several existence problems for t–designs and partial geometries. Randomized algorithms based on tabu search are employed to construct new Steiner 5–designs and large sets of combinatorial designs. In particular, partitions are found of the 4−subsets of a 16−set into 91 disjoint affine planes of order 4.

Steiner triple systems

A Pasch configuration in any incidence structure is a quadrilateral; that is, a set of six points and four blocks forming a quadrilateral. *Akbari, Khosrovshahi, Maysoori* and *Shariari* consider the size $f(n)$ of the largest set of triples on a set of size n containing no Pasch configuration. Upper and lower bounds for $f(n)$ found and applications of this result discussed.

Lindner, with many striking diagrams, leads us through the intricacies of embeddings for partial cycle systems. The embeddings of a partial 3−cycle system of order n in a 3−cycle system of order $3(2n + 1)$ and of a partial 5−cycle system of order n in a 5−cycle system of order $5(2n+1)$ are carefully explained and both pictures and examples turn a difficult piece of design theory into an elegant survey. Embedding partial even-cycle systems is much easier than embedding partial odd-cycle systems and the complete proof is given of the fact that a partial m−cycle system of order n can be embedded in an m−cycle system of order $2mn + 1$ when m is even.

A $2 - (v, k, \lambda)$ design or *balanced incomplete block design* (BIBD) is is called *minci* when the set $\mathcal{B}$ of blocks of can be partitioned into as many as possible maximum parallel classes and one other parallel class; many special types of designs are of this type. In particular, a *Rosa triple system* is a minci Steiner triple system with $v \equiv 2 \pmod 3$. *Ling* and *Colbourn* settle the existence of these triple systems completely.

Difference sets

A $(v, k, \lambda) - difference\ set$ in a group G of order v is a k-subset D of G, such that every element $g \neq 1$ of G has exactly λ representations $g = d_1 d_2^{-1}$ with $d_1, d_2 \in D$. As for designs, the parameter $n = k - \lambda$ is the *order* of the difference set. *Jungnickel* and *Schmidt* provide an update of the Jungnickel's 1992 survey on difference sets. Apart from reviewing the classical results, the authors describe new developments in the subject, some of whose techniques come from algebraic number theory and group characters, that have led to both the disproof of some long-standing conjectures and the construction of new families of difference sets. In recent years, research has focused on difference sets with $(v, n) > 1$, and all the new classes found are presented.

Latin squares

An intercalate in a Latin square of order n is a 2×2 subsquare. The maximum possible number is $\frac{1}{2}n^2(n - 1)$. This is achieved for $n = 2^k$, but it is unknown what happens for other n. *Danziger* and *Mendelsohn* show, among other results, that Latin squares of order n such that every cell is contained in an intercalate exist when $n \neq 1, 3, 5, 7$.

Diagram geometries

A string diagram of type $c^* - c$ is a rank three geometry. It has three types of objects: points, lines, and blocks, which satisfy the following axioms:

(a) any two distinct lines incident with a common point p are both incident with a unique common block;

(b) any two lines incident with a common block are incident with a unique common point;

(c) given a point p incident with a block B, there are exactly two lines incident with both;

(d) (The String Property) any point incident with a line of a block B is also incident with B.

All examples fall into these six classes. *Shult* surveys the current situation.

Generalized quadrangles

Let S be a generalized quadrangle of order (s, t) and let p be a point of S. An *elation about* p is either the identity collineation or a collineation of S which fixes each line incident with p and fixes no point not collinear with p. If S admits a group G of elations about p acting regularly on the points not collinear with p then S is an *elation generalized quadrangle* with *elation group* G and *base point* p. *O'Keefe* and *Penttila* survey some recent classification theorems for elation generalized quadrangles of order (q^2, q), q even, with particular emphasis on those involving subquadrangles of order q.

If x is a regular point of a generalized quadrangle $S = (P, B, I)$ of order $(s, t), s \neq 1$, then x defines a dual net with $t + 1$ points on any line and s lines through every point. The *Axiom of Veblen* states that if two lines intersect, then any two other lines meeting the first two and not passing through their meet also intersect. *Thas* and *Van Maldegehem* show that, if $s \neq t, s > 1, t > 1$, then S is isomorphic to a $T_3(O)$ of Tits if and only if S has a coregular point x such that, for each line L incident with x, the corresponding dual net satisfies the Axiom of Veblen. Translation generalized quadrangles are also considered.

Projective spaces

By considering the Veronese map from $PG(2, q)$ to $PG(5, q)$, *Bonisoli* and *Cossidente* construct a *cap*, that is a set of points no three of which are collinear, in $PG(5, q)$. It is still a mysterious problem to discover the largest size of cap in this space for $q > 3$.

A way of constructing a large *arc*, that is a set of points no three collinear, in $PG(2, q)$ with q a square is as an orbit of the $(q + \sqrt{q} + 1)$-th power of a Singer cycle. This produces a complete (maximal) arc of size $q - \sqrt{q} + 1$. For q even, this is as large as possible below the maximum size $q + 2$; for q odd, this is conjectured to be the next largest below the maximum size $q + 1$. In the even case, the associated algebraic curve has degree $\sqrt{q} + 1$ and is Hermitian; in the odd case, the curve has degree $2(\sqrt{q} + 1)$. In the latter case, *Cossidente* and *Korchmáros* show that the curve is birationally isomorphic to a Hermitian curve.

A *unital* is a $2 - (q^3 + 1, q + 1, 1)$ design, and a *Hermitian unital* is given by the rational points and non-tangent lines of a Hermitian curve. Buekenhout has given a construction of unitals in $PG(2, q^2)$ using the André representation of $PG(2, q^2)$ in the space $PG(4, q)$. Metz has shown that this construction produces Hermitian and non-Hermitian unitals. *Metsch* gives a geometric criterion in $PG(4, q)$ to decide whether the unital in $PG(2, q^2)$ is Hermitian or not.

A *spread* of $PG(3, q)$ is a set of $q^2 + 1$ lines which partition the points of the space. A *parallelism* or *packing* of $PG(3, q)$ is a collection of $q^2 + q + 1$ spreads which partition the lines of the space. A *regular* spread is one which contains the regulus determined by any three of its lines. A *subregular* spread of *index one* is a spread obtained from a regular spread by replacing a single regulus in the spread by its opposite. A *regular* parallelism is one consisting only of regular spreads. A *uniform* parallelism is one consisting entirely of spreads which are projectively equivalent. *Prince* shows that there are no regular parallelisms of $PG(3, 3)$ but that there are many parallelisms consisting entirely of subregular spreads of index one.

A double-five of planes is a set ψ of 35 points in $PG(5, 2)$ which admits two distinct decompositions $\psi = \alpha_1 \cup \alpha_2 \cup \alpha_3 \cup \alpha_4 \cup \alpha_5 = \beta_1 \cup \beta_2 \cup \beta_3 \cup \beta_4 \cup \beta_5$ into a set of five mutually skew planes such that $\alpha_r \cap \beta_r$ is a line, for each r, while $\alpha_r \cap \beta_s$ is a point, for $r \neq s$. *Shaw* delves into the properties of this surprisingly complex configuration. In particular the existence of an invariant symplectic form is demonstrated and some related duality properties are described.

On maximum size anti-Pasch sets of triples

S. Akbari   G. B. Khosrovshahi   Ch. Maysoori

S. Shahriari

Abstract

Given an n-set X, we denote the cardinality of a maximum size anti-Pasch (Pasch free) set of triples of X by $f(n)$. In this paper we provide lower and upper bounds for $f(n)$ and consequently we disprove a conjecture posed by Khosrovshahi at the fifteenth British Combinatorial Conference (BCC15).

1   Introduction

Let $X = \{1, \cdots, n\}$. We denote by $P_i(X)$ the set of all subsets of X of size i, and the elements of $P_3(X)$ will be called *triples*. Let $P_c = \{A_1, A_2, A_3, A_4\} \subset P_3(X)$ be a collection of four triples of X. If the union of these triples has size 6 and if the intersection of each pair of distinct triples in P_c has size one then P_c is called a *Pasch configuration*. It is easy to see that $P_c = \{abc, axy, bxz, cyz\}$ for some choice of $a, b, c, x, y, z \in X$. Thus a Pasch is exactly a Fano plane minus one point and the three lines incident with that point. A subset of $P_3(X)$ that does not contain a Pasch is called an *anti-Pasch*. We denote by $f(n)$ the cardinality of an anti-Pasch subset of $P_3(X)$ with maximum size. At the 15th British Combinatorial Conference, it was conjectured by Khosrovshahi [7] that $f(n) \leq \binom{n}{2}$. In this paper we present a counterexample to this conjecture. We will show that for $n \geq 7$

$$\frac{5}{4}\left(\binom{n}{2} - n\right) \leq f(n) \leq \frac{4}{7}\binom{n}{3}.$$

The lower bound disproves Khosrovshahi's Conjecture, and the very crude upper bound will help determine the exact value of $f(n)$ for small values of n. In fact, we will show that $f(6) = 14$ and $f(7) = 20$.

The existence of anti-Pasch configurations with additional properties has been the object of much study. Recall that a *Steiner triple system* of order v is a collection of triples on a set with v elements such that each unordered pair of elements is contained in exactly one triple in the collection. It has

7

been conjectured [1] that if $v \cong 1$ or $3 \pmod{6}$ and $v \neq 7, 13$ then an anti-Pasch Steiner triple system exists. Even though the conjecture remains open, infinite families of such systems have been found [1, 3, 5, 6, 9, 10]. The above conjecture is the refinement of a special case of a more general conjecture of Paul Erdös. Let $r \geq 3$ be an integer. A Steiner triple system is called *r-sparse* if every set of $r + 2$ points carries fewer than r triples. It is easy to see that a Steiner triple system is 4-sparse if and only if it is anti-Pasch, and an Steiner triple system is $r - 1$-sparse if it is r-sparse. Erdös [4] (see also [1, 2, 8]) asks whether for any $r \geq 3$ there is a v_r such that for all $v > v_r$ and $v \equiv 1$ or $3 \pmod{6}$ there exists an r-sparse Steiner triple system on v points. For an infinite family of 5-sparse Steiner triple systems as well as related results see [2]. For $r > 5$ very little is known about this seemingly very difficult question.

We conclude the introduction with some motivation for the conjecture of Khosrovshahi [7]. We construct an $\binom{n}{2} \times \binom{n}{3}$ matrix, $W(n)$, of zeros and ones with the rows indexed by the elements of $P_2(X)$, and the columns indexed by the elements of $P_3(X)$. If $A \in P_2(X)$ and $B \in P_3(X)$ then the (A, B) entry of the matrix is defined to be:

$$W(n)(A, B) = \begin{cases} 1 & \text{if } A \subseteq B, \\ 0 & \text{otherwise.} \end{cases}$$

It is known [4] that

$$\operatorname{rank} W(n) = \binom{n}{2}.$$

Therefore, any $\binom{n}{2} + 1$ columns of $W(n)$ must be linearly dependent.

Let $P_c = \{abc, axy, bxz, cyz\}$ be a Pasch configuration in $P_3(X)$ and define

$$\overline{P}_c = \{xyz, bcz, acy, abx\}.$$

$\overline{P}_c$ is also a Pasch configuration and is called the counterpart of P_c. Now given any Pasch configuration P_c, we define a $(0, 1, -1)$-vector

$$f = \left(f_1, \cdots, f_{\binom{n}{3}}\right)$$

as follows:

$$f_i = \begin{cases} 1 & \text{if } B_i \in P_c, \\ -1 & \text{if } B_i \in \overline{P}_c, \\ 0 & \text{otherwise,} \end{cases}$$

where B_i is the i-th indexed element of $P_3(X)$. It is very easy to check that $f \in \ker W(n)$. Since the linear dependence of any collection of the columns of $W(n)$ gives an element of the kernel, it was thus conjectured that any $\left(\binom{n}{2} + 1\right)$-subset of $P_3(X)$ would contain a Pasch and its counterpart. As was noted above our example will show that this conjecture is false.

2   Preliminaries

In this section we gather some further notation, as well as definitions, and some known simple facts.

Let $X = \{1, 2, \ldots, n\}$. For any $x \in X$, we denote by $x * *$, the set of all elements of $P_3(X)$ which contain x. Similarly, for any $\{x, y\} \in P_2(X)$, $xy*$ is the collection of triples that contain x and y. For any $k, 1 \leq k \leq n$, and $B \in P_k(X)$,

$$\Delta B = P_{k-1}(B)$$

is called the *shadow* of B.

Let $[7] = \{1, \cdots, 7\}$, then the set

$$\mathcal{F} = \{B_1, \cdots, B_7\} \subset P_3([7])$$

is called a *Fano plane* if for any $B_i, B_j \in \mathcal{F}$ ($i \neq j$), we have $|B_i \cap B_j| = 1$. It is known that the structure of a Fano plane is unique up to isomorphism [11] and is of the following form:

$$\{123, 145, 167, 246, 257, 347, 356\}.$$

It is also easy to show that there exist 30 distinct Fano planes on 7 points and for any $B \in P_3([7])$, there exist exactly 6 Fano planes containing B. It is straightforward to see that every 5 triples of a Fano plane contain a Pasch.

Now, suppose $n \geq 6$ and $X = \{1, \cdots, n\}$. Then we have the following easily verified facts [11]:

(i) For any $B \in P_3(X)$, there exist $(n-3)(n-4)(n-5)$ Pasches containing B;

(ii) There are all together $30\binom{n}{6}$ Pasches in $P_3(X)$;

(iii) Let $A, B \in P_3(X)$, with $|A \cap B| = 1$, then there are $2(n-5)$ Pasches containing both A and B.

3   Lower and Upper Bounds for $f(n)$

Recall that $f(n)$ is the maximum size of an anti-Pasch subset of $P_3(X)$. In this section we obtain some bounds for $f(n)$.

Theorem 1 *Let $n \geq 3$ be a positive integer, and let l be any integer with $1 \leq l \leq \lfloor \frac{n-1}{2} \rfloor$. Let $X = \{1, \cdots, n\}$, and define*

$$
\begin{aligned}
A_0 &= 1 * *, \\
A_i &= (2i)(2i+1) * \backslash A_0, \quad for \quad i = 1, 2, \cdots, l.
\end{aligned}
$$

Let $\mathcal{A} = \cup_{i=0}^{l} A_i$. Define $\mathcal{E} = \cup_{k=2}^{l} \cup_{j=1}^{k-1} E_{jk}$, where

$$E_{jk} = \{1(2j)(2k), 1(2j)(2k+1), 1(2j+1)(2k), 1(2j+1)(2k+1)\},$$

for $1 \leq j < k \leq l$. Then $\mathcal{A} \backslash \mathcal{E}$ is an anti-Pasch set with $\binom{n-1}{2} + l(n-3) - 4\binom{l}{2}$ elements.

Example. For $n = 9$ and $l = 2$ the above construction gives

$$\mathcal{A} = \cup_{i=0}^{2} A_i = 1 * * \cup 23 * \cup 45*;$$
$$\mathcal{E} = \{124, 125, 134, 135\}.$$

Thus $\mathcal{A} \backslash \mathcal{E}$ is an anti-Pasch set with $28 + 6 + 6 - 4 = 36$ triples.

Proof In a Pasch in $P_3(X)$ every element of X either does not occur or it occurs exactly twice. Likewise every pair of elements of X either are not contained in any of the triples or they are contained in exactly one of the triples. Thus if $\mathcal{A} \backslash \mathcal{E}$ contained a Pasch, this Pasch could have either zero or two triples from A_0 and at most one triple from any of the A_i for $i = 1, \ldots, l$. In addition, this Pasch could not have elements from four different A_i's since this would lead to the presence of at least 7 elements of X in the Pasch, and every Pasch uses only 6 of the elements of X. This leaves only the possibility that the Pasch would have two triples from A_0 and one triple from A_j and one from A_k where $1 \leq j < k \leq l$. Thus the elements of X that are contained in the triples of the Pasch are 1, $2j$, $2j + 1$, $2k$, $2k + 1$, and a sixth element a. Now the pair $1a$ can occur in at most one triple and thus at least one of the triples from A_0 will not contain a. Thus this triple will consist of 1, one element of $\{2j, 2j+1\}$ and one element of $\{2k, 2k+1\}$. This means that one of the triples has to come from E_{jk}, and these triples were eliminated in $\mathcal{A} \backslash \mathcal{E}$. Thus $\mathcal{A} \backslash \mathcal{E}$ is an anti-Pasch configuration. Clearly,

$$|A_0| = \binom{n-1}{2},$$
$$|A_i| = n - 3, \quad i = 1, 2, \cdots, l,$$
$$|\mathcal{E}| = 4\binom{l}{2}.$$

Therefore the size of $\mathcal{A} \backslash \mathcal{E}$ is as claimed and the proof is complete. $\square$

To find a large anti-Pasch configuration we choose l so as to maximize $|\mathcal{A} \backslash \mathcal{E}|$. For a fixed n the expression for $|\mathcal{A} \backslash \mathcal{E}|$ is a second degree polynomial in l with its maximum at $l = \frac{n-1}{4}$. Choosing $l = \lfloor \frac{n}{4} \rfloor$ (which is the closest integer to $\frac{n-1}{4}$) gives the largest possible value for $|\mathcal{A} \backslash \mathcal{E}|$ for an integer l. A straightforward calculation shows that, for $l = \lfloor \frac{n}{4} \rfloor$, we have $|\mathcal{A} \backslash \mathcal{E}| \geq$

$\frac{5n^2-14n+5}{8}$ and that this expression is in turn no smaller than $\frac{5}{4}\left(\binom{n}{2} - n\right)$. Thus we have proved:

Corollary 2 *Let X be a set of size n with $n \geq 3$. Let $f(n)$ be the maximum size of an anti-Pasch subset of $P_3(X)$. Then*

$$f(n) \geq \frac{5n^2 - 14n + 5}{8} \geq \frac{5}{4}(\binom{n}{2} - n). \qquad \square$$

It follows that for $n \geq 10$ we have an anti-Pasch configuration in $P_3(X)$ of size greater or equal to $\binom{n}{2} + 1$. This clearly provides a counterexample to Khosrovshahi's Conjecture. We now give a crude upper bound for $f(n)$.

Theorem 3 *Let X be a set of size n with $n \geq 7$, and let $f(n)$ be the maximum size of an anti-Pasch subset of $P_3(X)$. Then*

$$f(n) \leq \frac{4}{7}\binom{n}{3}.$$

Proof Suppose $\mathcal{A}$ is an anti-Pasch set of size $f(n)$, and let $\overline{\mathcal{A}}$ denote $P_3(X)\backslash\mathcal{A}$. Now, let

$$\Sigma = \{(B,\mathcal{F})|B \in \mathcal{F} \cap \overline{\mathcal{A}}, \mathcal{F} \text{ is a Fano plane in } P_3(X)\}.$$

We will count the elements of Σ in two ways. Note that there exists $\binom{n}{7} \times 30$ distinct Fano planes in $P_3(X)$. Let $\mathcal{F}$ be an arbitrary Fano plane in $P_3(X)$. Every five triples of a Fano plane contain a Pasch and thus since $\mathcal{A}$ is an anti-Pasch set we must have $|\mathcal{F} \cap \overline{\mathcal{A}}| \geq 3$. Thus

$$|\Sigma| \geq \binom{n}{7} \times 30 \times 3.$$

On the other hand, let $B \in \overline{\mathcal{A}}$ be an arbitrary triple, then there are $\binom{n-3}{4} \times 6$ distinct Fano planes containing B. Since $|\overline{\mathcal{A}}| = |P_3(X)\backslash\mathcal{A}| = \binom{n}{3} - f(n)$, it follows that

$$|\Sigma| = (\binom{n}{3} - f(n))\binom{n-3}{4} \times 6.$$

Solving for $f(n)$ we get $f(n) \leq \frac{4}{7}\binom{n}{3}$. $\square$

Proposition 4 *Let $g(n) = f(n)/\binom{n}{3}$. Then $g(n)$ is a non-increasing function. In particular, if there exists a real number c and a natural number n_0 such that $f(n_0) \leq c\binom{n_0}{3}$, then $f(n) \leq c\binom{n}{3}$ for any $n \geq n_0$.*

Proof Let $Y = \{1, 2, \ldots, n, n+1\}$, and let $\mathcal{A}$ be an anti-Pasch set of size $f(n+1)$ in $P_3(Y)$. For any i, $1 \le i \le n+1$, let $r_i = |i * * \cap \mathcal{A}|$ be the number of triples in $\mathcal{A}$ that contain i. Clearly, $\sum_{i=1}^{n+1} r_i = 3f(n+1)$. Therefore, there exist some j for which $r_j \le \frac{3f(n+1)}{n+1}$.

Now, $\mathcal{A}$ is an anti-Pasch set which implies that $\mathcal{B} = \mathcal{A}\backslash j * *$ is also an anti-Pasch set. Thus

$$f(n) \ge |\mathcal{B}| \ge f(n+1) - \frac{3f(n+1)}{n+1} = \frac{n-2}{n+1} f(n+1),$$

and, since $\frac{n+1}{n-2} = \frac{\binom{n+1}{3}}{\binom{n}{3}}$, we have

$$f(n+1) \le \frac{\binom{n+1}{3}}{\binom{n}{3}} f(n).$$

Therefore, $g(n+1) \le g(n)$. □

4   Small Values of n

Proposition 5 $f(6) = 14$.

Proof Let $X = \{1, 2, \ldots, 6\}$, and let $\mathcal{A} = 1 * * \cup \triangle 2345$, then $|\mathcal{A}| = 14$. Any Pasch could have at most two triples from $1 * *$ and one triple from $\triangle 2345$. Since a Pasch consists of 4 triples, $\mathcal{A}$ must be anti-Pasch, and $f(6) \ge 14$. Now suppose that $\mathcal{B}$ is any anti-Pasch set in $P_3(X)$ with more than 10 triples. $P_3(X)$ consists of 10 triples and their complements, and since $\mathcal{B}$ has more than 10 triples, it must contain a triple and its complement. Without loss of generality, assume that 123 and 456 are in $\mathcal{B}$. Organize the remaining 18 triples in $P_3(X)$ in two tables as follows:

145	146	156
245	246	256
345	346	356

124	134	234
125	135	235
126	136	236

The triple 123 together with any three triples from the first table such that no two are from the same row or column form a Pasch. Similarly, 456 forms a Pasch with any three triples from the second table such that no two are from the same row or column. Both 123 and 456 are in $\mathcal{B}$ and $\mathcal{B}$ is an anti-Pasch. Thus $\mathcal{B}$ must not contain at least three triples from each of the two tables. It follows that $|\mathcal{B}| \le 20 - 6 = 14$. So $f(6) \le 14$ and the proof is complete. □

Proposition 6 $f(7) = 20$.

Proof The upper bound in Proposition 3 implies that $f(7) \leq 20$. Now let

$$\mathcal{A} = (1 * * \cup 23 * \cup \Delta 2456) \backslash \{134, 135, 136\}.$$

We claim that $\mathcal{A}$ is an anti-Pasch set of size 20. Assume $\mathcal{A}$ contains a Pasch, P_c. P_c must have two triples from $1 * * \backslash \{134, 135, 136\}$, one triple from $23 * \backslash 1 * *$, and one triple from $\Delta 2456$. Now 3 must appear in two triples of P_c, one of these will be from $23 * \backslash 1 * *$, and the other from $1 * * \backslash \{134, 135, 136\}$. It follows that 137 is one of the triples in P_c. This implies that 7 must appear in two triples of P_c and only one of these can have 1. Thus 237 must be a triple in P_c. However 137 and 237 cannot be in the same Pasch. The contradiction proves the claim. □

Note. The example in Proposition 6 was constructed using a computer and by utilizing the fact that for $|X| = 7$, $P_3(X)$ can be partitioned into 2 disjoint Fano planes and a 2-$(7, 3, 3)$ design.

5 Some Open Problems

(i) Determine $f(8)$ and $f(9)$.

(ii) Determine the asymptotic behaviour of $f(n)$. In particular find $\lim_{n \to \infty} \frac{f(n)}{\binom{n}{3}}$ and $\lim_{n \to \infty} \frac{f(n)}{\binom{n}{2}}$.

(iii) Let X be a set with n elements, and let $T(n)$ be the smallest natural number such that we can partition the $\binom{n}{3}$ triples of $P_3(X)$ into sets $A_1, A_2, \cdots, A_{T(n)}$ such that every A_i is anti-Pasch. Our constructions show that $T(6) = T(7) = 2$. Find a good upper bound for $T(n)$.

Acknowledgement This research was carried out while the last author was visiting the Institute for Studies in Theoretical Physics and Mathematics (IPM) in Tehran and was supported by a TOKTEN award from the United Nations Development Programme.

References

[1] A. E. Brouwer, *Steiner triple systems without forbidden subconfigurations*, Mathematisch Centrum Amsterdam, ZW104/77, 1977.

[2] C. J. Colbourn, E. Mendelsohn, A. Rosa, and J. Širáň, Anti-mitre Steiner triple systems, *Graphs and Combin.* **10** (1994), 215-224.

[3] J. Doyen, Linear spaces and Steiner Systems, *Geometries and Groups*, Lecture Notes in Math., Vol. 893 Springer, Berlin, 1981, pp.30-42.

[4] P. Erdös, Problems and results in combinatorial analysis, *Creation in Mathematics*, **9** (1976), 25.

[5] M. J. Grannell, T. S. Griggs, and J. S. Phelan, A new look at an old construction for Steiner triple systems, *Ars Combin.* **25** (1988), 55-60.

[6] T. S. Griggs, J. Murphy, and J. S. Phelan, Anti-Pasch Steiner triple systems, *J. Combin. Inform. Systems Sci.* **15** (1990), 1-6.

[7] G.B. Khosrovshahi, Problem 15.15, *Problems from the 15th British Combinatorial Conference*, P. Cameron (ed.), 1995.

[8] H. Lefmann, K. T. Phelps, and V. Rödl, Extremal problems for triple systems, *J. Combin. Designs* **1** (1993), 379-394.

[9] R. M. Robinson, The structure of certain triple systems, *Math. Comput.* **29** (1975), 223-241.

[10] D. R. Stinson, and R. Wei, Some results on quadrilaterals in Steiner triple systems, *Discrete Math.* **105** (1992), 207-219.

[11] A. P. Street and D. J. Street, *The Combinatorics of Experimental Design*, Clarendon Press, Oxford, 1987.

S. Akbari, Institute for Studies in Theoretical Physics and Mathematics (IPM), and Sharif University of Technology, Tehran, Iran.
e-mail: s_akbari@rose.ipm.ac.ir

G.B. Khosrovshahi, Institute for Studies in Theoretical Physics and Mathematics (IPM), and University of Tehran, Tehran, Iran.
email: rezagbk@zagros.ipm.ac.ir

CH. Maysoori, Institute for Studies in Theoretical Physics and Mathematics (IPM), Tehran, Iran.
email: maysoori@rose.ipm.ac.ir

S. Shahriari, Department of Mathematics, Pomona College, Claremont CA 91711, USA.
email: sshahriari@pomona.edu

Some simple 7-designs

A. Betten   R. Laue   A. Wassermann

Abstract

Some simple 7-designs with small parameters are constructed with the aid of a computer. The smallest parameter set found is 7-(24, 8, 4). An automorphism group is prescribed for finding the designs and used for determining the isomorphism types. Further designs are derived from these designs by known construction processes.

1 Parameter Sets

Certain projective groups are 3-homogeneous and have a small number of orbits on k-subsets for moderately small k. They have therefore been a valuable tool in several geometric constructions. The first simple 6-designs were found by Magliveras and Leavitt using a prescribed automorphism group $P\Gamma L(2, 32)$, [13]. Later, further 6-designs were found having other projective automorphism groups, see [7], [15], [9]. For a recent survey on t-designs with large t see D. L. Kreher's contribution in [10]. The recipe used to construct these designs in principle also applies to the construction of simple 7-designs.

Theorem 1.1 *The following projective groups are automorphism groups of* t-(v, k, λ) *designs*:

I. $\mathrm{PSL}(2, 23)$ *of* 7-$(24, 8, \lambda)$, *where* $\lambda = 4,\ 5,\ 6,\ 7,\ 8$;

II. $\mathrm{PGL}(2, 23)$ *of* 7-$(24, 9, \lambda)$, *where* $\lambda = 40,\ 48,\ 64$;

III. $\mathrm{PGL}(2, 25)$ *of* 7-$(26, 8, 6)$;

IV. $\mathrm{P\Gamma L}(2, 25)$ *of* 7-$(26, 9, \lambda)$, $\lambda = 54,\ 63,\ 81$;

V. $\mathrm{P\Gamma L}(2, 32)$ *of* 7-$(33, 8, 10)$, [2].

The only 7-designs known before were those of Teirlinck [16] with $k = t + 1$ and astronomically large λ and v :

$$\lambda = (t + 1)!^{2t+1},\ v \equiv t \bmod \lambda.$$

Applying a construction from Tran van Trung[17], see also Kreher[9], yields further 7-designs from those of the theorem.

Corollary 1.2 *There exist simple 7-designs with the following parameter sets*:

VI. 7-$(25, 9, \lambda)$ *for* $\lambda = $ 45, 54, 72;

VII. 7-$(27, 9, 60)$.

Thus, there exist simple 7-designs for $v = 24$, 25, 26, 27, and 33 with some projective automorphism groups.

2 Methods

The designs are constructed by the Kramer-Mesner method [6]. This method assumes a prescribed group A of automorphisms of the desired t-(v, k, λ) designs. The group A is a permutation group on the underlying set V of v elements acting in the induced way on the set of all k-subsets of V. A design allows A as an automorphism group if and only if the set of blocks of the design consists of full k-orbits of A.

Therefore a collection of such k-orbits has to be chosen such that each t-subset T is contained in equally many blocks from these orbits. So for each k-orbit K^A the number $m(T, K^A)$ of members containing T is computed. If T is replaced by some T' from the t-orbit T^A these numbers remain unchanged. So it suffices to consider only one representative T from each t-orbit. There results a matrix M with a row for each t-orbit and a column for each k-orbit. Choosing k-orbits for a t-(v, k, λ) design means to multiply M by an appropriate 0/1-vector on the right such that a vector with constant entries λ results.

There have been different approaches to finding such 0/1 vectors. We have implemented a variant of the LLL-algorithm [12], see [18], which in comparison to Kreher and Radzizowski [8] has the new feature of considering λ as a variable. This helps find unsuspected values of λ. After applying the LLL-algorithm all solutions are determined by an exhaustive search. The Kramer-Mesner matrix is computed by a new version of B. Schmalz's Leiterspiel (snakes and ladders). Our computational system *DISCRETA* allows the choice of groups A from some predefined series. The user computes the matrices and solves the diophantine system of equations by pressing some buttons at a graphical user interface. Besides the LLL-solver we have also included in the system a clever backtrack-solver written by B.D. McKay [14] and a linear programming tool lp-solve [1]. McKay's solver, in particular, is frequently a valuable alternative to the LLL method. The system is written in C and uses a Motif package for the graphical surface. It can be obtained from the authors via ftp.

The following group theoretic results allow us to determine the isomorphism types of designs with prescribed automorphism groups in many important cases without isomorphism testing.

Theorem 2.1 *Let G be a group acting on a set Ω. Let A be a subgroup of G which is the full stabilizer of the points in a set $\Delta \subseteq \Omega$. Then two points of Δ may only lie in the same G-orbit if they lie in the same orbit of $N_G(A)$, the normalizer of A in G.*

If Δ in the theorem is the set of all points having stabilizer A then $N_G(A)$ acts on this set with orbits of length $|N_G(A)/A|$. Thus, the number of isomorphism types of designs having a prescribed full automorphism group A is obtained by dividing the total number of all designs having a prescribed full automorphism group A by the index of A in its normalizer taken in the full symmetric group on the underlying point set. If the group A is not the full automorphism group of some designs fixed by A then those designs must have a larger automorphism group. The principle of inclusion-exclusion allows to determine the number of isomorphism types with prescribed automorphism group in this situation. This is the method W. Burnside formalized with his table of marks [3] for general actions of finite groups, see also [15],[11] for constructive aspects of this approach. Since in many situations the subgroups which occur as stabilizers are not easy to determine, the following special situation is of interest.

Theorem 2.2 *Let G be a group acting on a set Ω. Let $\omega_1, \omega_2 \in \Omega$ and let P be a Sylow-p-subgroup of G fixing ω_1 and ω_2. Then if ω_1 and ω_2 are in the same orbit of G both points are already in the same orbit of $N_G(P)$, the normalizer of P in G.*

In the situation of the theorem for a subgroup A containing P no knowledge about the overgroups of A is needed to decide whether two points fixed by A lie in the same G-orbit. The only difficulty in formulating general counting formulas results from the fact that the set of points fixed by A usually is not closed under $N_G(P)$. However, it is sometimes possible to enlarge P and $N_G(P)$ so that the overgroup of $N_G(P)$ acts on the set of fixed points of the overgroup of P. Hence, for any prime p the projective group $\mathrm{PSL}(2,p)$ contains a Sylow-p-subgroup P of S_{p+1} and $\mathrm{PGL}(2,p)$ contains the normalizer of P. Therefore the following holds.

Corollary 2.3 *For any prime p all designs which admit $\mathrm{PGL}(2,p)$ as a group of automorphisms are pairwise non-isomorphic. All designs admitting $\mathrm{PSL}(2,p)$ but not $\mathrm{PGL}(2,p)$ as a group of automorphisms are grouped into isomorphic pairs under the action of $\mathrm{PGL}(2,p)/\mathrm{PSL}(2,p)$.*

By the preceding methods, the numbers of designs obtained for the cases I to V yield the following numbers of isomorphism types:

I. 1, 138, ≥ 590, ≥ 126, ≥ 65 for 7-$(24, 8, \lambda)$ and $\lambda = 4$, 5, 6, 7, 8;

II. 113, 5463, ≥ 15325 for 7-$(24, 9, \lambda)$, where $\lambda = 40$, 48, 64;

III. 7 for 7-$(26, 8, 6)$,

IV. 3989, 37932, ≥ 14 for 7-$(26, 9, \lambda)$, where $\lambda = 54$, 63, 81;

V. 4996426 for 7-$(33, 8, 10)$, [18].

The corollary also explains why in their investigations of Steiner 5-designs M.J. Grannel, T.S. Griggs and R.A. Mathon in a series of papers, see [4], always found two copies of each isomorphism type of Steiner systems with some prescribed automorphism group $PSL(2, p)$.

B.D. McKay [14] was the first to find more 7-$(33, 8, 10)$ designs different from those in [2]. He estimated the existence of about 5 million designs of type V, and this gave the impetus for the development of better equation solver for the Kramer-Mesner method, see [18]. In fact, there are 4996426 such designs which is suprisingly close to his estimate.

A detailed presentation of all results mentioned would be very space consuming. A moderate listing is planned to appear elsewhere together with some new 6-designs and material on deduced parameter sets. The details can be obtained from the authors, see also our WWW-pages. Here we include only one representative for the smallest value of λ in each of the basic cases I - IV.

3   Selected 7-Designs with Small λ

I: We use the following permutation representation of $PGL(2, 23)$, a group of order 12144. Generators are the following permutations:

$\alpha = (3\ 7\ 4\ 12\ 6\ 22\ 10\ 19\ 18\ 13\ 11\ 24\ 20\ 23\ 15\ 21\ 5\ 17\ 8\ 9\ 14\ 16\ )$,
$\beta = (3\ 16\ 14\ 9\ 8\ 17\ 5\ 21\ 15\ 23\ 20\ 24\ 11\ 13\ 18\ 19\ 10\ 22\ 6\ 12\ 4\ 7)$,
$\gamma = (2\ 3\ 4\ 5\ 6\ 7\ 8\ 9\ 10\ 11\ 12\ 13\ 14\ 15\ 16\ 17\ 18\ 19\ 20\ 21\ 22\ 23\ 24)$,
$\delta = (1\ 3\ 14\ 10\ 8\ 16\ 6\ 12\ 5\ 20\ 9\ 23\ 4\ 18\ 7\ 22\ 15\ 21\ 11\ 19\ 17\ 13\ 24)$.
The permutations β^2, γ, and δ generate $PSL(2, 23)$, a group of order 6072.

The 7-$(24, 8, 4)$ design consists of the orbits of the following 8-subsets, called *starter blocks*, under the action of $PSL(2, 23)$:

Starter Blocks								Orbit Length
17	18	19	20	21	22	23	24	6072
7	18	19	20	21	22	23	24	6072
5	1	19	20	21	22	23	24	3036
9	1	19	20	21	22	23	24	6072
5	4	19	20	21	22	23	24	6072
16	4	19	20	21	22	23	24	6072
10	8	19	20	21	22	23	24	3036
16	8	19	20	21	22	23	24	6072
10	3	1	20	21	22	23	24	3036
13	3	1	20	21	22	23	24	3036
17	3	1	20	21	22	23	24	6072
11	4	1	20	21	22	23	24	6072
16	4	1	20	21	22	23	24	6072
17	4	1	20	21	22	23	24	3036
12	5	1	20	21	22	23	24	6072
18	5	1	20	21	22	23	24	6072
9	7	1	20	21	22	23	24	6072
13	8	1	20	21	22	23	24	6072
8	5	3	20	21	22	23	24	6072
10	5	3	20	21	22	23	24	6072
8	6	3	20	21	22	23	24	6072
13	6	3	20	21	22	23	24	6072
13	7	3	20	21	22	23	24	6072
17	8	3	20	21	22	23	24	3036
13	10	3	20	21	22	23	24	3036
14	11	3	20	21	22	23	24	3036
16	14	3	20	21	22	23	24	6072
11	6	4	20	21	22	23	24	6072
18	6	4	20	21	22	23	24	6072
13	8	4	20	21	22	23	24	3036
18	13	4	20	21	22	23	24	3036
10	7	5	20	21	22	23	24	3036
11	8	5	20	21	22	23	24	3036
12	9	5	20	21	22	23	24	1518
16	10	5	20	21	22	23	24	3036
16	10	7	20	21	22	23	24	3036
17	14	12	20	21	22	23	24	759
18	14	6	4	21	22	23	24	759

II: One out of 113 isomorphism types of 7-$(24, 9, 40)$ designs has the following starter blocks for orbits under the action of $PGL(2, 23)$:

Starter Blocks									Orbit Length
16	17	18	19	20	21	22	23	24	6072
6	17	18	19	20	21	22	23	24	12144
8	17	18	19	20	21	22	23	24	12144
9	17	18	19	20	21	22	23	24	6072
8	1	18	19	20	21	22	23	24	12144
7	4	18	19	20	21	22	23	24	12144
8	4	18	19	20	21	22	23	24	12144
9	4	18	19	20	21	22	23	24	12144
12	5	18	19	20	21	22	23	24	4048
13	5	18	19	20	21	22	23	24	6072
9	8	18	19	20	21	22	23	24	12144
5	3	1	19	20	21	22	23	24	12144
12	3	1	19	20	21	22	23	24	6072
17	3	1	19	20	21	22	23	24	2024
9	4	1	19	20	21	22	23	24	12144
16	4	1	19	20	21	22	23	24	6072
11	5	1	19	20	21	22	23	24	12144
12	8	1	19	20	21	22	23	24	2024
10	9	1	19	20	21	22	23	24	12144
8	5	4	19	20	21	22	23	24	12144
10	5	4	19	20	21	22	23	24	12144
11	5	4	19	20	21	22	23	24	2024
16	5	4	19	20	21	22	23	24	12144
10	9	4	19	20	21	22	23	24	12144
16	10	4	19	20	21	22	23	24	6072
7	4	3	1	20	21	22	23	24	12144
13	4	3	1	20	21	22	23	24	12144
8	5	3	1	20	21	22	23	24	4048
11	5	3	1	20	21	22	23	24	12144
16	7	3	1	20	21	22	23	24	12144
13	12	3	1	20	21	22	23	24	12144
7	5	4	1	20	21	22	23	24	6072
9	5	4	1	20	21	22	23	24	12144
13	5	4	1	20	21	22	23	24	12144
9	8	4	1	20	21	22	23	24	12144
13	8	5	1	20	21	22	23	24	12144
14	8	5	1	20	21	22	23	24	12144
16	8	5	1	20	21	22	23	24	6072
13	8	5	3	20	21	22	23	24	6072
15	10	5	3	20	21	22	23	24	12144

III: We use the following permutation representation of $P\Gamma L(2, 25)$, a group of order 31200. Generators are the following permutations:

$$\alpha = (1\ 2\ 3\ 4\ 5\ 6\ 7\ 8\ 9\ 10\ 11\ 12\ 13\ 14\ 15\ 16\ 17\ 18\ 19\ 20\ 21\ 22\ 23\ 24),$$

$$\beta = (1\ 17\ 14\ 15\ 10)(2\ 5\ 13\ 22\ 3)(4\ 11\ 9\ 19\ 8)(6\ 18\ 12\ 25\ 24)$$
$$(7\ 21\ 23\ 16\ 20),$$

$$\gamma = (1\ 8\ 4\ 17\ 3)(2\ 21\ 22\ 19\ 11)(5\ 16\ 20\ 13\ 15)(6\ 12\ 26\ 24\ 18)$$
$$(7\ 10\ 9\ 14\ 23),$$

$$\delta = (1\ 5)(2\ 10)(3\ 15)(4\ 20)(7\ 11)(8\ 16)(9\ 21)(13\ 17)(14\ 22)(19\ 23).$$

The permutations α, β, and γ generate $PGL(2, 25)$, a group of order 15600.

One out of 7 isomorphism types of 7-$(26, 8, 6)$ designs has the following starter blocks for orbits under the action of $PGL(2, 25)$:

Starter Blocks								Orbit Length
2	20	21	22	23	24	25	26	15600
7	20	21	22	23	24	25	26	15600
8	20	21	22	23	24	25	26	15600
3	2	21	22	23	24	25	26	7800
4	2	21	22	23	24	25	26	15600
7	2	21	22	23	24	25	26	15600
9	2	21	22	23	24	25	26	15600
18	2	21	22	23	24	25	26	15600
10	3	21	22	23	24	25	26	15600
11	3	21	22	23	24	25	26	15600
6	4	21	22	23	24	25	26	15600
8	4	21	22	23	24	25	26	7800
9	4	21	22	23	24	25	26	15600
16	4	21	22	23	24	25	26	7800
11	5	21	22	23	24	25	26	15600
15	5	21	22	23	24	25	26	15600
16	5	21	22	23	24	25	26	7800
9	6	21	22	23	24	25	26	7800
14	6	21	22	23	24	25	26	15600
15	6	21	22	23	24	25	26	3900
9	7	21	22	23	24	25	26	15600
10	9	21	22	23	24	25	26	7800
11	10	21	22	23	24	25	26	1950
5	4	2	22	23	24	25	26	7800
10	4	2	22	23	24	25	26	7800
13	4	2	22	23	24	25	26	15600
18	4	2	22	23	24	25	26	15600
11	5	2	22	23	24	25	26	15600
17	5	2	22	23	24	25	26	15600
14	6	2	22	23	24	25	26	15600
16	6	2	22	23	24	25	26	15600
19	6	2	22	23	24	25	26	15600
16	8	2	22	23	24	25	26	7800
18	10	2	22	23	24	25	26	15600
19	10	2	22	23	24	25	26	7800
19	11	2	22	23	24	25	26	7800
5	4	3	22	23	24	25	26	3900
9	7	3	22	23	24	25	26	3900
14	9	3	22	23	24	25	26	7800
17	5	4	22	23	24	25	26	15600
12	8	4	22	23	24	25	26	7800
17	10	4	3	23	24	25	26	3900

IV: One out of 3989 isomorphism types of 7-$(26, 9, 54)$ designs consisting of the orbits of the following starter blocks under the action of $P\Gamma L(2, 25)$:

Starter Blocks									Orbit Length
18	19	20	21	22	23	24	25	26	15600
2	19	20	21	22	23	24	25	26	31200
4	19	20	21	22	23	24	25	26	31200
8	2	20	21	22	23	24	25	26	31200
15	2	20	21	22	23	24	25	26	15600
17	2	20	21	22	23	24	25	26	31200
8	3	20	21	22	23	24	25	26	15600
5	4	20	21	22	23	24	25	26	15600
6	4	20	21	22	23	24	25	26	15600
7	4	20	21	22	23	24	25	26	31200
8	4	20	21	22	23	24	25	26	31200
9	4	20	21	22	23	24	25	26	31200
10	4	20	21	22	23	24	25	26	15600
12	4	20	21	22	23	24	25	26	31200
13	5	20	21	22	23	24	25	26	15600
8	6	20	21	22	23	24	25	26	31200
10	6	20	21	22	23	24	25	26	15600
14	6	20	21	22	23	24	25	26	15600
9	8	20	21	22	23	24	25	26	31200
7	3	2	21	22	23	24	25	26	31200
9	3	2	21	22	23	24	25	26	31200
11	4	2	21	22	23	24	25	26	31200
13	4	2	21	22	23	24	25	26	15600
11	5	2	21	22	23	24	25	26	31200
8	6	2	21	22	23	24	25	26	31200
9	6	2	21	22	23	24	25	26	31200
14	7	2	21	22	23	24	25	26	31200
15	7	2	21	22	23	24	25	26	31200
15	8	2	21	22	23	24	25	26	15600
13	9	2	21	22	23	24	25	26	15600
14	9	2	21	22	23	24	25	26	15600
11	10	2	21	22	23	24	25	26	3900
15	13	2	21	22	23	24	25	26	31200
18	17	2	21	22	23	24	25	26	31200
17	12	3	21	22	23	24	25	26	7800
8	7	4	21	22	23	24	25	26	15600
17	7	4	21	22	23	24	25	26	31200
13	10	4	21	22	23	24	25	26	15600
8	7	5	21	22	23	24	25	26	31200
15	7	6	21	22	23	24	25	26	7800
19	10	6	2	22	23	24	25	26	15600
20	11	6	2	22	23	24	25	26	31200

References

[1] M.R.C.m. Berkelaar, lp-solve, a public domain MILP solver, freely available from ftp://ftp.es.ele.tue.nl/pub/lp_solve/

[2] A. Betten, A. Kerber, A. Kohnert, R. Laue, A. Wassermann, The discovery of simple 7-designs with automorphism group $P\Gamma L(2, 32)$, *AAECC Proceedings 1995*, Springer LNCS **948** (1995),131-145.

[3] W. Burnside, *Theory of Groups of Finite Order*, Second edition, Cambridge University Press, Cambridge, 1911 (Dover, 1955).

[4] C.J. Colbourn, R. Mathon, Steiner systems, *The CRC Handbook on Combinatorial Designs* C.J. Colbourn, J.H. Dinitz (eds.), CRC Press, Boca Raton, New York, London, Tokyo, 1996, pp. 66-75.

[5] B. Huppert, *Endliche Gruppen I*, Springer-Verlag, Berlin, 1967.

[6] E.S. Kramer and D.M. Mesner, t-designs on hypergraphs, *Discrete Math.* **15** (1976), 263-296.

[7] E.S. Kramer, S.S. Magliveras and D.W. Leavitt, Construction procedures for t-designs and the existence of new simple 6-designs, *Ann. Discrete Math.* **26** (1985), 247-274.

[8] D.L. Kreher, S.P. Radziszowski, Simple 5-$(28, 6, \lambda)$ designs from $PSL_2(27)$, *Ann. Discrete Math.* **37** (1987), 315-318.

[9] D.L. Kreher, An infinite family of (simple) 6-designs, *J. Combin. Des.* **1** (1993), 41-48.

[10] D.L. Kreher, t-designs, $t \geq 3$, *The CRC Handbook on Combinatorial Designs* C.J. Colbourn, J.H. Dinitz (eds.), CRC Press, Boca Raton, 1996, pp. 47-66.

[11] R. Laue, Construction of combinatorial objects – A tutorial, *Bayreuth. Math. Schr.* **43** (1993), 53-96.

[12] A.K. Lenstra, H.W. Lenstra Jr., and L. Lovász, Factoring polynomials with rational coefficients, *Math. Ann.* **261** (1982), 515-534.

[13] S. Magliveras and D.W. Leavitt, Simple 6-$(33, 8, 36)$ designs from $P\Gamma L_2(32)$, *Computational Group Theory*, M.D. Atkinson (ed.), Academic Press 1984, pp. 337-352.

[14] B.D. McKay, private communication.

[15] B. Schmalz, The t-designs with prescribed automorphism group, new simple 6-designs, *J. Combin. Des.* **1** (1993), 125-170.

[16] L. Teirlinck, Non trivial t-designs without repeated blocks exist for all t, *Discrete Math.* **65** (1987), 301-311.

[17] Tran van Trung, On the construction of t-designs and the existence of some infinite families of simple 5-designs, *Arch.Math.* **47** (1986), 187-192.

[18] A. Wassermann, Finding simple t-designs with enumeration techniques, in preparation.

A. Betten, Lehrstuhl II für Mathematik, Universität Bayreuth, D-95440 Bayreuth, Germany.
e-mail: Anton.Betten@uni-bayreuth.de

R. Laue, Lehrstuhl II für Mathematik, Universität Bayreuth, D-95440 Bayreuth, Germany.
e-mail: laue@btm2x2.mat.uni-bayreuth.de

A. Wassermann, Lehrstuhl Mathematik und ihre Didaktik, Universität Bayreuth, D-95440 Bayreuth, Germany.
e-mail: Alfred.Wassermann@uni-bayreuth.de

http://www.mathe2.uni-bayreuth.de/betten/DESIGN/d1.html

Inscribed bundles, Veronese surfaces and caps

A. Bonisoli   *A. Cossidente*

Abstract

The Veronese correspondence maps the set of all plane conics which are tangent to the sides of a given triangle in $PG(2,q)$, q odd, to a $(2q^2 - q + 2)$–cap in $PG(5,q)$ obtained as the complete intersection of three quadratic cones. This cap can also be represented as the union of two quadric Veroneseans sharing three conics pairwise meeting at one point. Some information about the (setwise) stabilizer of this cap in $PGL(6,q)$ is also given.

The *quadric Veronesean* of $PG(2,q)$ is the variety $\mathcal{V}$ consisting of all points $(x_0^2, x_1^2, x_2^2, x_0x_1, x_0x_2, x_1x_2)$ in $PG(5,q)$ as (x_0, x_1, x_2) varies over the points of $PG(2,q)$.

Definition *An* **inscribed bundle** *of* $PG(2,q)$ *consists of all conics of* $PG(2,q)$ *that are simultaneously tangent to the three sides of a triangle* T.

Observe that in our definition we do not require that a conic of an inscribed bundle be non–degenerate, cf. [2].

The generic conic γ of $PG(2,q)$, q odd, has equation

$$a_{00}x_0^2 + a_{11}x_1^2 + a_{22}x_2^2 + 2a_{01}x_0x_1 + 2a_{02}x_0x_2 + 2a_{12}x_1x_2 = 0$$

with coefficients a_{ij} in $GF(q)$. Denote by A the symmetric matrix associated to the conic γ, namely

$$A = \begin{pmatrix} a_{00} & a_{01} & a_{02} \\ a_{01} & a_{11} & a_{12} \\ a_{02} & a_{12} & a_{22} \end{pmatrix}.$$

The condition for the line ℓ with equation $u_0x_0 + u_1x_1 + u_2x_2 = 0$ to touch the conic γ is

$$A_{00}u_0^2 + A_{11}u_1^2 + A_{22}u_2^2 + 2A_{01}u_0u_1 + 2A_{02}u_0u_2 + 2A_{12}u_1u_2 = 0,$$

where A_{ij} is the cofactor of a_{ij} in A (see [1, p.106]). In the projective plane $PG(2,q)$, consider the canonical reference triangle T, whose vertices are the points $U_0 = (1, 0, 0)$, $U_1 = (0, 1, 0)$ and $U_2 = (0, 0, 1)$ and whose sides are the coordinate axes $x_0 = 0$, $x_1 = 0$ and $x_2 = 0$. By imposing that condition that the three sides of T be tangent to γ, we obtain the following relations:

$$A_{00} = a_{11}a_{22} - a_{12}^2 = 0, \quad A_{11} = a_{00}a_{22} - a_{02}^2 = 0, \quad A_{22} = a_{00}a_{11} - a_{01}^2 = 0. \quad (1)$$

The relations (1) thus define an inscribed bundle $\mathcal{B}$ of $PG(2, q)$. It is easy to see that the conics of $\mathcal{B}$ form a net.

Consider the projective space $PG(5, q)$, where X_{00}, X_{11}, X_{22}, X_{01}, X_{02}, X_{12} are homogeneous coordinates and map the conic γ of $PG(2, q)$ described by the symmetric matrix $A = (a_{ij})$ to the point $(a_{00}, a_{11}, a_{22}, a_{01}, a_{02}, a_{12})$ of $PG(5, q)$. For $i, j \in \{0, 1, 2\}$, $i \neq j$, define $F_{ij} = X_{ij}^2 - X_{ii}X_{jj}$.

Proposition 1 *The conics of $\mathcal{B}$ in the conic–point correspondence, are the points of a surface Z of order eight obtained by intersecting the three quadrics with equations $F_{01} = 0$, $F_{02} = 0$, $F_{12} = 0$.*

Proof. Let T be the triangle of $PG(2, q)$ defined by $x_0 x_1 x_2 = 0$. We have observed that the sides of T are tangent to the conic γ if and only if the conditions in (1) hold. The counterparts of these conditions in $PG(5, q)$ are precisely the equations $F_{ij} = 0$, each of which represents a quadratic cone in $PG(5, q)$ whose vertex is a plane. The intersection of the cones is a variety of order eight. □

Proposition 2 *The surface Z is the union of two Veronese surfaces $\mathcal{V}_2^4$ and $\overline{\mathcal{V}}_2^4$ of $PG(5, q)$. The intersection of $\mathcal{V}_2^4$ and $\overline{\mathcal{V}}_2^4$ is the union of three conics intersecting pairwise in one point. The points on one of these conics are the images of the plane conics given by the repeated lines through one of the vertices of T.*

Proof. The equations of Z are

$$X_{01}^2 - X_{00}X_{11} = 0, \quad X_{02}^2 - X_{00}X_{22} = 0, \quad X_{12}^2 - X_{11}X_{22} = 0. \quad (2)$$

It is easy to see that the first two equations are satisfied by putting

$$X_{00} = u^2, \ X_{11} = v^2, \ X_{22} = w^2, \ X_{01} = uv, \ X_{02} = uw,$$

where $u, v, w \in GF(q)$, $(u, v, w) \neq (0, 0, 0)$. Substitution in the third equation yields $X_{12}^2 - v^2 w^2 = 0$, and so $X_{12} = \pm vw$.

It follows that the points in the intersection of the cones (2) form two Veronese surfaces $\mathcal{V}_2^4$ and $\overline{\mathcal{V}}_2^4$ with parametric equations

$$X_{00} = u^2, \ X_{11} = v^2, \ X_{22} = w^2, \ X_{01} = uv, \ X_{02} = uw, \ X_{12} = \pm vw, \quad (3)$$

with $u, v, w \in GF(q)$, $(u, v, w) \neq (0, 0, 0)$. Note that setting $X_{01} = \pm uv$ or $X_{02} = \pm uw$ rather than $X_{12} = \pm vw$ simply yields a different parametric representation of the intersection of the same three cones.

Setting in turn $u = 0$, $v = 0$, $w = 0$ in (3), it is easily seen that the Veronese surfaces $\mathcal{V}_2^4$ and $\overline{\mathcal{V}}_2^4$ have the following three conics in common:

$$X_{02} = X_{12} = X_{22} = X_{01}^2 - X_{00}X_{11} = 0,$$
$$X_{01} = X_{11} = X_{12} = X_{02}^2 - X_{00}X_{22} = 0,$$
$$X_{00} = X_{01} = X_{02} = X_{12}^2 - X_{11}X_{22} = 0,$$

which lie in three distinct conic planes of the hypersurface $\mathcal{M}_4^3$ of the chords of $\mathcal{V}_2^4$, and so they meet pairwise in a point. This completes the proof. $\square$

Proposition 3 *The surface Z is a $(2q^2 - q + 2)$-cap of $PG(5, q)$.*

Proof. We know that each Veronese surface is a cap in $PG(5, q)$; see [4, Thm. 25.1.8]. So each one of the sets $\mathcal{V}_2^4$ and $\overline{\mathcal{V}}_2^4$ is a cap.

The discriminant of the conic corresponding to the generic point of $\overline{\mathcal{V}}_2^4$ has the value $-4u^2v^2w^2$, which is non–zero, unless $u = 0$, $v = 0$ or $w = 0$. These values of u, v, w yield precisely the conics in the intersection of $\mathcal{V}_2^4$ and $\overline{\mathcal{V}}_2^4$. Apart from these cases, the conics in $\overline{\mathcal{V}}_2^4$ are non–degenerate.

Suppose now that three points of Z are collinear. This implies that the line ℓ containing them is entirely contained in Z, since Z is the intersection of three quadratic cones.

Suppose that ℓ contains at least two points of $\mathcal{V}_2^4$, say P_1 and P_2. A third point P_3 on ℓ cannot be a point of $\mathcal{V}_2^4$, since $\mathcal{V}_2^4$ is a cap, and so P_3 lies on $\overline{\mathcal{V}}_2^4$. The line ℓ on Z should correspond to a pencil of conics in $\mathcal{B}$. This pencil is generated by the two repeated lines whose images are the points P_1 and P_2. It is easily seen that if q is odd a pencil of conics generated by two repeated lines further contains $(q-1)/2$ line pairs and $(q-1)/2$ single points; see for instance [3]. This means that P_3 cannot be a point on ℓ, as P_3 corresponds to a non–degenerate conic.

Suppose that at most one point of ℓ lies on $\mathcal{V}_2^4$. The remaining points lie then on $\overline{\mathcal{V}}_2^4$ and since the number of such points is at least $q \geq 3$, we should have three collinear points on $\overline{\mathcal{V}}_2^4$, which is not the case.

The proof is now complete. $\square$

We know from [4, §25.1] that each linear collineation of $PG(2, q)$ can be "lifted" to a linear collineation of $PG(5, q)$ preserving the quadric Veronesean $\mathcal{V}_2^4$, thus yielding a subgroup H of $PGL(6, q)$ fixing $\mathcal{V}_2^4$ (setwise). The group H is isomorphic copy of $PGL(3, q)$ and is actually the full (setwise) stabilizer of $\mathcal{V}_2^4$ in $PGL(6, q)$. We now want to give an explicit representation for H: when a matrix is viewed as a linear collineation then the points of the projective space are treated as row–vectors, with action given by matrix multiplication.

Proposition 4 *If* $\mathbf{y} \mapsto \mathbf{y}A$ *is a linear collineation of* $PG(2, q)$ *with* $A = (a_{ij})$, $i, j = 0, 1, 2$, *then the "lifted" linear collineation of* $PG(6, q)$ *fixing the quadric Veronesean is* $\mathbf{x} \mapsto \mathbf{x}B$ *where* B *is the following* 6×6 *matrix:*

$$
\begin{bmatrix}
a_{00}^2 & a_{01}^2 & a_{02}^2 & a_{00}a_{01} & a_{00}a_{02} & a_{01}a_{02} \\
a_{10}^2 & a_{11}^2 & a_{12}^2 & a_{10}a_{11} & a_{10}a_{12} & a_{11}a_{12} \\
a_{20}^2 & a_{21}^2 & a_{22}^2 & a_{20}a_{21} & a_{20}a_{22} & a_{21}a_{22} \\
2a_{00}a_{10} & 2a_{01}a_{11} & 2a_{02}a_{12} & a_{00}a_{11} + a_{10}a_{01} & a_{00}a_{12} + a_{10}a_{02} & a_{01}a_{12} + a_{11}a_{02} \\
2a_{00}a_{20} & 2a_{01}a_{21} & 2a_{02}a_{22} & a_{00}a_{21} + a_{20}a_{01} & a_{00}a_{22} + a_{20}a_{02} & a_{01}a_{22} + a_{21}a_{02} \\
2a_{10}a_{20} & 2a_{11}a_{21} & 2a_{12}a_{22} & a_{10}a_{21} + a_{11}a_{20} & a_{10}a_{22} + a_{20}a_{12} & a_{11}a_{22} + a_{21}a_{12}
\end{bmatrix}.
$$

Proof. The relations

$$
\begin{aligned}
(1, 0, 0) &\mapsto (a_{00}, a_{01}, a_{02}), \\
(0, 1, 0) &\mapsto (a_{10}, a_{11}, a_{12}), \\
(0, 0, 1) &\mapsto (a_{20}, a_{21}, a_{22}), \\
(1, 1, 0) &\mapsto (a_{00} + a_{10}, a_{01} + a_{11}, a_{02} + a_{12}), \\
(1, 0, 1) &\mapsto (a_{00} + a_{20}, a_{01} + a_{21}, a_{02} + a_{22}), \\
(0, 1, 1) &\mapsto (a_{10} + a_{20}, a_{11} + a_{21}, a_{12} + a_{22}), \\
(1, 1, 1) &\mapsto (a_{00} + a_{10} + a_{20}, a_{01} + a_{11} + a_{21}, a_{02} + a_{12} + a_{22}),
\end{aligned}
$$

in $PG(2, q)$ have the following counterparts in $PG(5, q)$:

$$
\begin{aligned}
(1, 0, 0, 0, 0, 0) &\mapsto (a_{00}{}^2, a_{01}{}^2, a_{02}{}^2, a_{00}a_{01}, a_{00}a_{02}, a_{01}a_{02}), \\
(0, 1, 0, 0, 0, 0) &\mapsto (a_{10}{}^2, a_{11}{}^2, a_{12}{}^2, a_{10}a_{11}, a_{10}a_{12}, a_{11}a_{12}), \\
(0, 0, 1, 0, 0, 0) &\mapsto (a_{20}{}^2, a_{21}{}^2, a_{22}{}^2, a_{20}a_{21}, a_{20}a_{22}, a_{21}a_{22}), \\
(1, 1, 0, 1, 0, 0) &\mapsto \\
&\quad ([a_{00} + a_{10}]^2, [a_{01} + a_{11}]^2, [a_{02} + a_{12}]^2, \\
&\quad\ [a_{00} + a_{10}][a_{01} + a_{11}], [a_{00} + a_{10}][a_{02} + a_{12}], [a_{01} + a_{11}][a_{02} + a_{12}]) \\
(1, 0, 1, 0, 1, 0) &\mapsto \\
&\quad ([a_{00} + a_{20}]^2, [a_{01} + a_{21}]^2, [a_{02} + a_{22}]^2, \\
&\quad\ [a_{00} + a_{20}][a_{01} + a_{21}], [a_{00} + a_{20}][a_{02} + a_{22}], [a_{01} + a_{21}][a_{02} + a_{22}]) \\
(0, 1, 1, 0, 0, 1) &\mapsto \\
&\quad ([a_{10} + a_{20}]^2, [a_{11} + a_{21}]^2, [a_{12} + a_{22}]^2, \\
&\quad\ [a_{10} + a_{20}][a_{11} + a_{21}], [a_{10} + a_{20}][a_{12} + a_{22}], [a_{11} + a_{21}][a_{12} + a_{22}]) \\
(1, 1, 1, 1, 1, 1) &\mapsto ([a_{00} + a_{10} + a_{20}]^2, [a_{01} + a_{11} + a_{21}]^2, [a_{02} + a_{12} + a_{22}]^2, \\
&\quad\ [a_{00} + a_{10} + a_{20}][a_{01} + a_{11} + a_{21}], [a_{00} + a_{10} + a_{20}][a_{02} + a_{12} + a_{22}], \\
&\quad\ [a_{01} + a_{11} + a_{21}][a_{02} + a_{12} + a_{22}]).
\end{aligned}
$$

We have

$$
\begin{aligned}
(1, 1, 1, 1, 1, 1) = \ & -(1, 0, 0, 0, 0, 0) - (0, 1, 0, 0, 0, 0) - (0, 0, 1, 0, 0, 0) \\
& +(1, 1, 0, 1, 0, 0) + (1, 0, 1, 0, 1, 0) + (0, 1, 1, 0, 0, 1),
\end{aligned}
$$

and so the seven points represented by the given row–vectors form a frame in $PG(5,q)$. If we now multiply the seven row–vectors by the matrix B we obtain precisely the above relations and the assertion follows.   □

Proposition 5 *The (setwise) stabilizer of Z in H, is the subgroup E obtained by "lifting" all monomial transformations of $PG(2,q)$.*

Proof.   Consider a transformation $g \in PGL(3,q)$ and its "lifted" counterpart $\hat{g}$ in H. We have $\hat{g}(Z) = Z$ if and only if $\hat{g}(\overline{\mathcal{V}}_2^4) = \overline{\mathcal{V}}_2^4$. Hence in particular g must permute the non–singular plane conics which are simultaneously tangent to the three sides of the reference triangle T. Is is easily seen that the three vertices of the triangle itself are the unique points of the plane which do not lie on any non–singular conic which is simultaneously tangent to all three sides of T.   □

Proposition 6 *The (setwise) stabilizer of Z in $PGL(6,q)$ contains the elementary abelian group N of order eight generated by the transformations*

$$(Y_1, Y_2, Y_3, Y_4, Y_5, Y_6) \mapsto (Y_1, Y_2, Y_3, -Y_4, Y_5, Y_6),$$
$$(Y_1, Y_2, Y_3, Y_4, Y_5, Y_6) \mapsto (Y_1, Y_2, Y_3, Y_4, -Y_5, Y_6),$$
$$(Y_1, Y_2, Y_3, Y_4, Y_5, Y_6) \mapsto (Y_1, Y_2, Y_3, Y_4, Y_5, -Y_6).$$

The subgroup $\langle E, N \rangle$ fixes Z (setwise) and has order $12 \cdot (q-1)^2$.

Proof.   Each transformation in N clearly exchanges $\mathcal{V}_2^4$ with $\overline{\mathcal{V}}_2^4$. The subgroup E has order $6 \cdot (q-1)^2$ and the subgroup $\langle E, N \rangle$ is the semidirect product of E by any one of the three involutions generating N.   □

Remark.   Computer tests performed for $q \leq 9$ have shown that the cap Z is generally far from being complete. Furthermore the group $\langle E, N \rangle$ is the full (setwise) stabilizer of the cap Z in $PGL(6,q)$ in these cases: we have not yet found a proof holding for all values of q. The software used in our tests was the Computer Algebra system MAGMA developed at the University of Sydney.

Acknowledgement.   This research was carried out within the activity of G.N.S.A.G.A. of the Italian C.N.R. with the support of the Italian Ministry for Research and Technology.

References

[1] H.F. Baker, *Principles of Geometry, Volume II*, Cambridge University Press, Cambridge, 1922 (Ungar, New York, 1991).

[2] R.D. Baker, J.M.N. Brown, G.L. Ebert and J.C. Fisher, Projective bundles, *Bull. Belg. Math. Soc.* **3** (1994), 329-336.

[3] J.W.P. Hirschfeld, Projective Geometries over Finite Fields, new edition (to appear).

[4] J.W.P. Hirschfeld and J.A. Thas, *General Galois Geometries*, Oxford University Press, Oxford, 1991.

A. Bonisoli, Dipartimento di Matematica, Università della Basilicata, via N.Sauro 85, 85100 Potenza, Italy.
e-mail: bonisoli@unibas.it

A. Cossidente, Dipartimento di Matematica, Università della Basilicata, via N.Sauro 85, 85100 Potenza, Italy.
e-mail: cossidente@unibas.it

Embedding partial geometries in Steiner designs

Andries E. Brouwer, Willem H. Haemers
Vladimir D. Tonchev

Abstract

We consider the following problem: given a partial geometry $\mathcal{P}$ with v points and k points on a line, can one add to the line set a set of k-subsets of points such that the extended family of k-subsets is a 2-$(v, k, 1)$ design (or a Steiner system $S(2, k, v)$). We give some necessary conditions for such embeddings and several examples. One of these is an embedding of the partial geometry $PQ^+(7, 2)$ into a 2-$(120, 8, 1)$ design.

1 Introduction

We consider the question whether, given a partial geometry $\mathcal{P} = (X, \mathcal{L})$, there is a Steiner 2-design $\mathcal{D} = (X, \mathcal{B})$ such that $\mathcal{L} \subseteq \mathcal{B}$. Clearly, the existence of such an embedding of $\mathcal{P}$ does not depend on the structure of $\mathcal{P}$, but only on its collinearity graph and line size.

Troughout $\mathcal{P} = (X, \mathcal{L})$ will denote a partial geometry with parameters s, t and α. The number v of points and the number l of lines of $\mathcal{P}$ are given by

$$v = (s + 1)(st + \alpha)/\alpha \ , \ l = (t + 1)(st + \alpha)/\alpha \ .$$

The collinearity graph (or point graph) is strongly regular having eigenvalues $s(t + 1)$, $s - \alpha$ and $-t - 1$ with multiplicities

$$1, \ f = \frac{st(s + 1)(t + 1)}{\alpha(s + t + 1 - \alpha)}, \ g = \frac{s(s + 1 - \alpha)(st + \alpha)}{\alpha(s + t + 1 - \alpha)},$$

respectively. For these and other results on partial geometries we refer to the survey paper by De Clerck and Van Maldeghem [3].

Suppose we have a collection $\mathcal{C}$ of $(s + 1)$-cocliques in the point graph of $\mathcal{P}$ that cover all non-collinear pairs of points exactly once. Then $\mathcal{P}$ is embedded in the Steiner 2-design $\mathcal{D} = (X, \mathcal{C} \cup \mathcal{L})$ with parameters:

$$v, \ k = s + 1, \ r = (st + t + \alpha)/\alpha, \ b = (st + \alpha)(st + t + \alpha)/\alpha^2.$$

Clearly a necessary condition for $\mathcal{P}$ to be embeddable is that r and b are integers. A quick inspection of the parameters of the known partial geometries shows that these divisibility conditions are satisfied very often. The following result excludes many more parameter sets.

Theorem 1.1 *Suppose $\mathcal{P}$ is embeddable in a Steiner 2-design $\mathcal{D}$ and suppose that $\mathcal{P}$ is not a Steiner 2-design or a net (i.e. $\alpha \neq s+1$ and $\alpha \neq t$). Then*

$$\alpha \leq \frac{t(s+1)}{t+s+1}.$$

If equality holds then $\alpha^f (t - \alpha)^g$ is the square of an integer.

This result is an immediate corollary of the following 'Fisher inequality'.

Theorem 1.2 *Let Γ be a strongly regular graph with parameters (v, k, λ, μ). If Γ is the collinearity graph of a partial linear space with l lines of size $s+1$, then either the given partial linear space is a partial geometry with parameters (s, t, α), or $l \geq v$. If $l = v$, then $\det(A + (k/s)I)$ is a square, where A is the adjacency matrix of Γ.*

Proof (of Theorem 1.2). Let N be the $v \times l$ point-line incidence matrix of the partial linear space, and A the adjacency matrix of Γ. Then $NN^{\top} = A + (t+1)I$, where $t+1 := k/s$ is the number of lines on each point. If $l = v$, then N is square, and $\det(A + (t+1)I) = (\det N)^2$. If $l < v$, then $NN^{\top}$ has rank less than v, so that A has eigenvalue $-t-1$, i.e., $(t+1)^2 - (\mu-\lambda)(t+1) + \mu - k = 0$. In this case, since $t+1$ divides k, it also divides μ, say $\mu = (t+1)\alpha$ for some nonnegative integer α, and we find $\lambda = s-1+\alpha t$, so that Γ has the parameters of the point graph of a pg(s, t, α), and since the lines are regular cliques, our partial linear space was in fact a partial geometry. $\square$

Proof (of Theorem 1.1). Apply Theorem 1.2 to the noncollinearity graph Γ of $\mathcal{P}$. We find either $b - l \geq v$, which reduces to $\alpha(s+t+1) \leq t(s+1)$, or Γ is the collinearity graph of a partial geometry $\mathcal{P}'$ with parameters (s', t', α'), where $s' = s$, $t' = s-\alpha$, $\alpha' = s-t$ and $(t-\alpha)(s+1-\alpha) = 0$. The adjacency matrix A of Γ has eigenvalues $st(s+1-\alpha)/\alpha$, $\alpha-1-s$, t with respective multiplicities 1, f, g, so if $b - l = v$ then $\alpha^f (s+t+1)^g$ is a square, and since $t^2 = (t-\alpha)(s+t+1)$ also $\alpha^f (t-\alpha)^g$ is a square. $\square$

A strongly regular graph is called *imprimitive* when it or its complement is a vertex disjoint union of cliques (i.e., when $\mu = 0$ or $\mu = k$ or there are no (non)edges at all). A union of m-cliques is the collinearity graph of a partial linear space with lines of size c if and only if a 2-$(m, c, 1)$ design exists. The complement of a union of n m-cliques is the collinearity graph of a partial linear space with lines of size c if and only if a group divisible design $GD(c, 1, m; nm)$ exists. Nothing nontrivial can be said here.

A partial geometry is called *improper* if $\alpha = 1$, s, $s + 1$, t or $t + 1$. For $\alpha = s$ or $\alpha = s + 1$ the point graph is imprimitive. Otherwise, if $\alpha = t + 1$, $\mathcal{P}$ is a dual Steiner system, which is not embeddable by Theorem 1.1. If $\alpha = t$, $\mathcal{P}$ is a net that we want to embed in a 2-$(k^2, k, 1)$ design, i.e. an affine plane. Therefore $\mathcal{P}$ is embeddable if and only if it is the union of some parallel classes of an affine plane.

Finally we consider the case $\alpha = 1$. Then $\mathcal{P}$ is a generalized quadrangle $GQ(s, t)$. In this case the divisibility conditions and the conditions of Theorem 1.1 are always fulfilled. Several of the known generalized quadrangles are constructed as a set of lines in a projective or affine space and hence are embeddable by construction, see Payne and Thas [6]. The smallest (with respect to v) open case is a possible embedding of $GQ(5, 3)$ in a 2-$(96, 6, 1)$ design.

One might ask whether an embedding will be unique. But when there is an embedding into some Steiner system with lots of subsystems, like a projective or affine space, then by twisting one or more subsystems one will in general get lots of embeddings. For example, in the smallest non-trivial case, that of the generalized quadrangle $GQ(2, 2)$, naturally embedded into $PG(3, 2)$ as the $Sp(4, 2)$ quadrangle, each plane contains three lines of the quadrangle and there are two ways of extending that set of three to a 2-(7,3,1) on that plane. Thus, a plane can be 'flipped'. Flipping one plane destroys all other planes except those that meet it in a line from the quadrangle. An arbitrary 2-(15,3,1) containing the lines of the quadrangle is obtained from $PG(3, 2)$ by flipping 0, 1, 2 or 3 planes on a given line of the quadrangle, and we find precisely four nonisomorphic 2-$(15, 3, 1)$ designs that contain $GQ(2, 2)$.

2 Proper partial geometries

Let us examine the known families of proper partial geometries for possible imbeddings.

First, we consider the class $\mathcal{S}(\mathcal{K})$. These geometries exist whenever there is a maximal arc $\mathcal{K}$ of degree d in a projective plane of order $q = dc$. The parameters are $s = d(c - 1)$, $t = c(d - 1)$, $\alpha = (c - 1)(d - 1)$. Substitution in Theorem 1.1 gives $d \leq 1 + c/(c - 1)^2$, which is satisfied only if $d = c = 2$. Then $\mathcal{P}$ is $GQ(2, 2)$, which has four embeddings, as we saw before. Note that the same conclusion holds for the dual geometries, since they belong to the same parameter family.

Next we consider the class $\mathcal{T}_2^*(\mathcal{K})$ with parameters $s = 2^h - 1$, $t = (2^h + 1)(2^m - 1)$, $\alpha = 2^m - 1$. They exist whenever m divides h and consist of all the points and a subset of the lines of $AG(3, 2^h)$, being all translates of the lines through the origin that correspond to a maximal arc $\mathcal{K}$ of degree 2^m in $PG(2, 2^h)$. So they all are embeddable. For the duals, all our necessary conditions are fulfilled, but we don't know whether any embeddings exists (except, of course, for the trivial case $m = h = 1$). For $m = 1$, $h = 2$ we get the open case $GQ(5, 3)$, that we mentioned earlier.

The partial geometries $PQ^+(4n-1, 2)$ were constructed by De Clerck, Dye and Thas [2] using a non-singular hyperbolic quadric Q in $PG(4n-1, 2)$ with a spread (i.e a partition of the point set of Q into maximal totally singular subspaces). The points of the partial geometry are the points of $PG(4n-1, 2)$ that are not on Q and the lines are the hyperplanes in the subspaces of the spread. A point x is on a line L if x lies in the polar space of L with respect to Q. The parameters of $PQ^+(4n-1, 2)$ are $s = 2^{2n-1} - 1$, $t = 2^{2n-1}$, $\alpha = 2^{2n-2}$.

We found, first by computer and later by hand, that for $n = 2$ this geometry indeed has an embedding. For an extensive discussion, see the next section.

Theorem 2.1 *The partial geometry $PQ^+(4n-1, 2)$ is embeddable in a Steiner 2-design if and only if $n \leq 2$.*

Proof First suppose the geometry is embeddable. Then the blocks of the embedding are cocliques of size $k = 2^{2n-1}$ in the point graph of $PQ^+(4n-1, 2)$, which is the orthogonality graph on the nonsingular points. The Gram matrix of the vectors spanning the points of any such coclique is $J - I$, which is nonsingular, and hence these vectors are linearly independent and their number cannot exceed the dimension of the space. That is, $2^{2n-1} \leq 4n$, so $n \leq 2$. The case $n = 1$ is trivial: $K_{3,3}$ can be extended to K_6. It remains to show embeddibility in case $n = 2$. That is, we have to construct a system of 8-cliques, one on each edge, for the nonorthogonality graph on the nonsingular points for $O_8^+(2)$. As follows: Pick a good system $\mathcal{O}$ of ovoids, one on each pair of nonorthogonal singular points, and pick a good totally singular 4-space V. For each ovoid O in $\mathcal{O}$ we find a unique point $O \cap V = \langle p \rangle$, and a *base* (basis consisting of 8 mutually nonorthogonal vectors) $B = \{a+p | \langle a \rangle \in O, \ a \neq p\}$. The set of 120 bases thus obtained is the required system of 8-cliques. For details on the choice of $\mathcal{O}$ and V (not any V will do), see the next section. $\square$

Recently, Mathon and Street [4] and De Clerck [1] have derived new partial geometries from $PQ^+(4n - 1, 2)$, but with the same parameters. We don't know if any of these admits an embedding. The non-existence argument for $n > 2$ doesn't work anymore, because these new geometries have other point graphs.

The parameter sets under consideration all meet the bound of Theorem 1.1, but the condition there is always fulfilled. For the related parameter sets $s = 2^{2m} - 1$, $t = 2^{2m}$ and $\alpha = 2^{2m-1}$, no geometry is known, but for $m > 1$ they may exist. The embedding however, can not exist by Theorem 1.1 (indeed, $f + g = v - 1$ is odd, so $2^{(2m-1)(f+g)}$ is not a square). For $n \neq 1$, the dual parameters never satisfy the conditions of Theorem 1.1.

The partial geometries $PQ^+(4n - 1, 3)$ (the construction method is due to Thas and is only known to work for $n = 1$) have parameters $s = 3^{2n-1} - 1$, $t = 3^{2n-1}$, $\alpha = 2 \cdot 3^{2n-1}$. These geometries and their duals have no embedding by Theorem 1.1.

Finally we consider the two known sporadic proper partial geometries. The one with parameters $s = 4$, $t = 17$ and $\alpha = 2$, constructed by the second author, does not satisfy the divisibility conditions, so has no embedding, but the dual may have one. The other one due to Van Lint and Schrijver with $s = t = 5$ and $\alpha = 2$ looks more interesting. An embedding would lead to a 2-$(81, 6, 1)$ design and such a design is not known to exist. By (incomplete) computer search we were able to extend quite far, but not far enough. Probably the embedding does not exist.

Remark The 120 points and the 120 blocks of the embedding of $PQ^+(7, 2)$ given in Theorem 2.1, form a (flag-transitive) partial linear space, with an incidence graph that is not a bipartite distance-regular graph of diameter 4 or 5, and yet, both the point and the block graph are primitive strongly regular graphs (in fact they are isomorphic). This seems to be a remarkable property. Examples with imprimitive strongly regular graphs are given by the elliptic semiplanes.

3   Triality, ovoids, spreads and bases

Let X be an 8-dimensional vector space over a field K, provided with a nondegenerate quadratic form Q of (maximal) Witt index 4. Let L be the collection of totally singular (t.s.) lines, and let Z_0, Z_1, Z_2 be the sets of singular points and of t.s. 4-spaces of the first and second kind, respectively. Put $Z = Z_0 \cup Z_1 \cup Z_2$. Natural incidence (symmetrized containment) defines a bipartite graph Γ on $Z \cup L$ with bipartition $\{Z, L\}$. This graph has automorphism group $G \simeq O_8^+(K).\mathrm{Sym}(3)$. The group G is transitive on Z and L and preserves $\{Z_0, Z_1, Z_2\}$. The subgroup $G_0 \simeq O_8^+(K)$ preserves the sets Z_0, Z_1, Z_2. The phenomenon that the three sets Z_0, Z_1, Z_2 can be permuted arbitrarily is called *triality*.

Let N_0 be the set of nonsingular points. We need to interpret these in terms of the graph Γ so that we can apply triality and also get sets N_1, N_2. One way of doing that is by representing a nonsingular point $\langle n \rangle$ by the reflection

$$r_{\langle n \rangle} \ : \ x \mapsto x - \frac{(x, n)}{Q(n)} n \ .$$

Let R_0 be this set of reflections. There is a 1-1 correspondence between N_0 and R_0. We have $R_0 \subset G$ and R_0 is closed under conjugation by G_0.2 $\simeq$ $PGO_8^+(K)$. It follows that we can find three sets R_0, R_1, R_2 of reflections under conjugation by G, where R_i consists of the reflections that preserve Z_i and interchange Z_j and Z_k for $\{i, j, k\} = \{0, 1, 2\}$.

Lemma 3.1 (cf. Tits [7]). *Let $r \in R_i$ and $s \in R_j$ with $i \neq j$. Then $(rs)^3 = 1$.*

Proof We may suppose $r \in R_0$, $s \in R_1$. Since $srsrs \in R_0$ it suffices to show that r and $srsrs$ fix the same singular points. Let $p \in Z_0$ be fixed by r, and put $W = sp$ and $V = rW$ so that $V \in Z_1$ and $W \in Z_2$. Now $\pi := V \cap W$ is a plane containing p and s fixes each line on p in π so that V and sV have at least a plane in common. But $sV \in Z_1$, so $V = sV$, i.e., $rsp = srsp$. □

So far the field was arbitrary, but from now on we take $K = \mathbf{F}_2$. The property that only holds in this case is: *If m, n are nonsingular vectors orthogonal to the t.s. plane π, then $(m, n) = 0$.* Indeed, $\pi^\perp$ is the union of three totally isotropic (t.i.) 4-spaces on π, of which two are t.s., so m and n are both contained in the third.

Now that $K = \mathbf{F}_2$, let us use $+$ between projective points instead of the spanning vectors, and write $\langle a \rangle + \langle b \rangle := \langle a + b \rangle$.

Let a *base* be a set of 8 mutually nonorthogonal nonsingular points. Let an *ovoid* be a set of 9 mutually nonorthogonal singular points. Let a *spread* be a set of 9 pairwise disjoint t.s. 4-spaces. All elements of a spread are of the same kind, and we talk about a *j-spread* when the spread is a subset of Z_j $(j = 1, 2)$. If we call ovoids 0-*spreads*, then i-spreads $(i = 0, 1, 2)$ correspond under triality.

If B is a base, then $b_0 := \sum_{b \in B} b$ is singular, and $O_B := \{b_0\} \cup \{b_0 + b | b \in B\}$ is an ovoid. Conversely, if O is an ovoid, and $p \in O$, then $B_{O,p} := \{a + p | a \in O, p \neq a\}$ is a base. Thus, we find a 9-1 correspondence between bases and ovoids.

Proposition 3.2 *Let S be a 1-spread. Then $\mathcal{O} := \{rS | r \in R_2\}$ is a system of ovoids, one on each pair of nonorthogonal singular points.*

These are the sets $\mathcal{O}$ called 'good' in the previous section.

Proof The numbers fit, so we have to show that no pair of noncollinear points is covered twice. Interchanging types 0 and 2, we have to show that no two disjoint 4-spaces W, W' are contained in both rS and $r'S$ for $r, r' \in R_0$. Let $r = r_m$ and $r' = r_n$. Then $m^\perp$ and $n^\perp$ meet W and W' in the same plane

π and π', respectively. Both m and n lie in the t.i. but not t.s. 4-spaces Y and Y' on π and π'. But then Y and Y' have the line l spanned by m and n in common, and l hits the disjoint planes π, π' in distinct points, so has at least 4 points, contradiction.   □

Proposition 3.3 *Take S and $\mathcal{O}$ as above, and fix an element $V \in S$. The set $\mathcal{B}$ of 120 bases $B_{O,p}$ with $O \in \mathcal{O}$ and $p = O \cap V$ is a system of 8-cliques, one on each edge, for the nonorthogonality graph on the 120 nonsingular points.*

Of course the existence of $\mathcal{B}$ is the whole point of this section (in fact, of this paper).

The above description does not show the symmetry between the 120 non-singular points and the 120 bases. A more symmetric description of the same configuration: Take a base B and put $O := O_B$. Let $\mathcal{S}_1 = \{rO | r \in R_2\}$ and $\mathcal{S}_2 = \{rO | r \in R_1\}$, and join $S \in \mathcal{S}_1$ to the 8 spreads $r_b S$ ($b \in B$) in $\mathcal{S}_2$.

Or, in terms of reflections: Take a base B and join $r \in R_2$ to the 8 reflections $r_b r r_b$ in R_1.

Proof We have to show that both descriptions are equivalent, and that they actually work. As to the former, interchanging types 0 and 1 we see that the spread S with fixed element V, the collection $\mathcal{O} = \{rS | r \in R_2\}$, an ovoid $O = rS$ in $\mathcal{O}$, the point $p = O \cap V$ and the set of 8 reflections fixing 7 points of O and interchanging p with the ninth point correspond to, respectively, the ovoid O with fixed element b_0, the collection $\mathcal{S}_1 = \{rO | r \in R_2\}$, a spread $S = rO$ in $\mathcal{S}_1$, the 4-space $rb_0 \in S$ containing b_0 and the set of 8 reflections $\{rr_b r | b \in B\}$. Since $rr_b r = r_b r r_b$, and $r_b O = O$, this shows that both descriptions are equivalent.

Since the bases by definition are 8-cliques in the nonorthogonality graph on the nonsingular points, and the numbers fit, we only have to check that no two bases $B_{O,p}$ have a pair in common, or, equivalently, that no two sets $\{rr_b r | b \in B\}$ and $\{sr_b s | b \in B\}$ have a pair in common (for $r, s \in R_2$). But if $rr_a r = sr_b s$ and $rr_c r = sr_d s$ ($a, b, c, d \in B$, $a \neq c$, $b \neq d$) then $r_a r r_a = r_b s r_b$, $r_c r r_c = r_d s r_d$, i.e., $r = r_a r_b s r_b r_a = r_a r_b r_d r_c r r_c r_d r_b r_a$. With $S = rO$ this means that $r_a r_b r_d r_c S = S$. But if a, b, c, d are all distinct, then $r_a r_b r_d r_c$ has order 5 (as is seen by its action on O) hence must fix some element $V \in S$. On the other hand, both $\langle a, b, c, d \rangle$ and $\langle a, b, c, d \rangle^{\perp}$ are elliptic quadrics, and $r_a r_b r_d r_c x = x + (x, c)c + (x, c+d)d + (x, b+d)b + (x, a+b)a$ while a, b, c, d and sums of two of them are not in Z_0, so when two of the inner products (x, c), $(x, c+d)$, $(x, b+d)$, $(x, a+b)$ vanish for $x \in V$, all do. But $V \cap \langle a, b, c, d \rangle^{\perp}$ does not contain a line, contradiction. If a, b, c, d are not all distinct, say $a = d$, then $r_a r_b r_d r_c = r_a r_b r_a r_c = r_e r_c$ for $e = r_a b = a + b$. But if $r_e r_c S = S$, then, since $r_e r_c$ fixes at least a line on each element of S, $r_e r_c$ must fix all elements of S. But $c \neq e$ and $r_e r_c x = x + (x, c)c + (x, e)e$, so we find that $(x, c) = (x, e)$ on each $V \in S$, so on all of X, so $c = e$, contradiction.   □

What about the automorphism group of these constructions? The 8 reflections r_b ($r \in B$) generate a Sym(9) and clearly this is the full stabilizer of O in G. The stabilizer of O in G_0 is Alt(9), and that is also the full group of the system $\mathcal{O}$. It follows that the full group of the system B is the stabilizer of b_0 in the previous, i.e., is Alt(8).

As mentioned earlier, De Clerck-Dye-Thas construct a partial geometry pg(7,8,4) on the nonsingular points by fixing a spread S and taking the sets of nonsingular points on the t.i. but not t.s. 4-spaces meeting some element of S in a plane. (This is the dual of the pg(8,7,4) obtained from $\mathcal{O}$. Indeed, the DDT system has point set R_0, and its lines are the planes in some element of S, where r is incident with π if $r\pi = \pi$. If we let S be a 1-spread, then these planes can be identified with the elements of Z_2 containing them, and interchanging types 0 and 2 we find the description of $\mathcal{O}$.)

If we join our system B to the set of lines of this partial geometry, we get a Steiner system $S(2, 8, 120)$ with automorphism group (at least) Alt(8), when both systems were constructed starting from the same spread S. (No doubt several nonisomorphic $S(2, 8, 120)$'s arise in this way, but we have not investigated the details. Several nonisomorphic $S(2, 8, 120)$'s were known already - obtained as the exterior lines and the points off a hyperoval in some projective plane of order 16, cf. [5].)

References

[1] F. De Clerck, New partial geometries derived from old ones, preprint (1996).

[2] F. De Clerck, R.H. Dye and J.A. Thas, An infinite class of partial geometries associated with the hyperbolic quadric in $PG(4n - 1, 2)$, *European J. Combin.* **1** (1980), 323-326.

[3] F. De Clerck and H. Van Maldeghem, Some classes of rank 2 geometries, Chapter 10 in Handbook of Incidence Geometries (F Buekenhout ed.), Elsevier, Amsterdam, 1995.

[4] R. Mathon and A.P. Street, Overlarge sets and partial geometries, J. Geometry, to appear.

[5] T. Penttila, G.F. Royle & M.K. Simpson, *Hyperovals in the known projective planes of order 16*, J. Combin. Designs **4** (1996) 59-65.

[6] S.E. Payne and J.A. Thas, *Finite Generalized Quadrangles*, Pitman, London, 1984.

[7] J. Tits, *Sur la trialité et les algèbres d'octaves*, Acad. Roy. Belg. Bull. Cl. Sci. (5) **44** (1958) 332-350. MR 21#2019

Andries E. Brouwer, Department of Mathematics, Eindhoven University of Technology, P.O. Box 513, 5600 MB Eindhoven, The Netherlands.
e-mail: aeb@cwi.nl

Willem H. Haemers, Department of Econometrics, Tilburg University, P.O. Box 90153, 5000 LE Tilburg, The Netherlands.
e-mail: haemers@kub.nl

Vladimir D. Tonchev, Department of Mathematical Sciences, Michigan Technological University, Houghton, Michigan 49931, USA.
e-mail: tonchev@mtu.edu

Finite geometry after Aschbacher's Theorem: PGL(n, q) from a Kleinian viewpoint

Peter J. Cameron

Abstract

Studying the geometry of a group G leads us to questions about its maximal subgroups and primitive permutation representations (the G-invariant relations and similar structures, the base size, recognition problems, and so on). Taking the point of view that finite projective geometry is the geometry of the groups PGL(n, q), Aschbacher's theorem gives us eight natural families of geometric objects, with greater or smaller degrees of familiarity. This paper presents some speculations on how the subject could develop from this point of view.

1 Introduction

The worldwide fellowship of finite geometers has some features in common with the Pythagorean brotherhood (as described by Jean Doyen at the conference). I hope that I don't suffer the fate of Hippasos for revealing the secret that, in the last decade or so, a number of initiates have questioned the direction of future research in the subject. If Pythagoras will not inspire me, I can invoke the help of two mathematicians of more recent times, whose names are in my title.

The central object of finite geometry is the projective space PG($n - 1, q$). Now, projective geometry has many different meanings. A topologist's real projective plane is the simplest closed non-orientable 2-manifold, obtained from the unit square by identifying each pair of opposite sides in opposite senses. A computer scientist uses it for the practical business of putting graphics on a screen.

Even to a finite geometer, there are several aspects. It can be regarded as a lattice, whose elements are the subspaces, ordered by inclusion. It is a building, whose chambers are the maximal flags of subspaces, two chambers i-adjacent if they agree in all dimensions except i. It may be a geometry in Buekenhout's sense, whose varieties are all the subspaces, and incidence is symmetrised inclusion. Probably, we are more used to thinking of it as a finite

43

set of points, equipped with various distinguished subsets called 'subspaces'. The translations between one viewpoint and another are not difficult; but choice of viewpoint influences how we think about the projective geometry, and how we choose to generalise it.

For example, the chamber system approach fits most naturally into the scheme of buildings, BN-pairs, algebraic groups, and so on. On the other hand, from the point of view of complexity, note that $PG(n-1, q)$ has roughly q^n points, $q^{n^2/4}$ subspaces of all dimensions, and $q^{n^2/2}$ maximal flags.

The theme of Felix Klein's *Erlanger Programm* is that geometry comes from groups. If we are given the group of a geometry, then interesting geometrical configurations are those stabilised by interesting subgroups of the group.

In this article, I want to draw attention to some implications of this viewpoint for the geometry $PG(n-1, q)$. We know that the full collineation group of the geometry is $P\Gamma L(n, q)$; for most purposes, we can work in the subgroup $PGL(n, q)$, or even in the covering matrix group $GL(n, q)$. The subgroups of these groups are now much better understood, and I want to look at the implications of this understanding for a Kleinian view of the geometry. Very little here is original, but I hope to show that this point of view leads to many interesting questions.

The main topics of geometry can be regarded as *axiomatisation* (or *recognition*), and the study of *configurations* or subsets. There is no sharp dividing line between the two, and many questions blend from one to the other.

For further information on permutation groups, see Wielandt [35] or Cameron [6]; for classical groups, Artin [1] or Dieudonné [11], or Kleidman and Liebeck [20] for their maximal subgroups; and for the finite simple groups, Gorenstein [14] or the ATLAS [9], or Carter [8] for the Lie-theoretic viewpoint.

2   Aschbacher's Theorem

Since the Classification of Finite Simple Groups was announced in 1980 (see Gorenstein [14]), one of the main areas of interest in finite group theory has been listing the maximal subgroups of almost simple groups. There are several sources for this interest. Some are considered later in this paper. One of the principal ones is the study of primitive permutation groups, along the lines suggested by the O'Nan–Scott Theorem [30]. This shows that the classification of primitive permutation groups can be 'reduced' (modulo various 'extension' problems) to two types: those groups which have an elementary abelian regular normal subgroup V (an n-dimensional vector space over $GF(p)$, where p is prime), in which the stabiliser of the origin is a primitive irreducible linear subgroup of $GL(n, p)$; and those groups which are almost simple, in which the point stabiliser is a maximal subgroup. This meant that it was necessary to study maximal subgroups of almost simple groups, and primitive irreducible linear groups.

Both of these problems point towards Aschbacher's Theorem [2]. Since the Classification, we know that 'most' simple groups are classical groups, of which PSL(n, q) is the prototype. Also, the first step towards finding all linear groups is to find the maximal ones.

The O'Nan–Scott Theorem was originally not about primitive permutation groups, but about maximal subgroups of symmetric and alternating groups. If such a subgroup is intransitive or imprimitive, then it is the full stabiliser of a subset or a partition, and these are easily described; so it is necessary only to examine the primitive maximal subgroups. Aschbacher's Theorem does a similar job for classical groups, though I will concentrate on the general linear groups.

The O'Nan–Scott Theorem states that a maximal subgroup of S_n or A_n either belongs to one of five easily described classes of 'large' subgroups (stabilisers of subsets or partitions, wreath products of symmetric groups in the product action, diagonal groups, and affine groups), or else is almost simple. Aschbacher's Theorem has a similar form. Though it applies to all types of classical groups, I will consider only the case of the (projective) linear groups. First, we describe the classes of 'large' subgroups of $G = \mathrm{PGL}(n, q)$ which occur. Let V be the underlying n-dimensional vector space over $\mathrm{GF}(q)$.

$\mathcal{C}_1$: Stabilisers of subspaces of V.

$\mathcal{C}_2$: Stabilisers of direct sum decompositions of V into subspaces of the same dimension.

$\mathcal{C}_3$: Stabilisers of extension fields of $\mathrm{GF}(q)$ of prime degree.

$\mathcal{C}_4$: Stabilisers of tensor product decompositions $V = V_1 \otimes V_2$.

$\mathcal{C}_5$: Linear groups over subfields of prime index.

$\mathcal{C}_6$: Normalisers of r-groups of symplectic type, where r is a prime different from p.

$\mathcal{C}_7$: Stabilisers of tensor product decompositions $V = \otimes_{i=1}^{t} V_i$, where the V_i all have the same dimension.

$\mathcal{C}_8$: Classical groups.

Theorem 2.1 (Aschbacher's Theorem) *Let H be a subgroup of $\mathrm{PGL}(n, q)$, not containing $\mathrm{PSL}(n, q)$. Then either*

(a) H is contained in a member of one of the classes C_1–C_8; or

(b) H is almost simple and is induced by an absolutely irreducible subgroup modulo scalars.

At first glance, this appears a very powerful result. On closer inspection, the theorem suggest at least the strategy of the proof. For example, if G is reducible, it fixes a subspace; if it is irreducible but not absolutely irreducible, then by Schur's Lemma it fixes an extension field; if it has a non-trivial normal subgroup, then Clifford's Theorem indicates how to decompose the space.

It next appears that the weakness of the theorem is that all the difficulties are swept into the ragbag case (b). Even such natural examples as $G = \mathrm{GL}(m, q)/\{\pm 1\}$, embedded in $\mathrm{GL}(n, q)$ (for $n = \binom{m}{2}$) by the action of $\mathrm{GL}(m, q)$ on the exterior square of its natural module, falls under case (b) for $n > 4$. (For $n = 4$, G preserves the Klein quadric, and is in C_8.) The gain is that, from the Classification of Finite Simple Groups [14], we can list the candidates for G; all we have to do is to determine their absolutely irreducible representations over $\mathrm{GF}(q)$. There remains the question of the maximality of the various subgroups; this is dealt with by Kleidman and Liebeck [20].

The classification of finite simple groups allows very strong general assertions to be made about the groups in case (b). For example, Liebeck [24] showed:

Theorem 2.2 Let H fall under case (b) in Aschbacher's Theorem. Then either H is S_m or A_m, with $m = n+1$ or $m = n+2$, or H has order at most q^{3n}.

In this theorem, the symmetric or alternating group acts on the set of words with coordinate sum zero, or (if the characteristic of the field divides m) the quotient of this space modulo constant words. For even moderate values of n, the bound q^{3n} is very small compared to the order of $\mathrm{PGL}(n, q)$ (approximately cq^{n^2-1}).

The knowledge of maximal subgroups of the general linear group given by Aschbacher's Theorem is important in other areas besides geometry. One of these is the computational investigation of matrix groups. The number of bits of information in a 100×100 matrix over $\mathrm{GF}(2)$ is approximately the same as for a permutation on 1146 points (this is just the assertion that $\mathrm{GL}(100, 2)$ has approximately the same order as S_{1146}). Permutation groups of this degree are now routinely handled by computer, and we might expect that groups of 100×100 matrices would be as easy. But the smallest permutation representation of $\mathrm{GL}(100, 2)$ has degree $2^{100} - 1 \approx 1.268 \times 10^{30}$, far too large for any foreseeable computer. So another method must be found.

One basic question is to decide whether a given set of $n \times n$ matrices over GF(q) generates at least SL(n, q). This will be the case if and only if there is no maximal subgroup which contains all the given matrices. To test this, the first step is to decide, for each of the Aschbacher classes, whether the matrices preserve a structure in that class. A classic result of this type is Parker's 'meat-axe' algorithm [28], which decides whether or not such a group is reducible (that is, fixes a subspace, and hence lies in a subgroup in the class $\mathcal{C}_1$). Holt, Leedham-Green, O'Brien and Rees [18], [23] are currently developing similar tests for the other classes. There remains the possibility that the group comes under alternative (b) in Aschbacher's Theorem. Since such groups are 'small', we might hope to handle them directly.

3 PGL(n, q) from a Kleinian viewpoint

Suppose we take the view that finite projective geometry is the geometry of the group PGL(n, q). What should we study? It is a truism that geometry is what geometers do; so I shall be descriptive rather than prescriptive, and suggest some properties which have been studied in various cases and may be interesting more generally.

Geometric objects will be elements of sets affording permutation representations of the group. Of course, any permutation representation is the disjoint union of transitive ones, so we should consider these first. The transitive representations (up to permutation isomorphism) correspond to conjugacy classes of subgroups of the group. However, there are still far too many of these, so we must select 'interesting' ones. Natural criteria for these include primitivity, small degree, and small permutation rank (or high degree of transitivity, if possible).

If the action of G on X is imprimitive, let Y be a system of imprimitivity. Then (as geometric objects) a member of X can be regarded as a member of Y (the block containing it) together with some additional structure or information (to specify it in the block). To keep individual objects as simple as possible, we first discard this structure. Now the primitivity of G on Y is equivalent to the maximality of the corresponding subgroup (the stabiliser of a point in Y).

For a simple example, consider the symmetric group S_n, acting on the set of ordered pairs (i, j) with $1 \le i, j \le n$ and $i \ne j$. The stabiliser of (i, j) is contained in the stabiliser of the unordered pair $\{i, j\}$; ands (i, j) is obtained just by 'ordering' $\{i, j\}$ in one of the two possible ways.

The other desiderata are more for convenience. We choose actions of small degree (corresponding to large maximal subgroups) in order not to have too many objects to look at; and actions of small rank in order that the number of different relationships in which two objects may stand is not too large. (The minimal G-invariant binary relations are the G-orbits on Ω^2, and their number is the rank of G.)

Of course, these conditions are not identical, but tend to push in the same direction. The transitive representation of smallest degree of a simple group is primitive; small degree and small rank often go together (but not always: the Higman–Sims simple group acts with rank 3 on 100 points, but 2-transitively on 176 points).

Given a collection of interesting transitive representations of G, we look for the binary (or higher-order) relations in each set preserved by G, and whether there are structures with geometric significance. For example, an invariant binary relation may be the collinearity graph of a partial linear space with lines of size greater than 2. If so, then several more questions arise: Is this partial linear space embeddable in a projective space (as a set of points together with all lines contained in this set)? Is it of some familiar type, such as a generalised polygon or a net? We can also ask whether one of the G-invariant graphs is distance-transitive. In more combinatorial terms, properties such as diameter, girth, connectivity, and expansion properties of these graphs can be examined.

Also, we may consider how many objects are required as a 'basis' for the space, so that their pointwise stabiliser in G is trivial. In principle at least, any element of the space is uniquely specified by its G-invariant relations to the basis elements. Of course, the simplest case is a vector space V, where a base for $\mathrm{GL}(V)$ is a basis for V in the usual sense, and any vector is uniquely a linear combination of basis vectors. For another example, consider the projective line over $\mathrm{GF}(q)$. The cross-ratio provides not only a description of the orbits of $\mathrm{PGL}(2, q)$ on 4-tuples, but also an indexing of the points relative to the base $(\infty, 0, 1)$.

We further consider the binary 'incidence' relations between elements of sets affording different representations of G. We can ask whether the resulting incidence structure belongs to one of the many special classes which have been studied (designs of various kinds, partial geometries, and so on).

If we take several permutation representations and study the binary relations, we obtain Buekenhout geometries of various kinds. It would also be possible to consider relations of higher arity, though I don't have any interesting examples.

The 'classical' example of the interplay between several types of maximal subgroups is the construction of the Witt designs and the binary Golay code from $\mathrm{PGL}(3, 4)$ (see Lüneburg [26]). The points of the 5-(24, 8, 1) large Witt design are the points of the projective plane together with three extra points (which are the cosets of $\mathrm{PSL}(3, 4)$ in $\mathrm{PGL}(3, 4)$). The blocks are the lines, hyperovals, Baer subplanes, and symmetric differences of two lines. Unitals give dodecads in the Golay code and hence carry the 5-(12, 6, 1) small Witt design.

A further problem concerns axiomatisation. Given one or several sets, and some relations or other structures on these sets, how do we recognise that we are looking at a geometry for a specified group?

For PGL(n, q), we take the interesting permutation representations to be those corresponding to subgroups in the Aschbacher classes. Most of our desirable properties will hold, and the structures should behave relatively uniformly for different n and q.

Needless to say, the same kind of analysis can be applied to all the classical groups (for which Aschbacher's Theorem holds), and also to the exceptional groups of Lie type, for which similar theorems have been proved. What about the symmetric and alternating groups? As noted earlier, the analogue of Aschbacher's theorem for these groups is the O'Nan–Scott theorem, which gives five classes of large maximal subgroups. The geometry of the first two of these (the maximal intransitive or imprimitive groups) is, from a different viewpoint, the combinatorics of subsets and partitions of a finite set, which has an extensive literature. Perhaps the other three classes (product actions of wreath products, diagonal groups, and affine groups) could be given a similar treatment.

The remainder of this paper, apart from the last section, consists of examples of the geometry of some of the Aschbacher classes taken from the literature.

4   Reducible groups

Classical projective geometry is more-or-less the geometry of the maximal reducible subgroups of PGL(n, q), since these are the stabilisers of subspaces of the projective space (points, lines, planes, ..., hyperplanes).

With finitely many exceptions, the faithful permutation representation of smallest degree is that on the points (or hyperplanes) of the projective space. This result goes back to the roots of the subject: one of Galois' discoveries [13] is that, among the groups PSL$(2, p)$ for p prime, the smallest-degree representation is the one on the $p + 1$ points except when $p = 2, 3, 5, 7, 11$. (In each of these cases, there is a representation of degree p, though it is not faithful for $p < 5$.) The smallest degrees of permutation representations of all classical groups were determined by Cooperstein [10].

More generally, Kantor [19] showed that a maximal subgroup of $G = $ PGL(n, q) with order greater than roughly the square root of $|G|$ is necessarily reducible. These results of Cooperstein and Kantor can be seen as precursors of Liebeck's theorem, asserting that subgroups of PGL(n, q) are either 'known' or small.

The subspace stabilisers play a very important role in the point of view which regards $\mathrm{PSL}(n, q)$ as a group of Lie type: they are the maximal parabolic subgroups. A subgroup is parabolic if it contains the *Borel subgroup* B of G (the normaliser of a Sylow p-subgroup, where p is the characteristic of $\mathrm{GF}(q)$). A Borel subgroup is the stabiliser of a maximal flag (set of mutually incident subspaces), and the parabolic subgroups containing it are precisely the stabilisers of sub-flags. There are two different but related points of view here. Diagram geometry takes the basic objects or 'varieties' to be all the subspaces (corresponding to the maximal parabolic subgroups), with two subspaces incident if one contains the other, that is, if their intersection is parabolic. The theory of buildings takes the basic objects as the chambers or maximal flags (corresponding to the Borel subgroup), and imposes on the set of chambers a number of partitions, one for each subset of the set of types. The finest non-trivial partitions correspond to the minimal parabolic subgroups. See Carter [8], or the chapters by Scharlau and Cohen in the *Handbook of Incidence Geometry* [3].

A general fact about parabolic representations of the groups of Lie type is that the behaviour of the G-invariant binary relations is independent of the field order. If a group G acts transitively on a set Ω, the orbits of G on Ω^2 (the minimal G-invariant binary relations) are in one-to-one correspondence with the double cosets HxH, where H is the stabiliser of a point. If G is a group of Lie type (more generally, a group with a BN-pair), these double cosets correspond to the double cosets in the corresponding parabolic representation of the Weyl group. Moreover, this correspondence preserves the decomposition into orbits of the composition of two such minimal relations. Also, the same holds for relations between two parabolic representations, using double cosets HxK, where H and K are the two parabolic subgroups.

These facts are easily seen directly in the case $G = \mathrm{PGL}(n, q)$, where the Weyl group is the symmetric group S_n. Corresponding to the representation of G on k-spaces is the representation of S_n on k-sets. The minimal binary relations for G have the form $\{(U_1, U_2) : \dim(U_1 \cap U_2) = i\}$, for $0 \le i \le k$; correspondingly, in the symmetric group we have relations $\{(X_1, X_2) : |X_1 \cap X_2| = i\}$. And so on.

Of course, we cannot expect anything so simple to occur in the other Aschbacher classes. If the subgroup H of G has order smaller than the square root of G (as do almost all the irreducible subgroups of $\mathrm{PGL}(n, q)$), then in the representation on the cosets of H, the rank (the number of H-orbits) is at least $|G|/|H|^2$; typically this will be an increasing function of q.

The group $\mathrm{PGL}(n, q)$ is 2-transitive on the set of points. If $n > 2$, its smallest orbit on triples is the ternary relation of *collinearity*; a *line* consists of two points together with all points collinear with them, and a *subspace* is a set of points containing the line through any two of its points. These subspaces are of course just those corresponding to the other subgroups in C_1. If, instead, we take the family of k-subspaces, for $0 < k < n - 2$, then the group is not 2-transitive, and it preserves a partial linear space, whose maximal singular subspaces give us the $(k - 1)$-subspaces and the $(k + 1)$-subspaces. So we can work up and down through the collection of subspaces.

Axiomatisation in the case of points and lines is classical; for other dimensions, see Sprague [31] and Tallini [32]. Also, the geometry of k-spaces can be embedded in $\bigwedge^k V$, the k-spaces $U = \langle v_1, \ldots, v_k \rangle$ being represented by the 1-space $\langle v_1 \wedge \cdots \wedge v_k \rangle$: the lines of the partial linear space are precisely the lines of the embedding projective space which are contained within the set of such points (any other line of the projective space meets the set in at most two points).

Alternatively, if we take two of these permutation representations (corresponding to subspaces of different dimension), then the smallest non-empty G-invariant relation between then is the usual incidence relation.

If $n = 2$, these methods give us no non-trivial geometry, since there is only one class of maximal reducible subgroups, and G is 3-transitive on the points of the projective line. The G-invariant quaternary relations are specified by prescribing values of the cross-ratio of the four points.

As far as I know, the size of a minimal base in each of these permutation representations has not been completely determined. For the action on points, everything is known: the smallest base has size n if $q = 2$, or $n + 1$ otherwise. For, if we take n points spanned by the vectors of a basis, their stabiliser consists of diagonal matrices. This is trivial for $q = 2$, since the only non-zero scalar is 1. For larger fields, we must also fix a point in general position.

At the other extreme, if $n = 2m$ is even, then there is a base of size 5 for the action on m-spaces, consisting of the row spaces of the $m \times 2m$ matrices (in block form) $(O\ I)$, $(I\ O)$, $(I\ I)$, $(I\ A)$, and $(I\ B)$, where O and I are the zero and identity matrices, and A and B generate an absolutely irreducible subgroup of $\mathrm{GL}(m, q)$. Now fixing the first and second spaces forces a matrix to be block-diagonal, with $m \times m$ blocks X and Y (say); fixing the third space forces $X = Y$; and fixing the fourth and fifth forces X to commute with A and B, and hence to be a scalar (so trivial in $\mathrm{PGL}(n, q)$).

In fact, we can say more:

Theorem 4.1 *(a) For fixed n and $q \to \infty$, almost all $(n+1)$-tuples of points are bases for $\mathrm{PGL}(n, q)$.*

(b) For fixed m and $q \to \infty$, almost all 5-tuples of m-spaces are bases for $\mathrm{PGL}(2m, q)$.

Problem. Complete these results for subspaces of other dimensions.

5   Direct sums and tensor products

The stabiliser of a direct sum decomposition of the vector space V into subspaces $V_1, \ldots, V_t$ (all of the same dimension) permutes these subspaces among themselves. It is irreducible but, by definition, imprimitive.

There is an interesting connection with permutation groups. One case in the O'Nan–Scott theorem consists of primitive groups G which have an abelian regular normal subgroup V, which is necessarily an elementary abelian p-group for some prime p. We can identify the set of points with V so that V acts on itself by translation. The stabiliser of the zero vector is then a linear group $H = G_0$. Any H-invariant subspace would be a block of imprimitivity for G; so the primitivity of G translates into irreducibility of H. Furthermore, if H is an imprimitive linear group (that is, if it preserves a direct sum decomposition of V), then G is contained in a wreath product with the product action, another case in the O'Nan–Scott theorem. (This means that G preserves a 'product structure' or 'Hamming graph'.) So it is often permissible to deal with this possibility in a different part of the analysis, and to assume that H is primitive.

Let G be the stabiliser of a direct sum decomposition. Then G has a normal subgroup which fixes all the subspaces in the decomposition; the factor group is a subgroup of the symmetric group S_t (permuting the subspaces). Of course, the subspaces $V_1, \ldots, V_t$ are independent (they span their direct sum).

In the case where $\dim(V_i) = 1$, the group G is important from the Lie-theoretic viewpoint: it is the normaliser of a maximal split torus T (induced by the group of diagonal matrices), and the quotient is the *Weyl group* of $\mathrm{PGL}(n, q)$, which is the symmetric group S_n in this case; see Carter [8].

In general, in the projective space $\mathrm{PG}(mt - 1, q)$ (where $\dim(V_i) = m$), G fixes a set S of points which is the union of t independent subspaces $\mathrm{PG}(m - 1, q)$. Some information about the relations between these objects, of the type described by Eisfeld for subfield geometries (see Section 7), could probably be established.

The problem of recognising when a matrix group is *imprimitive* (that is, preserves a direct sum decomposition of this type) has been solved by Holt *et al.* [18].

Questions about the rank, suborbits, or minimum base size for the permutation representations of this kind seem to be open.

Tensor products are inherently more complicated than direct sums. The direct sum $U \oplus W$ has subspaces naturally isomorphic to U and W, and can be recognised by the familiar internal characterisation of direct sums. In terms of the projective space this reads: any point outside the union of the two subspaces lies on a unique line meeting both of them. However, $V = U \otimes W$ does not contain copies of U and W invariant under the corresponding group.

One approach to the geometry, which has been exploited by Leedham-Green and O'Brien [22], [23] for matrix group algorithms, works as follows. Let $k = \dim(U)$, $l = \dim(W)$ (so that $\dim(V) = kl$). Now consider the subspaces $U \otimes X$, as X ranges over all subspaces of W. As partially ordered set, this family of subspaces is isomorphic to the subspace lattice of the projective geometry $\mathrm{PG}(l - 1, q)$ corresponding to W. Its atoms are $(k - 1)$-subspaces of $\mathrm{PG}(kl - 1, q)$. Leedham-Green and O'Brien show how, given generators of a group preserving such a tensor product, this lattice can be efficiently computed.

The smallest example of this case is given by the ruled (hyperbolic) quadric in $\mathrm{PG}(3, q)$. The 4-dimensional vector space is expressed as $U \otimes W$, where $\dim(U) = \dim(W) = 2$; the points on the quadric are those spanned by the rank 1 tensors $u \otimes w$, for $u \in U$ and $w \in W$.

The other tensor case, that where $V = \otimes_{i=1}^{t} V_i$, is even more recondite. The group G fixing such a decomposition has a normal subgroup N preserving the tensor factors, the quotient group being S_t. Geometrically, N preserves t sublattices of the projective space, each isomorphic to the projective space based on V_i.

6 Classical groups

These are the subgroups of G which preserve non-degenerate alternating or Hermitian forms or non-singular quadratic forms (the *symplectic, unitary* and *orthogonal* groups). Because we are considering only maximal subgroups, some cases do not need to be considered. In characteristic 2, an orthogonal group of odd degree fixes a point, namely the radical of the form, while an orthogonal group of even degree fixes the symplectic form obtained by polarisation and so is contained in the symplectic group.

This example illustrates the earlier remark about imprimitive permutation representations:

- if n is odd, an orthogonal geometry is specified by giving a point (the radical) together with a symplectic geometry on the quotient space;

- if n is even, an orthogonal geometry is specified by saying which of the totally isotropic subspaces for the symplectic geometry are totally singular.

The geometry of the classical groups is well-known, see for example the

Handbook of Incidence Geometry [3]. The usual description takes the varieties of the incidence geometry to be the subspaces which are totally isotropic or totally singular with respect to the form. There are familiar axiomatisations of these *polar spaces* (Tits, Buekenhout and Shult). No more need be said here.

The membership test for classical groups is reasonably straightforward. Suppose that the group G generated by a set of $n \times n$ matrices over $\mathrm{GF}(q)$ is irreducible. Then G is contained in the general symplectic group if and only if it fixes a 1-dimensional subspace in the space of alternating bilinear forms. (If there is a fixed 1-space of forms, the common radical of its members is fixed by G, and so is zero, since G is irreducible.) This space is the dual of $V \wedge V$, where $V = \mathrm{GF}(q)^n$, so the appropriate representation may be constructed and decomposed with standard computational tools. To test membership in orthogonal or unitary groups, we test similarly the spaces of symmetric bilinear or Hermitian forms.

The type of question which arises from our point of view is the following. Given two symplectic forms (say) on a vector space, what is the relationship between them? For example, what are the possible structures of the geometry of subspaces which are totally singular for both forms? The question on base size reads: What is the smallest set of symplectic forms with the property that the only linear maps preserving all of them (up to scalar factors) are scalar transformations?

The last question was investigated by Tracey Maund [27] in her (unpublished) thesis. She considered one particular relationship between two forms. Let $n = 2m$, and take a 2-dimensional vector space over $\mathrm{GF}(q^m)$. On restriction of scalars, this becomes an n-dimensional vector space over $\mathrm{GF}(q)$. Now all symplectic forms on $V(2, q^m)$ are equivalent under scalar multiplication. Take two such forms b and λb, where λ generates $\mathrm{GF}(q^m)$ over $\mathrm{GF}(q)$. Let Tr denote the trace map from $\mathrm{GF}(q^m)$ to $\mathrm{GF}(q)$. Then $B_1 = \mathrm{Tr}(b)$ and $B_2 = \mathrm{Tr}(\lambda b)$ are symplectic forms on $V(2m, q)$. The subgroup of $\mathrm{GL}(2m, q)$ fixing both these forms up to scalar factors is $\Gamma\mathrm{L}(2, q^m)$ (the extension of $\mathrm{GL}(2, q^m)$ by the group of automorphisms of $\mathrm{GF}(q^m)$ over $\mathrm{GF}(q)$). The maximal subspaces which are totally isotropic for both B_1 and B_2 form a Desarguesian spread in $V(2m, q)$. If n is large enough, it is possible to find a third symplectic form such that only scalars fix the three forms, so that there is a base of size 3 for this representation of $\mathrm{PGL}(n, q)$.

Problem. Do almost all triples constitute bases (say, as $q \to \infty$ and fixed n)?

The geometry of symplectic forms can be embedded in a projective space over $\mathrm{GF}(q)$, as follows. An alternating bilinear form on $V = \mathrm{GF}(q)^{2m}$ can be regarded as a linear map from $V \wedge V$ to $\mathrm{GF}(q)$. We dualise, and consider the vector space $V \wedge V$. Now the duals of non-degenerate forms are the elements of rank m, where the rank of an element of $V \wedge V$ is the smallest number of vectors of the form $v_1 \wedge v_2$ of which it is the sum. Since we are allowed to multiply forms by scalar factors, the geometry associated with this representation consists of all points of the projective space spanned by vectors of rank m in $V \wedge V$. This vector space also represents the 2-spaces of V, as the points spanned by vectors of rank 1.

One well-known instance is the case $m = 2$, where these two sets between them make up the whole projective space $PG(5, q)$. The points corresponding to rank 1 vectors comprise the Klein quadric, and those corresponding to symplectic forms to the points outside the quadric.

Another case which has been studied is the case $q = 2$, in connection with coding theory. Here, sets of alternating forms of maximum cardinality subject to the condition that the difference between any two is non-degenerate give rise to *Kerdock codes*. Moreover, the quadratic forms on $V(2m, 2)$ form a vector space which has the space of alternating bilinear forms as a homomorphic image. Much of the geometry and combinatorics of quadratic forms can be seen in this space; see Cameron and Seidel [7].

(The maximum cardinality of such a set on a $2m$-dimensional vector space is 2^{2m-1}. The Kerdock code consists of all the quadratic forms which polarise to bilinear forms in X. It is necessarily non-linear: indeed, a subspace of alternating forms all of whose non-zero members are non-degenerate has dimension at most m, as can easily be seen by applying the Chevalley–Warning theorem to the Pfaffian function on the space of skew-symmetric matrices representing the forms.)

Similar questions can be posed for the other types of classical groups, but have not received much attention yet from geometers. Gow [15] considered the permutation representation on non-degenerate Hermitian forms from a character-theoretic point of view, and showed that it is multiplicity-free.

7   Subfields and superfields

If $q = r^t$, the subgroup $PGL(n, r)$ of $PGL(n, q)$ acts on a projective geometry $PG(n - 1, r)$ 'embedded' in $PG(n - 1, q)$. Its subspaces are those spanned by vectors whose coordinates lie in the subfield $GF(r)$ of $GF(q)$. In the case $q = r^2$, this is a *Baer subgeometry*. In general, a line of the large geometry belongs to the subgeometry if two of its points do; so any line meets the subgeometry in $0, 1$ or $r + 1$ points. The Baer case is characterised by the fact that the case 0 does not occur; that is, every line meets the subgeometry which thus forms a minimal line-blocking set).

How are two such subgeometries related? Some relations have been studied. For example, if n and t are coprime (in particular, if n is odd and

$q = r^2$), a Singer cycle C for $\mathrm{PG}(n-1, q)$(a cyclic subgroup of $\mathrm{PGL}(n, q)$ acting transitively on the points) contains a subgroup which is a Singer cycle for $\mathrm{PG}(n-1, r)$; the images of this subgeometry partition the points of $\mathrm{PG}(n-1, q)$.

Ueberberg [33], [34] showed that, if two Baer subplanes Π_1 ands Π_2 of $\mathrm{PG}(2, r^2)$ are disjoint, then there is a unique Singer cycle for which the corresponding partition contains Π_1 and Π_2. So the geometry of Baer subplanes has the structure of a partial linear space with $r^2 - r + 1$ points per line. For $r = 2$, Ueberberg showed that conjugates of the Frobenius automorphism can be used to describe 'planes' in this geometry, each plane being a 3×3 grid. This rank 3 Buekenhout geometry, with diagram $C_2.c$, is a subgeometry of Ronan's geometry with the more general diagram $C_2.L$, consisting of the octads, trios and sextets related to the Golay code [29], [5]. Hence, Ueberberg's partial linear space is embeddable in the projective space $\mathrm{PG}(10, 2)$, corresponding to the (extended) Golay code modulo the all-1 vector.

Note that a Singer cycle belongs to the Aschbacher class $\mathcal{C}_3$ (superfield stabiliser). This is an example of a relationship between subfield and superfield geometries.

The relationships between these subfield geometries have been considered more generally by Jörg Eisfeld [12]. Let $q = r^t$, and let P be a $\mathrm{PG}(n-1, r)$ subgeometry of $\mathrm{PG}(n-1, q)$. For any point p, we define its 'distance' $d(p, P)$ from P to be the (geometric) dimension of the smallest subspace of P (that is, spanned by points of P) containing p. Then $0 \le d(p, P) \le t - 1$; and if $d(p, P) = t - 1$, the subspace in question is unique. Eisfeld constructed a 'large' set $\mathcal{M}$ of $\mathrm{PG}(n-1, q)$ subgeometries of $\mathrm{PG}(n-1, q)$ with the property that, for $P, P' \in \mathcal{M}$, the distance $d(p, P)$ is the same for all $p \in P'$. In particular, if t and n are coprime, the subgeometries of $\mathcal{M}$ partition the points of $\mathrm{PG}(n-1, q)$, and we obtain the Singer cycle partitions as before. At the other extreme, if $t = n$, there is a subset of $q - 1$ subgeometries which are pairwise at maximum distance $t - 1$.

Also, in the Baer case ($t = 2$), Gow [15] showed that the permutation character is multiplicity-free.

If $n = kl$, then $\mathrm{GF}(q^l)^k$ is identified with $\mathrm{GF}(q)^n$ by restricting scalars, and so $\mathrm{GL}(k, q^l)$ is a subgroup of $\mathrm{GL}(n, q)$. Moreover, the Galois automorphisms of $\mathrm{GF}(q^l)$ over $\mathrm{GF}(q)$ are also $\mathrm{GF}(q)$-linear and normalise $\mathrm{GL}(k, q^l)$. The points, lines, ... of $\mathrm{PG}(k-1, q^l)$ become l-spaces, $2l$-spaces, ... of the geometry over $\mathrm{GF}(q)$. In particular, the points give rise to a Desarguesian spread of l-spaces in $\mathrm{GF}(q)^n$.

8 Symplectic-type r-groups

I have little to say about this class of subgroups. But I would like to draw attention to a different but related context in which they arose recently, a surprising development in coding theory.

The Lee metric on $\mathbb{Z}_4$, defined by

$$d(x,y) = \min\{x - y, y - x\}$$

(where the arithmetic is mod 4) is isometric to the Hamming metric on $\mathbb{Z}_2^2$. Hence there is an isometry, the *Gray map*, from $\mathbb{Z}_4^n$ to $\mathbb{Z}_2^{2n}$. Hammons *et al.* [17] showed that some famous non-linear binary codes are the images of $\mathbb{Z}_4$-linear codes under the (non-linear) Gray map. These include the Kerdock codes discussed briefly in the last section, and also a modified version of the Preparata codes. Subsequently, Calderbank *et al.* [4] showed that the Gray map is closely connected with the real and complex geometry associated with extraspecial 2-groups. It may be that there are interesting finite analogues of this real and complex geometry.

9 Appendix: Very small base groups

Recall that, if G is a permutation group acting on a set Ω, a *base* for G is a sequence of points whose pointwise stabiliser is the identity. Apart from their role in geometry, bases are very important in computational permutation group theory, and the question of the minimal base size for a group has been considered by many authors. Often, efficient algorithms must treat groups differently according as they have 'large' or 'small' bases.

If G has a base of size c, then its images under distinct group elements are distinct, and so $|G| \le n^c$, where n is the degree of G. There is a feeling that primitive groups of polynomially bounded order should have bases of bounded size. This has been proved only in special cases, most of which exclude groups like $\mathrm{PGL}(n, q)$. The only general result I know, dealing with very small groups, is the following unpublished result of Cameron and Kantor.

Let $l(n)$ be a monotonic increasing function on the natural numbers satisfying $l(n!) = n$ for all n.

Theorem 9.1 *Let G be a primitive group of degree n, and suppose that the order of the point stabiliser in G is smaller than $\sqrt{l(n)}$. Then G has a base of size 2.*

The proof depends on the following result of some independent interest.

Theorem 9.2 *Let G be a primitive group of degree n. Then any element $g \in G$ which is not semiregular lies in a conjugacy class of size at least $l(n)$.*

Remark. The hypothesis excluding semiregular elements is necessary. For example, in the dihedral group of order $2p$ (where p is prime), elements of order p lie in conjugacy classes of size 2. Also, the result is not far from best possible, as can be seen by taking G to be the symmetric group S_m acting on the cosets of a very small maximal subgroup (so that n is nearly $m!$), and g to be a transposition in S_m (lying in a class of size $m(m-1)/2$).

Proof Let $|g^G| = m < l(n)$. Then $\bigcap_{h \in g^G} C_G(h) = N$ is a normal subgroup of G of index at most $m!$, hence less than n. So $N \neq 1$, and N is transitive. But N centralises g, so g is semiregular. $\qquad\qquad\qquad\qquad\qquad\square$

Proof of Theorem 9.1. Choose $\alpha \in \Omega$, and let $H = G_\alpha$. Count pairs (β, g) with $g \in G$, $g \neq 1$, $\beta g = \beta$. We have

$$
\begin{aligned}
n(|H| - 1) &= \sum_{g \in (H^{\sharp})^G} \mathrm{fix}(g) \\
&= \sum_i \mathrm{fix}(g_i) |g_i^G| \\
&= \sum_{h \in H^{\sharp}} \mathrm{fix}(h) |h^G| / |h^G \cap H| \\
&\geq \left(\sum_{h \in H^{\sharp}} \mathrm{fix}(h) \right) \left(\min_{h \in H^{\sharp}} |h^G| / (|H| - 1) \right).
\end{aligned}
$$

(Here $H^{\sharp} = H \setminus \{1\}$, and the sum in the second line is over a set of conjugacy class representatives of elements with fixed points.) So the number of choices of β for which (α, β) is *not* a base is at most

$$
\sum_{h \in H^{\sharp}} \mathrm{fix}(h) \leq n(|H| - 1)^2 / l(n) < n.
$$

$$\square$$

The proof in fact shows that, in a family of primitive groups with $|G_\alpha| = o(\sqrt{l(n)})$, asymptotically almost all pairs of points are bases.

In the case where $G = \mathrm{PGL}(n, q)$, rather than apply Theorem 9.1 directly, we use the fact that the smallest non-trivial conjugacy class has size roughly q^{2n}, together with the results of Hall, Liebeck and Seitz [16] on generation, and Landazuri and Seitz [21] on minimal degree. The conclusion is that, if $|H|$ is smaller than roughly q^n, there is a base of size 2. This deals with some of the smaller Aschbacher classes, depending on the relative sizes of n and q. Unfortunately, it doesn't deal with Aschbacher's case (b), since the bound is smaller than Liebeck's q^{3n}. It would be interesting to improve the technique to remove this defect.

References

[1] E. Artin, *Geometric Algebra*, Interscience, New York, 1957.

[2] M. Aschbacher, On the maximal subgroups of the finite classical groups, *Invent. Math.* **76** (1984), 469-514.

[3] F. Buekenhout (ed.), *Handbook of Incidence Geometry*, Elsevier, 1995.

[4] A. R. Calderbank, P. J. Cameron, W. M. Kantor and J. J. Seidel, $\mathbb{Z}_4$-Kerdock codes, orthogonal spreads, and extremal Euclidean line-sets, in preparation.

[5] P. J. Cameron, *Projective and Polar Spaces*, QMW Maths Notes **13**, Queen Mary and Westfield College, London, 1991.

[6] P. J. Cameron, Permutation groups, in *Handbook of Combinatorics* (ed. R. L. Graham, M. Grötschel and L. Lovász), Elsevier, 1995, pp. 611–645.

[7] P. J. Cameron and J. J. Seidel, Quadratic forms over GF(2), *Proc. Kon. Nederl. Akad. Wetensch.* (A) **76** (1973), 1–8.

[8] R. W. Carter, *Simple Groups of Lie Type*, Wiley, New York, 1972.

[9] J. H. Conway, R. T. Curtis, S. P. Norton, R. A. Parker and R. A. Wilson, *An ATLAS of Finite Groups*, Oxford Univ. Press, Oxford, 1985.

[10] B. N. Cooperstein, Minimal degree for a permutation representation of a classical group, *Israel J. Math.* **30** (1978), 213–235.

[11] J. Dieudonné, *La Géometrie des Groupes Classiques*, Springer, Berlin, 1963.

[12] J. Eisfeld, Some big sets of mutually disjoint subgeometries of $PG(d, q^t)$, preprint.

[13] É. Galois, Oeuvres mathématique: lettre à M. Auguste Chevalier, *J. Math. Pures Appl. (Liouville)* (1846), 400–415.

[14] D. Gorenstein, *Finite Simple Groups: An Introduction to their Classification*, Plenum Press, New York, 1982.

[15] R. Gow, Two multiplicity-free permutation representations of the general linear group $GL(n, q^2)$, *Math. Z.* **188** (1984), 45–54.

[16] J. I. Hall, M. W. Liebeck and G. M. Seitz, Generators for finite simple groups, with applications to linear groups, *Quart. J. Math. Oxford* (2) **43** (1992), 441–458.

[17] A. R. Hammons, Jr., P. V. Kumar, A. R. Calderbank, N. J. A. Sloane, and P. Solé, The $\mathbb{Z}_4$-linearity of Kerdock, Preparata, Goethals and related codes, *IEEE Trans. Inform. Theory*, to appear.

[18] D. F. Holt, C. R. Leedham-Green, E. A. O'Brien and S. E. Rees, Primitivity testing for matrix groups, *J. Algebra*, to appear.

[19] W. M. Kantor, Permutation representations of the finite classical groups of small degree or rank, *J. Algebra* **60** (1979), 158–168.

[20] P. B. Kleidman and M. W. Liebeck, *The Subgroup Structure of the Finite Classical Groups, London Math. Soc. Lecture Notes* **129**, Cambridge Univ. Press, Cambridge, 1990.

[21] V. Landazuri and G. M. Seitz, On the minimal degrees of projective representations of the finite Chevalley groups, *J. Algebra* **32** (1974), 418–443.

[22] C. R. Leedham-Green and E. A. O'Brien, Tensor products are projective geometries, *J. Algebra*, to appear.

[23] C. R. Leedham-Green and E. A. O'Brien, Recognising tensor products of matrix groups, preprint.

[24] M. W. Liebeck, On the orders of maximal subgroups of the finite classical groups, *Proc. London Math. Soc.* (3) **50** (1985), 426–446.

[25] M. W. Liebeck, C. E. Praeger and J. Saxl, On the O'Nan–Scott reduction theorem for finite primitive permutation groups, *J. Austral. Math. Soc.* **44** (1988), 389–396.

[26] H. Lüneburg, *Transitive Erweiterungen endlicher Permutationsgruppen*, *Lecture Notes in Math.* **84**, Springer, Berlin, 1969.

[27] T. Maund, D. Phil. thesis, Oxford, 1987.

[28] R. A. Parker and R. A. Wilson, The computer construction of matrix representations of finite groups over finite fields, *J. Symbolic Computation* **9** (1990), 583–590.

[29] M. A. Ronan, Locally truncated buildings and M_{24}, *Math. Z.* **180** (1982), 489–501.

[30] L. L. Scott, Representations in characteristic p, *Proc. Symp. Pure Math.* **37** (1980), 319–331.

[31] A. P. Sprague, Pasch's axiom and projective spaces, *Discrete Math.* **33** (1981), 79–87.

[32] G. Tallini, On a characterization of the Grassman manifold representing the lines in a projective space, pp. 354–358 in *Finite Geometries and Designs* (ed. P. J. Cameron, J. W. P. Hirschfeld and D. R. Hughes), *London Math. Soc. Lecture Notes* **49**, Cambridge University Press, Cambridge, 1981.

[33] J. Ueberberg, Projective spaces and dihedral groups, to appear.

[34] J. Ueberberg, A class of partial linear spaces related to $PGL(3, q^2)$, *Europ. J. Combinatorics*, to appear.

[35] H. Wielandt, *Finite Permutation Groups*, Academic Press, New York, 1964.

Peter J. Cameron, School of Mathematical Sciences, Queen Mary and Westfield College, Mile End Road, London E1 4NS, U.K.
e-mail p.j.cameron@qmw.ac.uk

The Hermitian function field arising from a cyclic arc in a Galois plane

A. Cossidente   *G. Korchmáros*

Abstract

It is shown that the function field of the algebraic envelope of the cyclic $(q-\sqrt{q}+1)$−arc in $PG(2,q)$, q square, is the Hermitian function field.

1 Introduction

An old but still open problem in Galois geometries is to determine the maximum size of a k−arc in $PG(2,q)$ which is not contained in a conic (for q even, in a hyperoval). The only known infinite family of large complete k−arcs consists of $(q-\sqrt{q}+1)$−arcs and it contains one such an arc for each square q. This family was originally found by Kestenband [7] who observed that two non−degenerate Hermitian curves can meet over $GF(q)$ in a $(q-\sqrt{q}+1)$−arc. The same family was independently obtained by Fisher, Hirschfeld and Thas [5], Boros and Szönyi [1] and Ebert [4] who showed that a point−orbit under a Singer cycle of size $(q-\sqrt{q}+1)$ is an arc. For this reason, such an arc is called the *cyclic* $(q-\sqrt{q}+1)-arc$ of $PG(2,q)$. Cossidente [2] showed that the existence of a cyclic $(q-\sqrt{q}+1)$−arc in $PG(2,q)$ depends on the fact that each $GF(q^3)$−rational point $P(x,y)$ of the Fermat curve $X^{q+\sqrt{q}+1} + Y^{q+\sqrt{q}+1} = 1$ is such that at least one of $x^{q+\sqrt{q}+1}$ and $y^{q+\sqrt{q}+1}$ belongs to $GF(q)$.

We are interested in the algebraic envelope Γ' of the cyclic $(q-\sqrt{q}+1)$−arc. To simplify the exposition we will refer to the corresponding algebraic curve Γ in the dual plane of $PG(2,q)$. By Segre's theorem [8] Γ has degree either $\sqrt{q}+1$ or $2(\sqrt{q}+1)$ according as $q = 2^h$ or $q = p^h$ and $p > 2$. In the former case, Γ turns out to be projectively equivalent to the Hermitian curve $XY^{\sqrt{q}} + X^{\sqrt{q}} + Y = 0$, see [10]. The odd order case has been investigated in a previous paper [3]. However the equation found for Γ did not allow us to determine the function field of Γ. This gap is filled here with the following result.

Theorem 1 *If $q > 9$, the function field of the algebraic envelope of the cyclic $(q - \sqrt{q} + 1)-$arc of $PG(2, q)$ is isomorphic to the function field of the Hermitian curve of $PG(2, q)$.*

2 Some remarks preliminary to the proof of Theorem 1

The existence of an algebraic envelope associated to a $k-$arc in $PG(2, q)$ comes from the fundamental theorem of Segre [8]; see also [6, Theorem 10.4.3].

Theorem 2 (B. Segre) *Let* K *be a k-arc in $PG(2, q)$.*
(a) *The $kt = k(q + 2 - k)$ unisecants through the points of* K *in $PG(2, q)$ lie on an algebraic envelope Γ' of degree t or $2t$, according as q is even or odd;*
(b) Γ' *contains no $2-$secant of* K *and so no pencil with vertex P in* K*;*
(c) K *is contained in a larger arc in $PG(2, q)$ if and only if Γ' has a linear component over $GF(q)$.*

In particular, for q odd,

(d) *the t unisecants of* K *through a point P of* K *each counts twice in the intersection of Γ' with the pencil Ψ_P of lines through P;*
(e) Γ' *may contain components of multiplicity at most two, but does not consist entirely of $2-$fold components;*
(f) *if $k \geq \frac{2}{3}(q + 2)$, then Γ' is uniquely determined by* K*.*

From now on we only consider the cyclic $(q - \sqrt{q} + 1)-$arc in $PG(2, q)$, and we also assume that $q = p^h$, where $p > 2$ is prime and h is even. We will denote this arc by K. In the dual plane, the algebraic envelope Γ' becomes an algebraic curve Γ. Since it is customary to consider curves rather then envelopes, we will work in the dual plane Π. The arc K is now viewed as a dual arc $\mathcal{K}$ in Π, that is, as a set of $(q - \sqrt{q} + 1)$ lines such that no three of them meet in a point. Properties (a)... (f) are easily reformulated for Γ.

The cyclic model of the finite desarguesian plane due to Singer [9] is certainly more suitable for the construction of K. However, it does not appear effective for investigating the algebraic curve Γ'. Our method is rather different; assume a projective frame in the cubic extension $PG(2, q^3)$ of $PG(2, q)$ in such a way that the Singer cycle transforms into diagonal form. It should be noted that plane Π no longer lies in canonical position; that is, it is not the subplane of $PG(2, q^3)$ coordinatized over $GF(q)$. Let $\mathcal{H}$ denote the Hermitian curve with affine equation

$$XY^{\sqrt{q}} + X^{\sqrt{q}} + Y = 0 \tag{1}$$

viewed as an algebraic curve in $PG(2, \mathcal{L})$ where $\mathcal{L}$ is the algebraic closure of $GF(q^3)$. Also, Γ will be regarded as an algebraic curve in $PG(2, \mathcal{L})$. To prove the theorem we have to exhibit a birational transformation from $\mathcal{H}$ to Γ.

3   The proof of Theorem 1

Let $\mathcal{F} = GF(q)$, where $q = p^h$ with h even and p an odd prime, and let $\mathcal{E} = GF(q^3)$ be a cubic extension of $\mathcal{F}$. We introduce the linear collineation of $PG(2, \mathcal{E})$:

$$\alpha : \begin{cases} \rho x_1' = b x_1 \\ \rho x_2' = b^{q+1} x_2 \\ \rho x_3' = x_3, \end{cases}$$

where $b \in \mathcal{R}$, and $\mathcal{R}$ denotes the set of all primitive $(q^2 + q + 1)$−th roots of unity over $\mathcal{E}$. Clearly, α has order $(q^2 + q + 1)$. Also, α fixes each vertex of the fundamental triangle $A_1 A_2 A_3$ of $PG(2, \mathcal{E})$.

The orbit of the point $E(1, 1, 1)$ under $\mathcal{A} = \langle \alpha \rangle$ is given by

$$\Pi = \{\alpha^i(E) : i = 0, 1 \ldots, q^2 + q\} = \{(c, c^{(q+1)}, 1) : c \in \mathcal{R}\}.$$

The orbit Π may be viewed as a subgeometry of $PG(2, \mathcal{E})$ induced by the lines meeting Π in at least two points. Actually, this subgeometry is a $PG(2, q)$.

Proposition 3 *The orbit Π is isomorphic to $PG(2, \mathcal{F})$. More precisely, Π is a projective subplane of $PG(2, \mathcal{E})$ lying in a non-classical position; that is $\Pi \neq PG(2, \mathcal{F})$. The lines of Π have equations $tX + t^{q+1}Y + 1 = 0$, with t running over $\mathcal{R}$, and form the line-orbit of $X + Y + 1 = 0$ under $\mathcal{A}$.*

Proof.   We begin by proving that the line $X + Y + 1 = 0$ meets Π in $q + 1$ points. To do this, we must check that if $u^{q+1} + u + 1 = 0$ and $u \in \mathcal{L}$, then u is a $(q^2 + q + 1)$−th root of unity. Clearly, we have $u^{q^2+q} + u^q + 1 = 0$, and hence also $u^{q^2+q+1} + u^{q+1} + u = 0$. From this $u^{q^2+q+1} = 1$ follows. Also we must check that the polynomial $f(Y) = Y^{q+1} + Y + 1$ has no repeated roots. This comes from the fact that $f(Y)$ and its derivative $f'(Y) = Y^q + 1$ have no common roots. Now, for any $(q^2 + q + 1)$−th root t of $\mathcal{E}$, let r_t denote the line $tX + t^{q+1}Y + 1 = 0$ in $PG(2, \mathcal{E})$. This line meets Π in the points $(c^i, c^{(q+1)i}, 1)$ satisfying the condition $(c^i t)^{q+1} + c^i t + 1 = 0$. Arguing as before, we see that there are exactly $q + 1$ such points. This shows that Π together with the lines r_t is a projective plane of order q. The last statement is easily checked by direct calculation.   □

Next, we establish some intersection properties of Π with conics and other curves of $PG(2, \mathcal{L})$.

Proposition 4 *The conic C with equation*

$$X^2 + Y^2 + 1 = 0 \tag{2}$$

and the algebraic curve $\mathcal{D}$ with equation

$$X = Y^{\sqrt{q}+1} \tag{3}$$

meet in $2(\sqrt{q}+1)$ distinct points, and half of these points lie on Π.

Proof. Substituting $Y^{\sqrt{q}+1}$ for X in (2) we obtain $Y^{2(\sqrt{q}+1)} + Y^2 + 1 = 0$. The latter equation has pairwise distinct roots since the polynomial $f(Y) = Y^{2(\sqrt{q}+1)} + Y^2 + 1$ and its derivative $f'(Y) = Y(Y^{2\sqrt{q}} + 1) = Y(Y^2 + 1)^{\sqrt{q}}$ have no common roots. We must to show that $\sqrt{q}+1$ of these roots have the property that the corresponding common point also lies on Π. Since $f(Y)^{\sqrt{q}}Y^2 = Y^{2(q+\sqrt{q}+1)} + Y^{2(\sqrt{q}+1)} + Y^2$ and $Y^{2(\sqrt{q}+1)} + Y^2 + 1 = 0$, each root of $F(Y)$ is also a root of the polynomial $g(Y) = Y^{2(q+\sqrt{q}+1)} - 1$. Since $f(Y) = f(-Y)$, and $g(Y) = g(-Y)$, half of the roots of $f(Y)$ are also roots of the polynomial $h(Y) = Y^{q+\sqrt{q}+1} - 1$. This shows that the corresponding $\sqrt{q}+1$ points lie on Π. $\qquad\square$

Proposition 5 *The conic C of equation (2) and the Hermitian curve $\mathcal{H}$ of equation (1) have $\sqrt{q}+1$ distinct points in common, and all lie in Π. Also, the curves C and $\mathcal{H}$ have the same tangent line at each of these points.*

Proof. In the proof of Proposition 4 $\sqrt{q}+1$ different points $P = (y^{\sqrt{q}+1}, y)$ in the interesection $C \cap \Pi$ were found, such that y is a root of both $f(Y)$ and $h(Y)$. Each of these points also lies on $\mathcal{H}$ since

$$y^{2\sqrt{q}+1} + y^{q+\sqrt{q}} + y = \frac{1}{y}(y^{q+\sqrt{q}+1} + y^{2(\sqrt{q}+1)} + y^2) = \frac{1}{y}(1 + y^{2(\sqrt{q}+1)} + y^2) = 0.$$

An easy calculation shows that the tangent line to $\mathcal{H}$ at the point $P = (y^{\sqrt{q}+1}, y)$ has slope equal to $-\sqrt{y}$. The same value is obtained when $\mathcal{H}$ is replaced by C. Hence C and $\mathcal{H}$ are tangent at each of the above $\sqrt{q}+1$ points. By the classical theorem of Bézout, no further common point of C and $\mathcal{H}$ exists, and this completes the proof of Proposition 5. $\qquad\square$

Now, consider the set of conics C_t with equation

$$tX^2 + t^{q+1}Y^2 + 1 = 0 \tag{4}$$

with $t \in \mathcal{R}$. Actually, this set consists of all conics which are images of the conic C of equation (2) under the group $\mathcal{A}$ of order $q - \sqrt{q} + 1$. Let $\mathcal{B}$ denote the subgroup of $\mathcal{A}$ of order $q - \sqrt{q} + 1$. Note that the orbit of C under $\mathcal{B}$ consists of the conics C_t where t ranges over the $(q - \sqrt{q}+1)$-roots of unity.

Proposition 6 *The conics C_t, with $t \in \mathcal{R}$, partition $\mathcal{H} \cap \Pi$ into $(q - \sqrt{q} + 1)$ pointsets each of size $\sqrt{q} + 1$.*

Proof.   First of all we check that $\mathcal{B}$ leaves $\mathcal{H}$ invariant. From

$$b^{q\sqrt{q}+\sqrt{q}+1}XY^{\sqrt{q}} + b^{\sqrt{q}}X^{\sqrt{q}} + b^{q+1}Y = 0$$

and dividing by $b^{\sqrt{q}}$, we get

$$b^{q-\sqrt{q}+1}XY^{\sqrt{q}} + X^{\sqrt{q}} + b^{q-\sqrt{q}+1}Y = 0;$$

since $b^{q\sqrt{q}+1} = b^{q-\sqrt{q}+1} = 1$, we are done. This shows that Proposition 5 remains valid when C is replaced by any conic C_t with t a $(q - \sqrt{q} + 1)$–th root of unity. From this we infer at first that no point on $\mathcal{H}$ lies on more than one of such conics. In fact, if both t and u are $(q - \sqrt{q} + 1)$–th roots of unity, and $P(x, y)$ is a point in $\mathcal{H} \cap C_t \cap C_u$, from Proposition 3 it follows that C_t and C_u must have the same tangent at $P(x, y)$. But, this can only occur when $-t^q = -u^q$, and hence $t = u$. From Proposition 5 we also infer that each conic C_t with t a $(q - \sqrt{q} + 1)$–th root of unity meets $\mathcal{H} \cap \Pi$ in $\sqrt{q} + 1$ points. Since $\mathcal{H} \cap \Pi$ has size $q\sqrt{q} + 1$, the assertion follows.   □

A relationship between conics C_t and lines r_t is given in the following result.

Proposition 7 *The quadratic transformation T defined by $X' = X^2$ and $Y' = Y^2$ induces a one-to-one correspondence on Π and maps the conic C_t in the line r_t.*

Proof.   The transformation T is one–to–one on Π since the mapping $x \rightarrow x^2$ is a bijection on the set of all $(q^2 + q + 1)$–th roots of unity. The second half of the assertion is easy to check.   □

Now we are in a position to show that the transformation T maps $\mathcal{H}$ in Γ. Let $T(\mathcal{H})$ denote the image of $\mathcal{H}$ under T. From Propositions 5, 6 and 7, it follows that each line r_t with t a $(q - \sqrt{q} + 1)$–th root of unity is a $(\sqrt{q} + 1)$–tangent to $T(\mathcal{H})$; that is, r_t is the tangent line to $T(\mathcal{H})$ at each of their common points. Since such lines form a dual arc, we see that $T(\mathcal{H})$ viewed on the dual plane is an algebraic envelope associated to the cyclic $(q - \sqrt{q} + 1)$–arc. According to (f) of Theorem 2, if $q > 9$ this algebraic envelope is uniquely determined.

To complete the proof of the theorem it must be checked that the transformation T maps $\mathcal{H}$ onto Γ birationally. Consider the linear system Φ of all conics with equation $\lambda X^2 + \mu Y^2 + \tau = 0$, with $\lambda, \mu, \tau \in \mathcal{L}$. We have to show that if $P(x, y)$ is a point of $\mathcal{H}$ in a generic position, and Φ_P is the linear subsystem of Φ consisting of all conics through $P = (x, y)$, then the conics of Φ_P have no further common point on $\mathcal{H}$. As $P = (x, y)$ is in generic position, both x and y must be different from 0. On the other hand, the conics of Φ_P

have exactly four common points on the plane $PG(2, \mathcal{L})$, namely, $P = (x, y)$, $P_1 = (x, -y)$, $P_2 = (-x, y)$ and $P_3 = (-x, -y)$. It is easily checked that no of the last three points lies on $\mathcal{H}$. Thus, the point $P(x, y)$ is the only common point of the conics of Φ_P which lies on $\mathcal{H}$. This completes the proof.

Acknowledgement. This research was carried out within the activity of G.N.S.A.G.A. of the Italian C.N.R. with the support of the Italian Ministry for Research and Technology.

References

[1] E. Boros and T. Szönyi, On the sharpness of a theorem of B. Segre, *Combinatorica* **6** (1986), 261-268.

[2] A. Cossidente, A new proof of the existence of $(q^2 - q + 1)$–arcs in $PG(2, q^2)$ *J. Geometry* **53** (1995), 37-40; Addendum, to appear.

[3] A. Cossidente and G. Korchmáros, The algebraic envelope associated to a complete arc, *Recent Progress in Geometry*, Rendiconti del Circolo Matematico di Palermo, to appear.

[4] G.L. Ebert, Partitioning projective geometries into caps, *Canad. J. Math.* **37** (1985), 1163-1175.

[5] J.C. Fisher, J.W.P. Hirschfeld and J.A. Thas, Complete arcs in planes of square orders, *Ann. Discrete Math.* **30** (1986), 243-250.

[6] J.W.P. Hirschfeld, *Projective Geometries over Finite Fields*, Oxford University Press, Oxford, 1979.

[7] B.C. Kestenband, A family of complete arcs in finite projective planes, *Colloq. Math.* **17** (1989), 59-67.

[8] B. Segre, Introduction to Galois geometries, *Atti Accad. Naz. Lincei Mem.* **8** (1967), 133-236.

[9] J. Singer, A theorem in finite projective geometry and some applications to number theory, *Trans. Amer. Math. Soc.* **43** (1938), 377-385.

[10] J.A. Thas, Complete arcs and algebraic curves in $PG(2, q)$, *J. Algebra* **106** (1987), 457-464.

A. Cossidente, Dipartimento di Matematica, Università della Basilicata, via N.Sauro 85, 85100 Potenza, Italy.
e-mail: ca015sci@unibas.it

G. Korchmáros, Dipartimento di Matematica, Università della Basilicata, via N.Sauro 85, 85100 Potenza, Italy.
e-mail:kg159sci@unibas.it

Intercalates everywhere

P. Danziger   *E. Mendelsohn*

This paper is dedicated to the memory of Peter Rodney $Z''L$. (1965 – 1995)

Abstract

The counting of the number of Intercalates, 2×2 subsquares, possible in a latin square of side n is in general a hard problem. N_2–Free latin squares, those for which there are no intercalates, are known to exist for $n \neq 1, 2, 4$. N_2–complete latin squares, those which have the property that they have the maximum number of N_2's possible, $\frac{n}{2} \cdot \binom{n}{2}$, must be isotopic to Z_2^k and thus of side 2^k. The maximum for $n \neq 2^k$ is in general unknown. We propose an intermediate possibility, that of N_2–ubiquitous. A latin square is N_2 ubiquitous if and only if every cell a_{ij} is contained in some 2×2 subsquare. We show these exist for $n \neq 1, 3, 5, 7$. It is also determined for which n, C–ubiquitous latin squares exist for every partial latin square, C with four cells. We also enumerate the number of times each 4-cell configuration can appear in a latin square and show that this number depends only on n and the number of intercalates.

1 Introduction

For basic results and terminology on latin squares, the reader is referred to "Latin Squares and Their Applications" [DK74] or "The CRC Handbook of Combinatorial Designs" [CD]. A partial latin square $\mathcal{P}$ is a subset of $P \times P \times P$ called cells, where P is a finite set, with the property that the projections $\Pi_{12}, \Pi_{13}, \Pi_{23}$, where $\Pi_{ij}(x_1, x_2, x_3) = (x_i, x_j)$, are one to one from $\mathcal{P}$ to $P \times P$. When these projections are onto, $\mathcal{P}$ is a latin square. Analogously to the definition in [GRR] a partial latin square on a small number of cells is called a *configuration*.

The cell (r, c, e) is said to have entry e in row r and column c and is sometimes denoted by the following array.

$$\begin{array}{c|c} * & c \\ \hline r & e \end{array}$$

We may also use quasigroup notation $r * c = e$. We will often represent partial latin squares by arrays of cells. Usually the specific row and column are unimportant and are omitted, rather it is the relation between the cells which is of interest. Thus the 2×2 subsquare would be represented by the array

a	b
b	a

Two other concepts are needed The first is the concept of a homotopism (we would like to use the term homotopy but it is overused in other fields). A homotopism Φ from $\mathcal{P}$ to $\mathcal{Q}$ is a set of three functions ϕ_r, ϕ_c and ϕ_e so that $(r, c, e) \in \mathcal{P} \leftrightarrow (\phi_r(r), \phi_c(c), \phi_e(e)) \in \mathcal{Q}$. A homotopism is an inclusion if and only if ϕ_r, ϕ_c and ϕ_e are one to one. If ϕ_r, ϕ_c and ϕ_e are bijections the homotopism is called an isotopy. We say that $\mathcal{P}$ *appears* in $\mathcal{Q}$ if there is an inclusion $\Phi : \mathcal{P} \longrightarrow \mathcal{Q}$, in quasigroup notation this means that $\phi_r(x) * \phi_c(y) = \phi_e(x \cdot y)$.

For example consider the latin square

1	2	3	4	5
2	1	5	3	4
3	4	1	5	2
4	5	2	1	3
5	3	4	2	1

the element 1 appears in a 2×2 subsquare of the form

1	a
a	1

with every other element a, but the cell $(2, 3, 5)$ is in no appearance of a 2×2 subsquare.

The use of small partial latin squares in studying the structure of latin squares is an important tool. Two of the common such partial squares are the 2×2 subsquare or intercalate, denoted by N_2 (we use these terms interchangably), and the transversal. The folklore theorem that a latin square of side n has an orthogonal mate if and only if it has n disjoint transversals is an example of this analysis. The Evans [Sm] conjecture states that for any n and k with $n > k$ any k-cell configuration may be completed to a latin square of side n. This is a theorem about the embeddability of certain configurations in a complete latin square. There is a rich literature on embeddability of partial special latin squares, for example see Lindner and Rogers [LR].

The 2×2 subsquare has been used as a fundamental tool. The number of 2×2 subsquares possible in a latin square of side n is still unknown. In fact the maximum number is still unknown for $n \neq 2^k$. Cameron [C] has proved that the elementary counting bound $\frac{n}{2} \cdot \binom{n}{2}$ is achieved only when $n = 2^k$. Some further results are obtained in Heinrich and Wallis [HW]. On the other hand the minimum number of intercalates, 0, can be realized in a latin square of side n for every $n \neq 2, 4$, see [AM, H80, KLR, M].

In section 2 we shall enumerate all the partial latin squares on four or fewer cells. In section 3 we shall deal with the concept of ubiquity:

Definition 1.1 *A configuration, $\mathcal{P}$, is ubiquitous in a latin square $\mathcal{L}$ if every cell of the latin square is contained in the image of an inclusion $\Phi : \mathcal{P} \longrightarrow \mathcal{L}$. (an appearance of $\mathcal{P}$ in $\mathcal{L}$).*

In other words each cell is contained in a copy of $\mathcal{P}$. This definition is motivated from its usefulness in totally symmetric idempotent latin squares [MR] and our Theorem 3.1 that if a configuration appears in the Cayley table of a group then it is ubiquitous. The result we obtain is

Theorem 1.2 *For every four or fewer cell configuration and every $n > 7$, there is a latin square of side n which is ubiquitous with respect to that configuration.*

We shall obtain relationships between the number of four cell configurations and show an analog of the main theorem of Grannell Griggs and Mendelsohn [GGM], i.e. that the number of copies of any four cell configuration is determined from the number of copies of two configurations.

Theorem 1.3 *A basis for the number of four cell configurations is the four cell partial row and the number of 2×2 subsquares.*

2   Enumeration

In order to simplify the enumeration and the problem of ubiquity we need the concept of conjugate or parastrophe [S] of a partial latin square. Let $\pi \in S_3$ and let $\mathcal{P}$ be a partial latin square, then $\mathcal{P}_\pi$ is the partial latin square with cells $(x_{\pi(1)}, x_{\pi(2)}, x_{\pi(3)})$ where $(x_1, x_2, x_3) \in \mathcal{P}$. The set of all distinct conjugates of $\mathcal{P}$ is called the main class containing $\mathcal{P}$. By distinct we mean that there is no permutation of rows, columns or entries which will map $\mathcal{P}$ to $\mathcal{P}_\pi$, i.e. distinct up to isotopy.

Clearly any configuration $\mathcal{P}$ has up to isotopy 1, 2, 3, or 6 conjugates. For example:

N_2 has 1 conjugate.

The 5 cell configuration

b	c	
a		b
		a

has 2 conjugates, this configuration is the smallest configuration with this property.

The 3–unipotent configuration

a		
	a	
		a

has 3 conjugates – itself, the partial row and the partial column.

The configuration

a	b	c
d		

has 6 conjugates.

We can enumerate the configurations by main class. In order to facilitate this we will define the following invariant. The partition of the number of rows (the projection of $\mathcal{P}$ on its first coordinate) will indicate how many cells have the same first entry. We use a conjugate definition on the columns and the entries. We separate the partitions by a "|", thus every partial latin square is represented by a triple $X \mid Y \mid Z$, where X, Y and Z represent partitions of the number of rows, columns and entries respectively. We note that for $n \geq 4$ this invariant is no longer sharp (see Table 4).

For example the partial latin square

a	b	c
d		

has invariant $3 + 1 \mid 2 + 1 + 1 \mid 1 + 1 + 1 + 1$, so that there are three elements in one row and one in another, two elements in the same column and two columns with one element each, and each entry is distinct.

The following lemma verifies that the number of configurations of a given size grows quickly.

Lemma 2.1 *Let $\mathcal{Q}$ be the set of all distinct partitions of n, each of the form $p = \{n_1, \ldots, n_k\}$, with $n_k > n_{k-1} > \ldots > n_2 > n_1$. The number of non isotopic n cell configurations using exactly n columns, at most n rows and at most n entries exceeds $\Sigma_{p \in \mathcal{Q}} 2^k \cdot n_1$.*

Table 1: The One Cell Configuration

Partition	Example	Number	Conjugates
$1 \mid 1 \mid 1$	a	n^2	1

Proof For each $p \in \mathcal{Q}$, $p = \{n_1, \ldots, n_k\}$, with $n_k > n_{k-1} > \ldots > n_2 > n_1$. We will construct $n_1 2^k$ non isotopic partial latin squares with n cells. The partial latin squares each have n columns, the partition p gives the number of cells in common rows. For each $t \leq n_1$ choose a subset $U \subseteq \{n_1, \ldots, n_k\}$. If $n_i \in U$ fill t cells of n_i with entries from 1 to t, fill all other cells with distinct entries.

$$\boxed{1\,2\ldots t\, a_1\, a_2 \ldots a_{n_k-t}}$$
$$\boxed{b_1\, b_2\ \ldots\ b_{n_{k-1}}}$$
$$\ddots$$
$$\boxed{1\,2\ldots t\, c_1\, c_2 \ldots c_{n_1-t}}$$

$$n_1, n_k \in U, \ n_{k-1} \notin U$$

For each choice of p and each choice of t the resulting configurations have different invariants, and so are not isotopic. Now let p and t be fixed. Each choice of U gives a different distribution of the number of common elements in rows. □

We note that this bound is well below the actual number of non isotopic n cell configurations.

In order to enumerate the configurations we will take an arbitrary ordering on the partitions of n and enumerate the configurations whose invariant is $X \mid Y \mid Z$ where $X \geq Y \geq Z$. We note that if X, Y and Z are distinct partitions of n then the number of isotopes in the main class is 6, otherwise the conjugates must be checked.

We now enumerate up to main class all fewer than four cell configurations in a latin square of side n. The following tables give, for each configuration, the invariant type, a canonical example, the number of that configuration as given and the number of conjugates. The total number of each main class is the number of configurations times the number of conjugates. Here we also use the symbol N_2 to denote the number of 2×2 subsquares.

Table 2: The Two Cell Configurations
The partition order is $(2) > (1+1)$.

Type	Partition	Example	Number	Conjugates
A_1	$2 \mid 1+1 \mid 1+1$	$a \quad b$	$n\binom{n}{2}$	3
A_2	$1+1 \mid 1+1 \mid 1+1$	$\begin{array}{c}a\\ \quad b\end{array}$	$n\binom{n}{2}(n-2)$	1

Table 3: The Three Cell Configurations
The partition order is $(3) > (2+1) > (1+1+1)$. We label these partitions by the number of common pairs in each to get $3 > 1 > 0$.

Type	Partition	Example	Number	Conjugates
B_1	$3 \mid 0 \mid 0$	$a \quad b \quad c$	$n\binom{n}{3}$	3
B_2	$1 \mid 1 \mid 1$	$\begin{array}{c}a \quad b\\ a\end{array}$	$2n\binom{n}{2}$	1
B_3	$1 \mid 1 \mid 0$	$\begin{array}{c}a \quad b\\ c\end{array}$	$2n\binom{n}{2}(n-2)$	3
B_4	$1 \mid 0 \mid 0$	$\begin{array}{c}a \quad b\\ \quad c\end{array}$	$n\binom{n}{2}(n-2)(n-3)$	3
B_5	$0 \mid 0 \mid 0$	$\begin{array}{c}a\\ b\\ \quad c\end{array}$	$\frac{1}{3}n\binom{n}{2}(n-2)$ $((n-4)^2+2(n-3))$	1

Table 4: The Four Cell Configurations

The invariant is no longer sharp. C_8, C_9; C_{12}, C_{13}, C_{14}; C_{15}, C_{16} have the same invariants.

The partition order is $(4) > (3+1) > (2+2) > (2+1+1) > (1+1+1+1)$.

We label these partitions by the number of common pairs in each to get $6 > 3 > 2 > 1 > 0$.

Type	Partition	Example	Number	Conjugates
C_1	6|0|0	$a \quad b \quad c \quad d$	$n\binom{n}{4}$	3
C_2	3|1|1	$a \quad b \quad c$ b	$6n\binom{n}{3}$	3
C_3	3|1|0	$a \quad b \quad c$ d	$3n\binom{n}{3}(n-3)$	6
C_4	3|0|0	$a \quad b \quad c$ $\qquad\qquad d$	$n\binom{n}{3}(n-3)(n-4)$	3
C_5	2|2|2	$a \quad b$ $b \quad a$	N_2	1
C_6	2|2|1	$a \quad b$ $c \quad a$	$n\binom{n}{2} - 2N_2$	3
C_7	2|2|0	$a \quad b$ $c \quad d$	$\frac{1}{2}n\binom{n}{2}(n-3) + N_2$	3
C_8	2|1|1	$a \quad b$ $\quad c \quad b$	$2n\binom{n}{2}(n-3) + 4N_2$	3
C_9	2|1|1	$a \quad b$ $\quad c \quad a$	$n\binom{n}{2}(n-2)$	3
C_{10}	2|1|0	$a \quad b$ $\quad c \quad d$	$n\binom{n}{2}(n-3)^2 - 2N_2$	6
C_{11}	2|0|0	$a \quad b$ $\qquad c \quad d$	$\frac{1}{4}n\binom{n}{2}(n-3)^2(n-4) + N_2$	3
C_{12}	1|1|1	$a \quad b$ $\quad c$ $\qquad a$	$2n\binom{n}{2}(n-2)(n-3)$	3
C_{13}	1|1|1	$a \quad b$ $\quad c$ $\qquad b$	$2n\binom{n}{2}((n-3)^2 + (n-2))$ $-4N_2$	1
C_{14}	1|1|1	$a \quad b$ $\quad a$ $\qquad c$	$2n\binom{n}{2}(n-3)^2 - 4N_2$	1

Four Cell Configurations (continued)

Type	Partition	Example	Number	Conjugates
C_{15}	$1 \mid 1 \mid 0$	a b c d	$2n\binom{n}{2}(n-3)^3 + 4N_2$	3
C_{16}	$1 \mid 1 \mid 0$	a b c d	$\frac{1}{2}n\binom{n}{2}(n-2)(n-3)(n-4)$	3
C_{17}	$1 \mid 0 \mid 0$	a b c d	$\frac{1}{2}n\binom{n}{2}(n-3)$ $(n^3 - 10n^2 + 34n - 38)$ $-2N_2$	3
C_{18}	$0 \mid 0 \mid 0$	a b c d	$\frac{1}{12}n\binom{n}{2}(n-3)$ $(n^4 - 14n^3 + 76n^2$ $-188n + 174)$ $+N_2$	1

We give a derivation for the enumeration of C_{10} by way of example. The derivations for the other configurations are similar. C_{10} has 6 conjugates and is given by

$$C_{10}$$

a	b	
	c	d

We will enumerate the conjugate given.

Lemma 2.2 *The number of C_{10} configurations appearing in a latin square of side n is $n\binom{n}{2}(n-3)^2 - 2N_2$.*

Proof

To begin with we choose all possible 4 cell squares, there are three possibilities, either the square is a C_5, C_6 or a C_7 configuration. The intercalate, C_5, cannot give rise to a C_{10}.

$$C_6 \qquad C_7$$

a	b
b	c

a	b
e	c

Each appearance of a C_6 gives rise to 2 possibilities for a C_{10}, depending which of the three cells are chosen, for each of these there are $n-3$ choices for d. Each appearance of a C_7 gives rise to 4 possibilities for a C_{10}, depending which of the three cells are chosen, for each of these there are $n - 4$ choices for d. Thus if $N(X)$ represents the number of appearances of configuration X we have

$$
\begin{aligned}
2N(C_{10}) &= 2(n-3)N(C_6) + 4(n-4)N(C_7) \\
&= 2(n-3)(n\binom{n}{2} - 2N_2) + 4(n-4)(\tfrac{1}{2}n\binom{n}{2}(n-3) + N_2) \\
&= 2n\binom{n}{2}(n-3)^2 - 4N_2
\end{aligned}
$$

Since the four original cells can be taken in two ways, we have in fact counted every C_{10} twice, □

We can see that with this sort of argument the enumeration of conjugates depends on the conjugates of previous configurations, down to C_5 (N_2), C_6 and C_7. It is straightforward but tedious to check that the enumeration of the conjugates of these configurations are all equal. Hence we may conclude that the number of times a 4–cell configuration appears is the same for each of the conjugates of that configuration.

3 Ubiquity

We begin our discussion of ubiquity by considering the case when $\mathcal{L}$ is the Cayley table of a group, in this case we may derive the following general theorem.

Theorem 3.1 *If a latin square $\mathcal{L}$ is the Caley table of a group, G, and $\mathcal{P}$ is a configuration which appears in $\mathcal{L}$ then $\mathcal{P}$ is ubiquitous in $\mathcal{L}$.*

Proof We shall use $x \cdot y$ to indicate multiplication in $\mathcal{P}$ and $x * y$ to indicate multiplication in $\mathcal{L}$. $\mathcal{P}$ appears in $\mathcal{L}$ means that there exists a one to one map $\Phi : \mathcal{P} \longrightarrow \mathcal{L}$. such that for any $(x,y,z) \in \mathcal{P}$, $\Phi(x,y,z) = (\phi(x), \psi(y), \theta(z)) \in \mathcal{L}$ satisfies $\phi(x)*\psi(y) = \theta(x \cdot y)$. We must show that for any $(u,v,w) \in \mathcal{L}$ there exists a one to one map $\Phi_{(u,v,w)} : \mathcal{P} \longrightarrow \mathcal{L}$ such that $\Phi_{(u,v,w)}(x,y,z) = (u,v,w)$ for some $(x,y,z) \in \mathcal{P}$.

Fix $(a,b,c) \in \mathcal{P}$ and for each $(x,y,z) \in \mathcal{P}$ define the map $\Phi_{(u,v,w)}(x,y,z) = (\phi_{(u,v,w)}(x), \psi_{(u,v,w)}(y), \theta_{(u,v,w)}(z))$ by

$$
\begin{aligned}
\phi_{(u,v,w)}(x) &= u * [\phi(a)]^{-1} * \phi(x), \\
\psi_{(u,v,w)}(y) &= \psi(y) * [\psi(b)]^{-1} * v \\
\theta_{(u,v,w)}(z) &= u * [\phi(a)]^{-1} * \phi(x) * \psi(y) * [\psi(b)]^{-1} * v.
\end{aligned}
$$

We have used the associativity of G implicitly in the definition of $\theta_{(u,v,w)}$, obviously $\theta_{(u,v,w)}(x \cdot y) = \phi_{(u,v,w)}(x) * \psi_{(u,v,w)}(y)$ and $\Phi_{(u,v,w)}(a,b,c) = (u,v,w)$. Since G is a group and Φ is one to one $\Phi_{(u,v,w)}$ is one to one.

It remains to show that $\theta_{(u,v,w)}$ is well defined Suppose that $z = x\cdot y = x'\cdot y'$, then

$$\theta_{(u,v,w)}(x \cdot y)$$
$$= u * [\phi(a)]^{-1} * \phi(x) * \psi(y) * [\psi(b)]^{-1} * v$$
$$= u * [\phi(a)]^{-1} * \theta(x \cdot y) * [\psi(b)]^{-1} * v$$
$$= u * [\phi(a)]^{-1} * \theta(x' \cdot y') * [\psi(b)]^{-1} * v$$
$$= u * [\phi(a)]^{-1} * \phi(x') * \psi(y') * [\psi(b)]^{-1} * v$$
$$= \theta_{(u,v,w)}(x' \cdot y').$$

$\square$

We now continue our discussion of ubiquity by considering the case of partial transversals. We denote by T_k the k cell partial transversal.

Theorem 3.2 *For every $n > 2$, there is a latin square which is T_k ubiquitous for every $k \leq n$.*

Proof We first note that if a square is T_n ubiquitous then it is T_k ubiquitous for all $k \leq n$. It is well known that a latin square of side n has an orthogonal mate if and only if it has n disjoint transversals. Thus any latin square with an orthogonal mate is T_k ubiquitous for every $k \leq n$. Further it is well known that for every $n \neq 2, 6$ there is such a square. Finally we note that the following square of side 6 is T_6 ubiquitous.

1	2	3	4	5	6
5	3	6	1	4	2
4	1	5	2	6	3
2	4	1	6	3	5
6	5	4	3	2	1
3	6	2	5	1	4

$\square$

We now consider the case of N_2 ubiquitous latin squares. We will ultimately give a construction which results in such a square for every $n > 7$.

Definition 3.3 *An N_2 ubiquitous latin square of side n, denoted $U(n)$, is a latin square in which every cell appears in a 2×2 subsquare.*

Theorem 3.4 *If there exists a latin square of side n and a $U(m)$ then there exists a $U(mn)$.*

Proof We use a standard multiplicative technique to create the square of side mn and show that the result is N_2 ubiquitous. If the latin square of side n has entries from a set X and the $U(m)$ has entries from a set I, we create a square with entries in $X \times I$. We denote entries by x_i where $x \in X$ and $i \in I$. The cell with row x_i, and column y_j, has entry z_k, where (x, y, z) is a cell in the latin square and (i, j, k) is a cell in the U(m). Each cell (i, j) in the U(m) is in an N_2 with rows i, i' and columns j, j', say. This means that each cell (x_i, y_j) is in an N_2 with rows x_i, $x_{i'}$ and columns y_j, $y_{j'}$.   □

Theorem 3.5 *There exists a U(n) for all n even.*

Proof Trivially the 2×2 square, $\mathbb{Z}_2$, is N_2 ubiquitous. Let $n = 2m$ and let L be a latin square of side m, we may now appeal to the previous theorem to get the result.   □

Definition 3.6 *Given a latin square with a transversal and a subsquare, we say that the subsquare is a* transversal avoiding subsquare *if it contains no element of the transversal.*

Definition 3.7 *A Partially Ubiquitous Transverse latin square of side n, PUT(n), is a latin square with a transversal such that every element not in the transversal is in at least one transversal avoiding N_2.*

Theorem 3.8 *If there exists a PUT(n) then there exists a U($n+1$).*

Proof We prolong [Be] the transversal, that is we start with the PUT(n) and add an extra row and column. For each transversal entry with row i, column j and entry e place a new element ∞ in the i, j position, put entry e in the $i, n+1$ and $n+1, j$ cells. Finally place ∞ in the $n+1, n+1$ position.

Now each non transversal cell is in one of the original transversal avoiding N_2. Each of the transversal cells, and those in the new row and column are in an N_2 of the following form.

$$
\begin{array}{c|ccc}
 & j & & n+1 \\
\hline
j & \infty & \cdots & e \\
 & \vdots & \ddots & \vdots \\
n+1 & e & \cdots & \infty
\end{array}
$$

          □

We in fact will use Theorem 3.10 to prove the existence of a PUT(n), however for completeness we include the following more general constructive technique, based on pairwise balanced designs, see [CD] for definitions relating to pairwise balanced designs.

Theorem 3.9 *If there exists a PBD(v, K) and for each $k \in K$ there exists a U(k) then there exists a PUT(v).*

Proof We will define our new square on the pointset of the PBD, X. We first define all elements to be idempotent, i.e. (x, x, x), and take these cells as the transversal. For each block of the PBD, B, with size k we construct a $U(k)$ on the points of B. To find the entry of the new square in row x column y we consider the block in the PBD containing x and y, B say, suppose B has size k. If the $U(k)$ on B has x, y entry z we take (x, y, z) as the entry in the new square.

To see that the resulting square is a PUT(v) consider a cell (x, y, z), where $x \neq y$. We wish to show that this is in a transversal avoiding N_2. Let B be the block containing x and y. The cell (x, y, z) in the $U(k)$ over B occurs in an N_2, let this N_2 have rows x and x' and columns y and y'. Now since x, x', y and y' all appear in B the entries in the new square will be the same as those in the $U(k)$ over B, which proves the result. $\square$

At this point we would normally use Wilson's recursive construction (see [CD]) with some base cases. However the search for base cases yielded the following direct construction. We now present our main construction for a PUT(v).

Theorem 3.10 *For all even v, with $v \geq 8$, there exists a PUT(v).*

Proof We construct the PUT(v) by using the latin square obtained from the multiplication table of the dihedral group D_n, where $n = v/2$. We label elements of the group by x^i and ax^i for $i = 0$ to $n - 1$ and all exponential arithmetic is assumed to be modulo n. We identify four quadrants within the square, in the top left those cells of the form (x^i, x^j, x^{i+j}), top right those of the form (x^i, ax^j, ax^{n-i+j}), bottom left of the form (ax^i, x^j, ax^{i+j}) and bottom right those of the form (ax^i, ax^j, x^{n-i+j}). We note that in this square every cell is in exactly n N_2's, with one cell from each of the four quadrants.

We first deal with the case where n is even. In this case D_n has a transversal as follows.

$$
\begin{array}{ll}
(x^i, x^i, x^{2i}) & \text{for } i = 0 \text{ to } \frac{n}{2} - 1 \\
(x^{n-(i+1)}, ax^i, ax^{2i+1}) & \text{for } i = 0 \text{ to } \frac{n}{2} - 1 \\
(ax^i, x^i, ax^{2i}) & \text{for } i = \frac{n}{2} \text{ to } n - 1 \\
(ax^{n-(i+1)}, ax^i, x^{2i+1}) & \text{for } i = \frac{n}{2} \text{ to } n - 1
\end{array}
$$

Given an arbitrary cell, it may be in up to three cells involving transversal elements, one in the same row, one in the same column, one with the same entry. Since each cell was in n N_2's to begin with, each cell will be in at least $n - 3$ transversal avoiding N_2's. This establishes the result for $n \geq 4$, n even.

We now consider the case when $n \geq 7$ is odd. We first identify the following partial transversal.

$$
\begin{array}{ll}
(x^i, x^i, x^{2i}) & \text{for } i = 0 \text{ to } \frac{n-1}{2} \\
(x^{n-(i+1)}, ax^i, ax^{2i+1}) & \text{for } i = 0 \text{ to } \frac{n-3}{2} \\
(ax^{i+1}, x^i, ax^{2i+1}) & \text{for } i = \frac{n+1}{2} \text{ to } n-2 \\
(ax^{n-i}, ax^i, x^{2i}) & \text{for } i = \frac{n+1}{2} \text{ to } n-1 \\
(a, x^{n-1}, ax^{n-1}) &
\end{array}
$$

This leaves the cell $(ax^{\frac{n+1}{2}}, ax^{\frac{n-1}{2}}, x^{n-1})$ unused and the element a uncovered. To complete the construction we introduce the idea of a *switch* of an N_2 (see [CD] p. 474). Given an N_2 we interchange the positions of the two elements.

a	b
b	a

$\longrightarrow$

b	a
a	b

We now wish to switch an a into the $(ax^{\frac{n+1}{2}}, ax^{\frac{n-1}{2}})$ cell, we cannot do this directly since it would involve switching a transversal element, instead we do a series of switches to get a into the required position. As stage 1 we switch the following subsquares.

$$
\begin{array}{c|ccc}
 & x^{\frac{n+1}{2}} & & ax^{\frac{n-1}{2}} \\
\hline
x^{\frac{n-1}{2}} & x^0 & \cdots & a \\
 & \vdots & \ddots & \vdots \\
ax^{\frac{n-1}{2}} & a & \cdots & x^0
\end{array}
$$

$$
\begin{array}{c|ccc}
 & x^{\frac{n-1}{2}} & & ax^{\frac{n-3}{2}} \\
\hline
x^{\frac{n-3}{2}} & x^{n-2} & \cdots & a \\
 & \vdots & \ddots & \vdots \\
ax^{\frac{n+1}{2}} & a & \cdots & x^{n-2}
\end{array}
$$

This gives

$$
\begin{array}{c|cc}
 & ax^{\frac{n-3}{2}} & ax^{\frac{n-1}{2}} \\
\hline
ax^{\frac{n-1}{2}} & x^{n-2} & a \\
ax^{\frac{n+1}{2}} & a & x^{n-2}
\end{array}
$$

We now switch this subsquare and take $(ax^{\frac{n+1}{2}}, ax^{\frac{n-1}{2}}, a)$ as the last element of the transversal. We must now ensure that all of the other switched elements are in transversal avoiding N_2. In order to do this we perform a further sequence of four of switches, which we refer to as stage 2 switches, as follows.

$$
\begin{array}{c|ccc}
 & x^{\frac{n-3}{2}} & & ax^{\frac{n-5}{2}} \\
\hline
x^{\frac{n-1}{2}} & x^{n-2} & \cdots & ax^{n-2} \\
 & \vdots & \ddots & \vdots \\
ax^{\frac{n-1}{2}} & ax^{n-2} & \cdots & x^{n-2}
\end{array}
$$

$$
\begin{array}{c|cc}
 & x^{\frac{n+3}{2}} & ax^{\frac{n+1}{2}} \\
\hline
x^{\frac{n-3}{2}} & x^0 & \cdots & ax^2 \\
 & \vdots & \ddots & \vdots \\
ax^{\frac{n+1}{2}} & ax^2 & \cdots & x^0
\end{array}
$$

$$
\begin{array}{c|cc}
 & x^{\frac{n+5}{2}} & ax^{\frac{n+3}{2}} \\
\hline
x^{\frac{n+3}{2}} & x^4 & \cdots & a \\
 & \vdots & \ddots & \vdots \\
ax^{\frac{n-5}{2}} & a & \cdots & x^4
\end{array}
$$

$$
\begin{array}{c|cc}
 & x^{\frac{n+3}{2}} & ax^{\frac{n-5}{2}} \\
\hline
x^{\frac{n-5}{2}} & x^{n-1} & \cdots & a \\
 & \vdots & \ddots & \vdots \\
ax^{\frac{n-3}{2}} & a & \cdots & x^{n-1}
\end{array}
$$

We note that none of these switches involve transversal elements, and so these switched elements are all still in a transversal avoiding N_2 subsquare, namely the one which was switched.

This results in the subarray given below. We can see that each of the stage 1 switched elements is in at least one transversal avoiding N_2.

	$x^{\frac{n-3}{2}}$	$x^{\frac{n-1}{2}}$	$x^{\frac{n+1}{2}}$	$x^{\frac{n+3}{2}}$	$x^{\frac{n+5}{2}}$	...	$ax^{\frac{n-5}{2}}$	$ax^{\frac{n-3}{2}}$	$ax^{\frac{n-1}{2}}$	$ax^{\frac{n+1}{2}}$	$ax^{\frac{n+3}{2}}$	...	ax^{n-1}
$x^{\frac{n-5}{2}}$	x^{n-4}	x^{n-3}	x^{n-2}	[a]	x^0	...	x^{n-1}	[ax]	ax^2	ax^3	ax^4	...	$ax^{\frac{n+3}{2}}$
$x^{\frac{n-3}{2}}$	$\mathbf{x^{n-3}}$	[a]	x^{n-1}	ax^2	[x]	...	ax^{n-1}	x^{n-2}	ax	x^0	ax^3	...	$ax^{\frac{n+1}{2}}$
$x^{\frac{n-1}{2}}$	ax^{n-2}	$\mathbf{x^{n-1}}$	[a]	x	$[x^2]$	...	x^{n-2}	ax^{n-1}	x^0	ax	ax^2	...	$ax^{\frac{n-1}{2}}$
$x^{\frac{n+1}{2}}$	x^{n-1}	x^0	x	x^2	x^3	...	ax^{n-3}	$\mathbf{ax^{n-2}}$	ax^{n-1}	a	ax	...	$ax^{\frac{n-3}{2}}$
$x^{\frac{n+3}{2}}$	x^0	[x]	$[x^2]$	x^3	[a]	...	$\mathbf{ax^{n-4}}$	ax^{n-3}	ax^{n-2}	ax^{n-1}	x^4	...	$ax^{\frac{n-5}{2}}$
	⋮					⋱			⋮				⋱
a			...					$[x^{\frac{n-3}{2}}]$	$[x^{\frac{n-1}{2}}]$			...	$[x^{n-1}]$
	⋮					⋱			⋮				⋱
$ax^{\frac{n-5}{2}}$	ax^{n-4}	ax^{n-3}	ax^{n-2}	ax^{n-1}	x^4	...	x^0	x	x^2	x^3	a	...	$x^{\frac{n+3}{2}}$
$ax^{\frac{n-3}{2}}$	ax^{n-3}	ax^{n-2}	ax^{n-1}	x^{n-1}	ax	...	a	x^0	x	x^2	$\mathbf{x^3}$	...	$x^{\frac{n+1}{2}}$
$ax^{\frac{n-1}{2}}$	x^{n-2}	ax^{n-1}	x^0	[ax]	ax^2	...	x^{n-2}	[a]	$[x^{n-1}]$	$\mathbf{x}$	x^2	...	$x^{\frac{n-1}{2}}$
$ax^{\frac{n+1}{2}}$	ax^{n-1}	x^{n-2}	ax	x^0	ax^3	...	x^{n-3}	$[x^{n-1}]$	$\underline{a}$	x^0	x	...	$x^{\frac{n-3}{2}}$
$ax^{\frac{n+3}{2}}$	a	ax	$\mathbf{ax^2}$	ax^3	ax^4	...	x^{n-2}	x^{n-1}	x^0	x	x^2	...	$x^{\frac{n-5}{2}}$
$ax^{\frac{n+5}{2}}$	ax	ax^2	ax^3	$\mathbf{ax^4}$	ax^5	...	x^{n-3}	x^{n-2}	x^{n-1}	x^0	x	...	$x^{\frac{n-7}{2}}$
	⋮					⋱			⋮				⋱

The boldface elements represent members of the transversal, the framed elements indicate the new subsquares which contain the stage 1 switch elements.

We now consider those cells not involved in any switch. Consider an arbitrary cell not involved in any switch. As above it may be involved in at most three N_2's which contain transversal elements. Each switch may destroy at most one subsquare containing the cell. This is because the cell may have at most one of a common row, column or entry with the switched square. Subsquares contain one cell from each quadrant, so if the subsquare has a common row, one of the columns must be in the same quadrant as the original square, and so does not affect any subsquares it may be in. A conjugate argument holds for subsquares with a common column or entry.

There are 7 switches, thus each non switched cell is in at least $n - 10$ transversal avoiding N_2 subsquares, this proves the result for $n \geq 11$. In fact the form of the switches is such that no cell is affected by more than three of the switches, this gives the result for $n \geq 7$.

The case of $n = 5$ is an exception, but may be constructed in a similar manner. The result is given in the table below.

$\mathbf{x^0}$	x	x^2	x^3	x^4	a	ax	ax^2	ax^3	ax^4
ax^3	$\mathbf{x^2}$	x^3	x^4	x^0	ax^4	a	ax	ax^2	x
x^2	x^3	$\mathbf{x^4}$	x^0	x	ax^3	ax^4	a	ax	ax^2
x^3	ax^2	x^0	x	x^2	x^4	$\mathbf{ax^3}$	ax^4	a	ax
a	ax^4	x	x^2	x^3	$\mathbf{ax}$	ax^2	ax^3	x^0	x^4
x^4	ax	ax^2	ax^3	ax^4	x^0	x	x^2	x^3	$\mathbf{a}$
ax	x^4	ax^3	$\mathbf{ax^4}$	a	ax^2	x^0	x	x^2	x^3
ax^2	ax^3	ax^4	a	ax	x^3	x^4	x^0	$\mathbf{x}$	x^2
x	x^0	a	ax	$\mathbf{ax^2}$	x^2	x^3	x^4	ax^4	ax^3
ax^4	a	ax	ax^2	ax^3	x	x^2	$\mathbf{x^3}$	x^4	x^0

Boldface elements are the transversal. □

We may now put together theorems 3.5, 3.8 and 3.10 to conclude the following.

Theorem 3.11 *There exists an N_2 ubiquitous latin square for all $n > 7$.*

Finally we consider ubiquity for the rest of the four cell configurations. Configurations C_1 to C_4, C_9, C_{12} and C_{16} are what are commonly referred to as *constant cell* configurations [DGGM], these are configurations whose number of appearances is constant for every cell, that is dependent solely on n. Since the enumeration of any constant cell configuration depends only on n the counting arguments cannot depend on any particular positioning of the cells, and so we may conclude the following.

Theorem 3.12 *If any constant k cell configuration appears, then it is ubiquitous in every latin square of side n with $n \geq k$.*

We note in passing that this implies that all of the 1, 2 and 3 cell configurations are ubiquitous in every latin square of side 3 or more, further the constant 4-cell configurations are ubiquitous for every latin square of side n, with $n > 4$

For the rest of the four cell configurations the counting argument (for the conjugates given in the tables) begins "Pick a pair of cells, (a, b) in the same row", this is the common factor of $n\binom{n}{2}$ in the enumerations. The argument then continues "pick a different row r". It then bifurcates depending whether $a\,b$ forms an N_2 with the elements in r. For configurations C_7, C_8, C_{11} and C_{15} the exisence of an N_2 increases the count. Since a and b where chosen arbitrarily we can see that these configurations are ubiquitous.

This leaves the configurations $C_6, C_{10}, C_{13}, C_{14}$ and C_{17}. It is evident that these configurations will be ubiquitous in an N_2 subsquare free latin square, if they appear. In [AM, H80, KLR, M] it is shown that latin squares with no intercalates exist for every $n \neq 2$ or 4, and so there are latin squares which are ubiquitous with respect to these configurations for all other values of n. This leads to our final theorem.

Theorem 3.13 *For each four cell configurations C_k, $k = 1 \ldots 18$ and for all $n > 7$ there is a latin square of side n which is ubiquitous with respect to C_k.*

4   Conclusions and Conjectures

The existence of ubiquitous 4-line configurations in latin squares poses some interesting further questions. The first is related to Cayley's Theorem and Theorem 3.1, we conjecture the converse of Theorem 3.1.

Conjecture 4.1 *If a latin square L has the property that all configurations which appear in L are ubiquitous in L, then L is isotopic to a group.*

This seems to be related to whether a block transitive 3-GDD of type n^3 must come from a latin square isotopic to a group.

In [MR] there is a discussion of the concept of simultaneous ubiquity. A latin square L is $\mathcal{C}$-ubiquitous for a collection of configurations $\mathcal{C}$ if it ubiquitous for all configurations $C \in \mathcal{C}$. For example the preceding conjecture can be reworded as "If C_L is the set of all configurations that appear in L then if L is C_L ubiquitous then it is isotopic to a group." A latin square is n-ubiquitous if it is $\mathcal{C}$-Ubiquitous where $\mathcal{C}$ is the set of all n-line configurations. We note that in simultaneous ubiquity the simplification of using main classes and conjugates is not as easily available. A latin square may be ubiquitous for a configuration and not for one of its conjugates.

Conjecture 4.2 *For every n there is an N such that if $k \geq N$ there is a latin square of side k which is n-ubiquitous.*

We note that the Cayley table of D_n is 4-ubiquitous, since every cell is in at least n N_2's as well as being in an $n \times n$ subsquare isotopic to C_n. The $U(n)$'s with n odd constructed in Theorem 3.11 are very close to being 4-ubiquitous, the only "nonubiquity" is that the (n, n) cell appears in no C_6. It appears that the number of C_6's relative to the number of intercalates is crucial in constructing such a square. Indeed from the discussion leading to Theorem 3.13 it seems likely that if a latin square is both C_5 and C_6 ubiquitous then it is 4-ubiquitous.

Given an n cell configuration C with a designated subconfiguration D one can generalize the concept of ubiquitous to that of (C, D)-ubiquitous. A latin square L is (C, D)-ubiquitous if whenever D appears in L it can be extended in L to an appearance of C, the cells of D are called *designated* cells. Kotzig et al [KLR] investigated this for N_2 with two diagonal cells designated. They showed that a latin square which is (C, D)-ubiquitous with respect to this configuration is isotopic to $\mathbf{Z}_2^n$ for some n.

This sort of investigation can be very useful in determining the underlying structure. For example the following is an adaptation of the quadrangle criterion.

Lemma 4.3 *If a quasigroup is (C, D) ubiquitous with respect to the configurations C and D below then it is isotopic to a group.*

	x		y
	z		w
x	y		
z	w		

	x		y
	z		
x	y		
z			

$$C \qquad\qquad D$$

This can be seen by noting that any quasigroup is isotopic to a quasigroup with an identity, e say, and considering the following subarray.

	e	b	c	bc
e		b		bc
a		ab		$a(bc)$
b	b		bc	
ab	ab		$(ab)c$	

However we believe that the use of designated cells is vital to this property and conjecture the following.

Conjecture 4.4 *There is no configuration C whose ubiquity in a quasigroup is equivalent to the associative law.*

Another generalization is that of *strong ubiquity* [MR]. A latin square is strongly ubiquitous with respect to the configuration C if every cell of the latin square appears at least once in each position of C. We note that by a similar argument which led to Theorem 3.12 we can conclude that all constant k-cell configurations are strongly ubiquitous in every latin square with side greater than k.

Finally this concept can be generalized in several different directions depending on which generalization of latin square one wishes to look at. One could ask the same questions about 3-GDD's of type n^g (latin squares being the case of n^3) or about transversal designs $TD(n, k)$ (latin squares being the case $k = 3$). In [LM] the idea of the conjugates of a $TD(n, 4)$ was first discussed. The conjugates of a configuration are obtained by simply applying S_4 to the positions, but the situation is complicated by the fact that not every subgroup of S_4 is the invariant group of the conjugates of some $TD(4, n)$.

Acknowledgement The authors acknowledge the support of NSERC (research grants #OGP0170220 and #OGP0077681).

References

[AM] L.D. Andersen, and E. Mendelsohn, A direct construction for latin squares without proper subsquares, *Ann. Discrete Math.* **15** (1982), 27-53.

[Be] V. D. Belousov, Extensions of quasigroups, *Bull. Acad. Ştince RSS Moldoven* **8** (1967), 3-24.

[B] A.E. Brouwer, Steiner triple systems without forbidden sub-configurations, *Math. Centrum Amsterdam*, ZW104/77.

[BJL] Beth, T. Jungnickel, D. and Lenz, H., *Design theory*, Bibl. Inst., Mannheim, 1985.

[C] P. J. Cameron, Parallelism of complete designs, *London Math. Soc. Lecture Notes* **23**, Cambridge University Press.

[CD] C.J. Colbourn and J.H. Dinitz, (editors), *CRC Handbook of Combinatorial Designs*, CRC Press, Boca Raton, 1996.

[CMRS] C.J. Colbourn, E. Mendelsohn, A. Rosa and J. Širáň, Anti-mitre Steiner triple systems, *Graphs Combinat.* **10** (1994), 215-224.

[DGGM] P. Danziger, M. J. Grannell, T. S. Griggs and E. Mendelsohn, Five line configurations in Steiner triple systems, To Appear, Utilitas Math.

[D]   R.H.F. Denniston, Remarks on latin squares with no subsquares of order two, *Utilitas Math.* **13** (1978), 299-302.

[DK74] J. Dénes and A.D. Keedwell, *Latin Squares and Their Applications*, English Universities Press, London, 1974.

[DK91] J. Dénes and A.D. Keedwell, *Latin Squares: New Developments in the Theory and Applications*, North-Holland, Amsterdam, 1991.

[DS]   J.H. Dinitz and D.R. Stinson, *Contemporary Design Theory — A Collection of Surveys*, John Wiley & Sons, New York, 1992.

[GGM] M. J. Granell, T. S. Griggs and E. Mendelsohn, A small basis for four line configurations in Steiner triple systems, *J. C. D.* **3** (1995), 51-59.

[GMR] T.S. Griggs, E. Mendelsohn, and A. Rosa, Simultaneous decompositions of Steiner triple systems, *Ars Combinat.* **37** (1994), 157-173.

[GR]   T.S. Griggs, and A. Rosa, Avoidance in triple systems, *Acta Math. Univ. Comen.* **63** (1994), 117-131.

[GRR] T.S. Griggs, M.J. de Resmini, and Rosa, A., Decomposing Steiner triple systems into four-line configurations, *Ann. Discrete Math.* **52** (1992), 215-226.

[H80] K. Heinrich, Latin squares with no proper subsquares, *J. Combin. Theory, Ser. A* (1980), 346-353.

[H91] K. Heinrich, Latin squares with and without subsquares of prescribed type, *Latin Squares: New Developments in the Theory and Applications*, J. Dénes and A.D. Keedwell (eds.), North-Holland, Amsterdam, 1991, pp.101-148.

[HW]   K. Heinrich and W.D. Wallis, The maximum number of intercalates in a latin square, in *Combinatorial Math. VIII* (Proc. Eighth Austral. Conf., Geelong, 1980), Lecture Notes in Math. No. 884, Springer Verlag, Berlin (1981), pp.221-233.

[HKZ] A.M. Hobbs, A. Kotzig and J. Zaks, Latin squares with high homogeneity, Proc. Thirteenth S.E. Conf. on Combinatorics, Graph Theory and Computing. *Congressus Numerantum* **35** (1982), 333-345.

[KLR] A. Kotzig, C.C. Lindner and A. Rosa, Latin squares with no subsquares of order two and disjoint Steiner triple systems, *Utilitas Math.* **7** (1975), 287-294.

[LM] C.C. Lindner and E. Mendelsohn, On the conjugates of an $n^2 \times 4$ orthogonal array, *Discrete Math.* **20** (1977), 123-132.

[LR] C.C. Lindner and C.A. Roger, Decomposition into Cycles II: Cycle Systems in [DS], pp.325-370.

[M] M. McLeish, On the existence of latin squares with no subsquares of order two, *Utilitas Math.* **8** (1975), 41-53.

[MR] E. Mendelsohn and A. Rosa *Ubiquitous configurations in Steiner triple systems, J. C. D.* To Appear.

[OP] D. Owens, and J. Preese, Aspects of complete sets of 9×9 pairwise orthogonal latin squares, (1996), pre-print.

[S] A. Sade, Quasigroupes parastrophiques. Expressions et identités, *Math. Nachr.* **20**(1959), 73-106.

[Sm] B. Smetaniuk, A new construction on latin squares — I: A proof of the Evans conjecture, *Ars Combin.* **11** (1981), 155-172.

[SW] D.R. Stinson, and Y.J. Wei, Some results on quadrilaterals in Steiner triple systems, *Discrete Math.* **105** (1992), 207-219.

P. Danziger, Department of Mathematics, Ryerson Polytechnic University, Toronto, Canada.
email: danziger@scs.ryerson.ca

E. Mendelsohn, Department of Mathematics, University of Toronto, Toronto, Canada.
email: mendelso@math.toronto.edu

Difference sets: an update

Dieter Jungnickel   *Bernhard Schmidt*

Abstract

In the last few years there has been rapid progress in the theory of
difference sets. This is a survey of these fascinating new developments.

1   Introduction

This paper is an update of the survey of the first author [Jungnickel (1992)].
Recent surveys on related topics are Ma (1994) (partial difference sets) and
Pott (1996) (relative difference sets). For the connections to coding theory,
we refer the reader to Assmus, Key (1992, 1992a, to appear) and Pott (1992).
A comprehensive introduction to difference sets can be found in Beth, Jung-
nickel, Lenz (1986) and Jungnickel (1992).

For the convenience of the reader, we recall the basic definition. A (v, k, λ)-
difference set in a group G of order v is a k-subset D of G, such that every
element $g \neq 1$ of G has exactly λ representations $g = d_1 d_2^{-1}$ with $d_1, d_2 \in D$.
The parameter $n = k - \lambda$ is called the order of the difference set. We say
that D is abelian, cyclic etc. if G has this property.

The known families of difference sets can be subdivided into three classes:
difference sets with Singer parameters, cyclotomic difference sets
and **difference sets with $(\mathbf{v}, \mathbf{n}) > 1$**. The difference sets with Singer parameters include the classical Singer difference sets and the Gordon-Mills-Welch
series. By cyclotomic difference sets we mean the Paley series consisting of
the quadratic residues in $GF(q)$, $q \equiv 3 \bmod 4$, the families using higher order residues and also the twin prime power series. The families of difference
sets with $(v, n) > 1$ are the Hadamard difference sets, the McFarland and
Spence family and two new series, one found by Davis/Jedwab and one found
by Chen. The construction methods for these three classes of difference sets
are completely different: The Singer difference sets are cyclic and can be
obtained from the action of a cyclic group of linear transformations on the
one-dimensional subspaces of a finite field (viewed as a vector space over a
suitable subfield), while cyclotomic difference sets live in elementary abelian
groups (or the direct product of two such groups) and are unions of cosets
of <u>multiplicative subgroups</u> of finite fields. The class of difference sets with

$(v, n) > 1$ is by far the richest; only recently, in a major work of Davis and Jedwab (1996), it has been discovered that all these difference sets are in fact very similar. In their paper, Davis and Jedwab give a recursive construction which covers all abelian groups known to contain a difference set with $(v, n) > 1$ (a modification is needed to include Chen's series). The best way to describe their construction is in terms of (abelian) characters: The difference set is built up from smaller pieces, which are in some sense orthogonal to each other with respect to the character group.

In the **existence theory** of difference sets, usually the following kind of problem is considered. Given a parameter series (for instance the Singer series), which (abelian) groups can contain a difference set with these parameters? Here the group order is prescribed, and the task is to find necessary and sufficient conditions on the group structure for the existence of a difference set. This turns out to be an extremely difficult problem; up to now, it has been solved for only two infinite parameter series, namely for difference sets in abelian 2-groups [Davis (1991), Kraemer (1993)] and (almost) for McFarland difference sets under the self-conjugacy assumption [McFarland (1973), Ma, Schmidt (1995a, submitted)].

The existence theory for Singer and cyclotomic difference sets on one side and difference sets with $(v, n) > 1$ on the other side is clearly separated. While the main tool for the study of the Singer and cyclotomic difference sets is the multiplier theorem, almost all results on difference sets with $(v, n) > 1$ are exponent bounds and rely on the character theoretic approach introduced by Turyn (1965). Most of the nonexistence results presented in this survey are of the latter type, since the research focussed on difference sets with $(v, n) > 1$ in the last few years.

In this context, we encounter a notion coming from algebraic number theory again and again. A prime p is called **self-conjugate** modulo a positive integer m, if there exists j, such that $p^j \equiv -1 \mod m'$, where m' is the p-free part of m. If we study difference sets in an abelian group G, we usually say that the self-conjugacy assumption is satisfied, if every prime divisor of the order n is self-conjugate modulo $\exp(G)$.

2   Difference Sets with $(v, n) > 1$

There are five known families of difference sets with $(v, n) > 1$, namely the Hadamard difference sets (called Menon difference sets in Jungnickel (1992)), the McFarland and the Spence family, a series similar to the Spence difference sets discovered by Davis, Jedwab (1996), and a series generalizing Hadamard difference set found by Chen.

A striking fact that should be mentioned in such a survey is that all known abelian difference sets with $(v, n) > 1$ satisfy a common condition which might be called the **character divisibility property**: We say that an (v, k, λ)-difference set D of square order $n = k - \lambda$ in an abelian group G satisfies the character divisibility property if the character value $\chi(D)$ is divisible by $\sqrt{n}$ for all nontrivial characters χ of G. Although every known abelian difference set with $(v, n) > 1$ has this property we feel that it would not be a good idea to turn this observation into a conjecture as has been done in similar situations. Instead, we pose the following

Research Problem: Construct difference sets with $(v, n) > 1$ that do not have the character divisibility property.

In our exposition of the latest results, we begin with the Hadamard difference sets (HDSs) which form by far the richest and most important family. In parts of the sections on HDSs we have drawn from the survey Davis, Jedwab (1996a). For more details we refer the reader to this article which deals exclusively with HDSs.

2.1   Abelian HDSs

By a **Hadamard difference set (HDS)**, we mean a difference set with parameters
$$(v, k, \lambda) = (4N^2, 2N^2 - N, N^2 - N).$$
In the last few years, there has been rapid progress in the theory of HDSs. Perhaps this is best demonstrated by the following outdated conjecture which is usually (but erroneously) attributed to McFarland and was wiped out completely by the new results.

Conjecture 2.1 *If there is an abelian HDS with $v = 4N^2$, then $N = 2^r 3^s$ for some integers r, s.*

The reader should compare this conjecture with Theorem 2.4!

We first come to the new constructions and then summarize the recent nonexistence results. We call a difference set D in a group G **reversible** if $\{d^{-1} : d \in D\} = D$.

1) After working hard for about ten years, Xia (1992) found a construction for reversible HDSs in all groups
$$\mathbf{Z}_2 \times \mathbf{Z}_2 \times \mathbf{Z}_{p_1}^4 \times \cdots \times \mathbf{Z}_{p_t}^4,$$
where each p_i is a prime with $p_i \equiv 3 \bmod 4$ We note that this also yields HDSs in $\mathbf{Z}_4 \times \mathbf{Z}_{p_1}^4 \times \cdots \times \mathbf{Z}_{p_t}^4$ (which are not reversible). Xia's sensational construction was the first result disproving Conjecture 2.1.

The proof of the correctness of Xia's construction was considerably simplified by Xiang and Chen (1996).

2) Using a recursive construction Jedwab (1992) showed that HDSs exist in all groups

$$H \times \mathbf{Z}_{s_1} \times \cdots \times \mathbf{Z}_{s_r}$$

(where H is an abelian 2-group of square order with $exp(H) \leq 2\sqrt{|H|}$) if there exists a binary supplementary quadruple (BSQ) (see Jedwab's paper for the definition) of size $s_1 \times \cdots \times s_r$.

A new and much clearer way of viewing this construction is presented in Davis, Jedwab (1996, Cor. 6.4). Because of its importance, we explain this approach in some detail. Davis and Jedwab introduce the notion of a **covering extended building set (covering EBS)**. An $(a, m, h, \pm)$ covering EBS in an abelian group G is a family $\{D_1, ..., D_h\}$ of subsets of G with the following properties.

a) $|D_1| = a \pm m$ and $|D_i| = a$ for $i = 2, ..., h$.

b) For every nonprincipal character χ of G there is exactly one i with $|\chi(D_i)| = m$ and $\chi(D_j) = 0$ if $j \neq i$.

Once a covering EBS is known, it is easy to construct difference sets corresponding to this covering EBS, as is shown in Davis, Jedwab (1996, Theorem 2.4):

Theorem 2.2 *Suppose there exists an $(a, m, h, \pm)$ covering EBS in an abelian group G. Then there exists an $(h|G|, ah \pm m, ah \pm m - m^2)$-difference set in any abelian group containing G as a subgroup of index h.*

The reason why covering EBSs are so useful is that there is a very powerful recursive construction for these objects. Using this method, Davis and Jedwab obtain a unified construction covering all abelian groups which are known to contain a difference set with $(v, n) > 1$.

3) Arasu, Davis, Jedwab, Sehgal (1993) constructed a BSQ of size $3^b \times 3^b$ for all $b \geq 1$. In the terminology of Davis and Jedwab this amounts to a $(3^b(3^b - 1)/2, 3^b, 4, +)$ covering EBS in $\mathbf{Z}_{3^b}^2$.

4) The most recent constructions yield reversible HDSs in $\mathbf{Z}_2 \times \mathbf{Z}_2 \times \mathbf{Z}_p^4$ for all odd primes p (note that this also gives HDSs in $\mathbf{Z}_4 \times \mathbf{Z}_p^4$); in the setting of Davis and Jedwab this amounts to $(p^2(p^2 - 1)/2, p^2, 4, +)$ covering EBSs in $\mathbf{Z}_p^4$. This important new development began with the discovery of a reversible HDS in $\mathbf{Z}_2 \times \mathbf{Z}_2 \times \mathbf{Z}_5^4$ by van Eupen and Tonchev (preprint) who found this difference set by a computer search. Wilson, Xiang (submitted, Theorem 2.2) gave a very general construction method for reversible HDSs in the groups $\mathbf{Z}_2^2 \times \mathbf{Z}_p^4$. They showed that the construction of an HDS in $H \times \mathbf{Z}_p^4$, where $H \cong \mathbf{Z}_4$ or $\mathbf{Z}_2^2$, can be reduced to the construction of a spread S in $PG(3, p)$ and two projective two-weight codes which are connected with S by certain intersection properties. We note that Xia's construction, whose original proof had been very involved, is an easy corollary to this result. By their method,

Wilson and Xiang obtained HDSs in $\mathbf{Z}_2 \times \mathbf{Z}_2 \times \mathbf{Z}_p^4$, $p = 13, 17$ (with the help of a computer search), and exponentially many inequivalent reversible HDSs in $\mathbf{Z}_2 \times \mathbf{Z}_2 \times \mathbf{Z}_p^4$ for $p \equiv 3$ mod 4.

In an earlier version of this survey, we wrote that "it seems very likely that HDSs in $H \times \mathbf{Z}_p^4$, $H \cong \mathbf{Z}_4$ or $\mathbf{Z}_2^2$, exist for all primes $p \equiv 1$ mod 4" and that "a construction should probably use Theorem 2.2 of Wilson and Xiang". In the meantime, exactly this was done by Chen (submitted) in a brilliant work. And he did even more than this: he generalized his construction to get the following new series of difference sets.

Theorem 2.3 *Let r, s, t be any positive integers, and let $q = 3^r$ or $q = p^{2s}$ for any odd prime p. Then there exists a difference set with parameters*

$$(v, k, \lambda) = (4q^{2t}\frac{q^{2t}-1}{q^2-1}, q^{2t-1}[\frac{2(q^{2t}-1)}{q+1}+1], q^{2t-1}(q-1)\frac{q^{2t-1}+1}{q+1})$$

in $K \times V$, where K is any abelian group of order $4\frac{q^{2t}-1}{q^2-1}$ and V is the elementary abelian group of order q^{2t}.

The state of knowledge about abelian HDSs on the existence side is summarized in the next theorem. No other abelian groups are known to contain an HDS. The best method to understand this result is described in Davis, Jedwab (1996, section 6); one has to apply their recursive procedure to the covering EBSs mentioned above to get the following.

Theorem 2.4 *Let H be an abelian group of order 2^{2a+2} ($a \geq 0$) with $exp(H) \leq 2^{a+2}$, let $b_1, ..., b_r$ be positive integers, and let $p_1, ..., p_t$ be (not necessarily distinct) odd primes. Then the group*

$$H \times \mathbf{Z}_{3^{b_1}}^2 \times \cdots \times \mathbf{Z}_{3^{b_r}}^2 \times \mathbf{Z}_{p_1}^4 \times \cdots \times \mathbf{Z}_{p_t}^4$$

contains an HDS. Here $r = 0$ or $t = 0$ is allowed and is interpreted in the obvious way.

On the nonexistence side, all new results rely on the character theoretic approach. In this section, we only mention the results which need the self-conjugacy assumption. More nonexistence results on HDSs can be found in Section 3.

Chan, Ma and Siu (1994) found necessary conditions for groups of p-rank two (p odd) to contain an HDS. This result was generalized by Arasu, Davis and Jedwab (1995) who proved part (a) of the following theorem. Part (b) was independently obtained by Davis, Jedwab (submitted (a)) and Ma, Schmidt (1995).

Theorem 2.5 *Let G be an abelian group with Sylow p-subgroup P of order p^{2a} (p odd) containing an HDS. Write $G = H \times P$. Assume that p is self-conjugate modulo $exp(G)$. Then the following hold.*

(a) $exp(P) \leq p^a$.
Furthermore, if $exp(P) = p^a$, then $P \cong \mathbf{Z}_{p^a} \times \mathbf{Z}_{p^a}$.
(b) If $P \cong \mathbf{Z}_{p^a} \times \mathbf{Z}_{p^a}$, then also each of the groups $H \times \mathbf{Z}_{p^b} \times \mathbf{Z}_{p^b}$ with $b < a$ contains an HDS.

Ray-Chaudhuri and Xiang (to appear (a)) obtained the following partial generalization of the result of Mann and McFarland (1973).

Theorem 2.6 *There are no HDSs in abelian groups $\mathbf{Z}_2 \times \mathbf{Z}_2 \times P$, where $|P| = p^{2a}$, a is odd and p is a prime congruent to 1 mod 4.*

2.2 Nonabelian HDSs

The recent research on nonabelian HDSs focussed on 2-groups and groups of order $4p^2$ (p an odd prime). Davis (1992) showed that Kraemer's method which settled the existence question for HDSs in abelian 2-groups can be modified to construct HDSs in nonabelian 2-groups. This result was generalized to non-2-groups by Meisner (1992, 1996a) who gave a recursive construction of nonabelian HDSs using relative difference sets. A research effort initiated by Dillon (1990) led to the decision of the existence question for HDSs for all 267 groups of order 64. Constructions were found for 258 of these groups and nonexistence was proved for 8. The last remaining case, the so-called modular group of exponent 32, was settled by Liebler and Smith (1993) who found a construction for HDSs in this group with the help of a representation theoretic sieve. Their construction was extended by Davis and Smith (1994) who proved that there exists a group of order 2^{2a+2} and exponent 2^{a+3} containing an HDS for every $a \geq 2$. Even higher exponents where achieved by a recent construction of Davis, Iiams (submitted). They showed that there is an HDS in a nonabelian group of order 2^{4d+2} and exponent 2^{3d+2} for every $d > 0$. By comparison with Turyn's exponent bound for abelian HDSs, this is not exactly what one would have expected!

Liebler (1993) used arguments of McFarland (1989) and techniques from representation theory to prove the following result.

Theorem 2.7 *If p is an odd prime and the group*

$$G = \langle x, y, z | x^p = y^p = z^4 = 1, yx = xy, z^{-1}xz = x^{-1}, z^{-1}yz = y^{-1} \rangle$$

contains an HDS, then $p = 3$.

Iiams (1995) extended Liebler's work to other groups of order $4p^2$ and obtained the following.

Theorem 2.8 *Let $p \geq 5$ be a prime and let G be a group of order $4p^2$ containing an HDS. Then G has an irreducible complex representation of degree 4 or $G \cong \langle x, y, z | x^p = y^p = z^4 = 1, \ xy = yx, \ xz = zx, \ zyz^{-1} = y^{-1} \rangle$ and $p \equiv 1 \bmod 4$.*

We note that Theorem 2.8 excludes 10 of the 16 isomorphism classes of groups of order $4p^2$ in the case $p \equiv 1 \bmod 4$ and 11 of the 12 isomorphism classes in the case $p \equiv 3 \bmod 4$. For both $p \equiv 1$ and 3 mod 4, McFarland (1989) had already excluded four of these classes (the abelian) and Liebler (1993) had excluded one nonabelian class, see Theorem 2.7.

There has also been an interesting discovery of a single nonabelian HDS by Smith (1995), namely an HDS in a group of order 100. McFarland (1989) had shown that an HDS in an abelian group of this order cannot exist. Smith's HDS gives the first example of a parameter triple (v, k, λ) such that a non-abelian but no abelian (v, k, λ)-difference set exists.

Finally, we note that the existence of abelian reversible HDSs implies the existence of certain nonabelian HDSs; for instance, a reversible HDS in $\mathbf{Z}_2^2 \times \mathbf{Z}_p^4$ leads to HDSs in all semi-direct products of $\mathbf{Z}_2^2$ and $\mathbf{Z}_p^4$. For related material, see Meisner (1992, 1996, 1996a).

2.3 McFarland difference sets

A **McFarland difference set** is a difference set with parameters

$$
\begin{aligned}
v &= q^{d+1}[1 + (q^{d+1} - 1)/(q - 1)], \\
k &= q^d(q^{d+1} - 1)/(q - 1), \\
\lambda &= q^d(q^d - 1)/(q - 1),
\end{aligned}
$$

where $q = p^f$ is a prime power and d is a positive integer. A series of difference sets with these parameters was constructed by McFarland (1973) in his important paper. We will assume $(p, f) \neq (2, 1)$, as this is the case of Hadamard difference sets in 2-groups.

In the last years there has been a lot of progress in the existence theory of McFarland difference sets. Under the self-conjugacy assumption the existence problem for abelian McFarland difference sets has been solved almost completely by the work of Ma and Schmidt (1995a, submitted). A new construction in the case $q = 4$ is due to Davis and Jedwab (1996). It shows that the exponent bound of Ma and Schmidt (submitted) is necessary and sufficient in this case.

Let us first look at some previous results. Arasu, Sehgal (1995) constructed a McFarland difference set with $q = 4$ and $d = 1$ in $\mathbf{Z}_2 \times \mathbf{Z}_4^2 \times \mathbf{Z}_3$, a group which is not covered by McFarland's original construction. Arasu, Sehgal (1995a) gave a nonexistence proof for two special cases of McFarland difference sets. In the case $d = 1$, Arasu, Davis, Jedwab, Ma and McFarland (1996) slightly improved the exponent bounds that can be obtained by the arguments of Turyn (1965). Finally, Ma and Schmidt (1995a, submitted) found the following exponent bounds which are best possible and almost completely solve the existence problem for abelian McFarland difference sets under the self-conjugacy assumption.

Theorem 2.9 *Assume that there exists a McFarland difference D set in an abelian group G of order $v = q^{d+1}[1 + (q^{d+1} - 1)/(q - 1)]$, where $q = p^f$ and p is self-conjugate modulo $exp(G)$. Let P be the Sylow p-subgroup of G. Then the following hold.*

(a) If p is odd, then P is elementary abelian.

(b) If $p = 2$ and $f \geq 2$, then $exp(P) \leq 4$.

Note that by the construction of McFarland (1973) condition (a) of Theorem 2.9 is also sufficient. As mentioned above, the construction of Davis and Jedwab (1996) shows that condition (b) is also sufficient for $f = 2$; for $p = 2$ and $f > 2$ there are still a lot of open cases, and it is an interesting question if the methods of Davis, Jedwab (1996) and Ma, Schmidt (1995a) may be combined to solve this problem. It should be mentioned that Theorem 2.9 remains true if the self-conjugacy condition is replaced by the weaker assumption that D has the character divisibility property. This implies that, if the self-conjugacy condition does not hold, constructions of putative difference sets in groups exceeding the exponent bounds of Theorem 2.9 have to be extremely involved. It is a very important question if such constructions are possible.

In the case $f = 1$ of Theorem 2.9 (a) it is possible to determine **all** McFarland difference sets with the given parameters. Ma and Schmidt (1997) proved the following.

Theorem 2.10 *If $f = 1$ in the situation of Theorem 2.9 (a), then D is one of the difference sets constructed by McFarland.*

2.4   The Davis–Jedwab series

After Spence's work [Spence (1977)] who constructed a series of difference sets with parameters

$$
\begin{aligned}
v &= 3^{d+1}(3^{d+1} - 1)/2, \\
k &= 3^d(3^{d+1} + 1)/2, \\
\lambda &= 3^d(3^d + 1)/2,
\end{aligned}
$$

the first discovery of a new parameter series of difference sets is due to Davis and Jedwab (1996). The new parameters are

$$
v = 2^{2d+4}(2^{2d+2} - 1)/3,
$$

$$k = 2^{2d+1}(2^{2d+3} + 1)/3,$$
$$\lambda = 2^{2d+1}(2^{2d+1} + 1)/3,$$

where d is a positive integer. Davis and Jedwab (1996) constructed difference sets with these parameters in all abelian groups of the given order v which have a Sylow 2-subgroup P of exponent at most 4, with the single exception of $d = 1$ and $P \cong \mathbf{Z}_4^3$. This construction is a part of a unifying construction including all abelian groups known to contain a McFarland or Spence difference set, see Corollary 5.3 of Davis, Jedwab (1996). The method is the same as explained in Section 2.1: A recursive construction for covering EBSs combined with Theorem 2.2.

Previously, Ma and Schmidt (1995a) had proved that in the case $d = 1$, that is $(v, k, \lambda) = (320, 88, 24)$, an abelian group containing such a difference set <u>must have</u> a Sylow 2-subgroup of exponent at most 4. Schmidt (preprint) generalized this result and obtained the following. We call a difference set with the above parameters a **Davis-Jedwab difference set**.

Theorem 2.11 *Let G be an abelian group of order $2^{2d+4}(2^{2d+2} - 1)/3$ with Sylow 2-subgroup P. With the possible exception of $d = 1$ and $P \cong \mathbf{Z}_4^3$, a Davis-Jedwab difference set in G <u>that has the character divisibility property</u> exists if and only if $\exp(P) \leq 4$.*

Roughly speaking, this result shows that Davis and Jedwab did a very good job and constructed everything which is possible without using extremely complicated character sums.

3   Difference sets without self-conjugacy

The best way to understand the importance of the self-conjugacy assumption is in terms of (abelian) characters. Let D be a k-subset of an abelian group G. If D is viewed as an element of the group ring $\mathbf{Z}G$, then D is a (v, k, λ)-difference set in G if and only if

$$\chi(D)\overline{\chi(D)} = n \qquad\qquad (3.1)$$

for all nonprincipal characters χ of G, where $n = k - \lambda$ is the order of D. This approach to the study of difference sets was introduced in the classical paper of Turyn (1965). For simplicity, let us assume that n is a square, say $n = u^2$. It can be shown [see Turyn (1965)] that if n is self-conjugate modulo $\exp(G)$, then (3.1) implies that $\chi(D) = u\xi$, where ξ is a root of unity (let us call such solutions **trivial solutions**). This means that $\chi(D)$ can be determined explicitly from (3.1) under the self-conjugacy assumption. Together with the fact that $\chi(D)$ is the image of a subset of G this gives rise to necessary conditions for difference sets. This is the reason why difference sets are quite well understood today, if self-conjugacy is assumed.

Much less is known about difference sets without self-conjugacy, since in this case it is much more difficult to determine $\chi(D)$ from (3.1). How complicated matters can become is best demonstrated by McFarland's work [McFarland (1989)] on HDSs in abelian groups of order $4p^2$. He needed 70 pages to show that no such difference set exists if p is a prime $\equiv 1$ mod 4 (where self-conjugacy does not hold); the proof of the same result for primes $p \equiv 3$ mod 4, $p > 3$, (where self-conjugacy holds) only takes one page [see Mann, McFarland (1973)]!

In this context, it should be mentioned that, from the point of view of the character approach via equation (3.1), Ryser's conjecture (a (v, k, λ)-difference set with $(v, n) > 1$ cannot be cyclic) and Lander's conjecture (if a (v, k, λ)-difference set exists in an abelian group G and p is a prime divisor of (v, n), then the Sylow p-subgroup of G cannot be cyclic) are very dubious. The reason for this is that almost all "evidence" for these conjectures comes from cases where (3.1) has only the trivial solutions. If (3.1) has further solutions, then the situation changes considerably and very little is known about this. We remark that these important problems are closely connected to the question whether there exists a McFarland difference set exceeding the exponent bounds of Theorem 2.9.

A new approach to avoid the self-conjugacy assumption is due to Chan (1993); he showed that in some special cases equation (3.1) has only the trivial solutions, although the self-conjugacy assumption does not hold. This resulted in necessary conditions for the existence of HDSs in groups of the form $\mathbf{Z}_{2pq} \times \mathbf{Z}_{2pq}$ resp. $\mathbf{Z}_{2p} \times \mathbf{Z}_{2p} \times H$, where p, q are distinct prime numbers and H is an abelian q-group. In particular, he showed that an HDS in $\mathbf{Z}_{6p} \times \mathbf{Z}_{6p}$ can only exist if $p = 3$ or $p = 13$. Applications of Chan's method to divisible difference sets can be found in Arasu, Pott (1996).

The most difficult and most interesting problems arise in the cases where equation (3.1) has other solutions than the trivial ones. An example where equation (3.1) has three essentially distinct types of solutions, namely the case of abelian McFarland difference sets with $q = 9$ and $d = 1$ (i.e. $(891, 90, 9)$-difference sets) is studied in the remarkable work of Arasu and Ma (in preparation). Avoiding the explicit determination of these solutions, they prove that such a difference set can only exist if the exponent of the underlying group is 33 (by McFarland's construction [McFarland (1973)], this condition is also sufficient).

Some theorems which are very useful for the study of difference sets without self-conjugacy can be found in Ma's important work [Ma (to appear)] on relative $(n, n, n, 1)$-difference sets.

Another approach to difference sets without self-conjugacy was chosen by

Schmidt (in preparation). He uses properties of the decomposition group of the prime ideal divisors of the order of the difference set together with arguments similar to those of McFarland (1989, section 4) to find restrictions on the solutions of (3.1). To give a flavor of these results, we mention the following special case. By ξ_t we denote a primitive complex t-th root of unity.

Theorem 3.1 *Let $d = p^a m$, where p is an odd prime and $m > 0$ is an odd integer relatively prime to p. If $X \in \mathbf{Z}[\xi_d]$ satisfies*

$$X\overline{X} = p,$$

then with suitable j either $\xi_d^j X \in \mathbf{Z}[\xi_m]$ or $X = \pm \xi_d^j Y$, where Y is a generalized Gauss sum (see Ireland, Rosen (1990)).

With the help of results similar to Theorem 3.1 it is often possible to find all solutions of equation (3.1). One of the most interesting applications concerns the following well-known conjecture on Hadamard matrices, see Jungnickel (1992, section 12).

Conjecture 3.2 *There is no circulant Hadamard matrix of order $m > 4$ (or, equivalently, there is no cyclic Hadamard difference set with $N > 1$).*

Here m and N are connected via $m = 4N^2$. Since Turyn's classical work [Turyn (1965)] it has been known that Conjecture 3.2 is true for $m < 12,100$. Since then there have been a lot of incorrect claims on this subject (and also on Conjecture 3.4 below), see Lin, Wallis (1993). We restrict our attention to the few results which have a chance to be correct. Schmidt (in preparation) extends the bound of $12,100$ for which Conjecture 3.2 is known to be true and also proves some general nonexistence theorems on Hadamard difference sets relying on the approach mentioned above and the sub-difference set method due to McFarland (1990).

A notion closely connected to cyclic HDSs is **Barker sequences**. These are finite sequences $a_1, ..., a_v$ of ones and minus ones, such that the so-called aperiodic autocorrelation

$$c_j = \sum_{i=1}^{v-j} a_i a_{i+j}$$

takes only the values 0 and ± 1 for $j = 1, ..., v - 1$. The only known examples of Barker sequences have length $v \in \{2, 3, 4, 5, 7, 11, 13\}$. It is conjectured that these are all possible values of v.

Conjecture 3.3 *There is no Barker sequence of length $v > 13$.*

The following facts are well-known, see Jungnickel (1992, Section 12).

Theorem 3.4 *There is no Barker sequence of odd length $v > 13$. If there exists a Barker sequence of even length v, then $v = 4N^2$ for some N and there exists an HDS in the cyclic group of order $4N^2$.*

Hence Conjecture 3.3 is weaker than Conjecture 3.2. There is one important theorem on Barker sequences due to Eliahou, Kervaire, Saffari (1990) which is not known to be true for cyclic HDSs:

Theorem 3.5 *There is no Barker sequence of length $4N^2$ if N has a prime divisor congruent to 3 mod 4.*

Using this theorem together with the results of Turyn (1965), Eliahou, Kervaire (1992) showed that Conjecture 3.3 is true for $v < 1,898,884$. Some details of this paper were discussed in Jedwab, Lloyd (1992), Broughton (1994) and Eliahou, Kervaire (1994).

The results of Schmidt (in preparation) give some further restrictions on the length of Barker sequences which do not follow from Turyn's work or Theorem 3.5.

4   Miscellanea

4.1   Difference Sets with multiplier -1

A well-known conjecture of McFarland states that, up to a single exception, all abelian difference sets with multiplier -1 must be HDSs. This conjecture has been the main research problem in this field for many years. Recently, Cao (private communication) claimed to have a proof of McFarland's conjecture (we have not seen the paper yet). Hence the following might be true.

Theorem 4.1 *Let D be a (v, k, λ)-difference set with multiplier -1 (w.l.o.g. assume $k < v/2$ by complementation). Then either $(v, k, \lambda) = (4000, 775, 150)$ or D is a Hadamard difference set.*

Ma (1991) had shown that the proof of Theorem 4.1 can be reduced to the proof of two number theoretic conjectures on solutions of certain diophantine equations, see Jungnickel (1992, section 13). Le, Xiang (1996) could verify the first of these two conjectures. The key to their proof is the observation that a solution violating Ma's conjecture would lead to a fundamental solution of Pell's equation. Finally, Cao (private communication) claimed to have a proof of both of Ma's conjectures completing the proof of Theorem 4.1.

Concerning Hadamard difference sets with multiplier -1, Xiang (submitted) proved that no such difference sets exist in $\mathbf{Z}_2^2 \times \mathbf{Z}_9^2$ and $\mathbf{Z}_4^2 \times \mathbf{Z}_3^2$. This settled the last two open cases with $N < 10$, see Ma (1990).

The new constructions for difference sets with multiplier -1 were already mentioned in Section 2.2; the constructions of Xia (1992), van Eupen, Tonchev (preprint), Wilson, Xiang (submitted) and Chen (submitted) all provide reversible HDSs. Dillon (1990) constructed reversible HDSs in all groups $\mathbf{Z}_{2^t} \times \mathbf{Z}_{2^t}$. Putting all these examples together with the trivial HDS in $\mathbf{Z}_4$ and Turyn's reversible HDS in $\mathbf{Z}_2^2 \times \mathbf{Z}_3^2$ into the recursive constructions of Menon (1962) and Turyn (1984), one obtains the following theorem. No other abelian groups are known to contain a reversible HDS.

Theorem 4.2 *There exist reversible HDS in G and $G \times \mathbf{Z}_2^2 \times \mathbf{Z}_3^{2a} \times \mathbf{Z}_{p_1}^4 \times \cdots \times \mathbf{Z}_{p_s}^4$ for all groups $G = \mathbf{Z}_4^b \times \mathbf{Z}_{2^{c_1}}^2 \times \cdots \times \mathbf{Z}_{2^{c_r}}^2$, where the p_i are (not necessarily distinct) odd primes with and $a, b, c_1, ..., c_r$ are nonnegative integers.*

4.2 Skew Paley-Hadamard difference sets

By **Paley-Hadamard difference sets** we mean difference sets with parameters $(v, k, \lambda) = (4n-1, 2n-1, n-1)$. These were called just Hadamard difference sets in Jungnickel (1992); we have switched to "Paley-Hadamard" to avoid confusion with the HDSs from Sections 2.1 and 2.2. A Paley-Hadamard difference set in a group G is called a **skew Paley-Hadamard difference set** if G is the disjoint union of D, $D^{(-1)} := \{d^{-1} : d \in D\}$ and the identity element. It is well known (see Jungnickel (1992, section 9)) that a skew Paley-Hadamard difference set in an abelian group of order v can only exist if $v = p^m \equiv 3 \bmod 4$ for some prime p and some positive integer m. The following is a longstanding open conjecture.

Conjecture 4.3 *If there exists an abelian skew Paley-Hadamard difference set in a group G of order $v = p^m \equiv 3 \bmod 4$, then G must be elementary abelian.*

Chen, Xiang, Sehgal (1994) made some progress towards Conjecture 4.3. They proved the following result which, in particular, shows that Conjecture 4.3 is true for $m \leq 5$.

Theorem 4.4 *Let G be an abelian p-group, where p is a prime with $p \equiv 3 \bmod 4$, and write $|G| = p^m$, $exp(G) = p^s$. If G admits a skew Paley-Hadamard difference set and $s \geq 2$, then $s \leq (m+1)/4$.*

4.3 Dihedral difference sets

Leung, Ma, Wong (1992) derived strong necessary conditions for the existence of difference sets in dihedral groups lending support to the following conjecture.

Conjecture 4.5 *No nontrivial difference sets exist in dihedral groups.*

Using a computer search, Leung, Ma and Wong verified Conjecture 4.5 for all parameter triples (v, k, λ) with $k - \lambda \leq 10^6$, except five undecided cases. In this context, we mention the following observation of Schmidt (submitted). Special cases of this result were previously obtained by Fan, Ma and Siu (1985) for dihedral difference sets and by Shiu (1996) for arbitrary difference sets.

Theorem 4.6 *There is no nontrivial symmetric (v, k, λ)-design with $v = 2p^m$ for any odd prime p and any positive integer m. In particular, there is no nontrivial difference set in any group of order $2p^m$.*

4.4   Multipliers

There have been some attempts to make progress towards Hall's multiplier conjecture, see Qiu (1993, 1994, 1995, 1995a, 1996, submitted (a), submitted (b)); let us first recall this conjecture.

Conjecture 4.7 *If D is a (v, k, λ)-difference set in an abelian group and t is a divisor of $n = k - \lambda$ relatively prime to v, then t is a multiplier of D.*

Maybe the most interesting result in this direction is the following obtained by Muzychuk (submitted). It finishes the case $n = 2p^a$ which was first considered by Turyn (1964) and Mann, Zaremba (1969), but not completely settled.

Theorem 4.8 *Let D be a (v, k, λ)-difference set in an abelian group, where $n = 2p^a$ for some odd prime p and $(p, |G|) = 1$. Then p is a multiplier of D.*

A unified theorem containing most of the numerous variations of Hall's multiplier theorem is due to Arasu and Xiang (1995). There are also some new results concerning the structure of multiplier groups. Xiang (1994) used techniques from algebraic number theory to get restrictions on the numerical multiplier group of difference sets. Xiang and Chen (1995) obtained the following upper bound for the size of the multiplier group of a cyclic difference set.

Theorem 4.9 *The multiplier group M of a cyclic (v, k, λ)-difference set D has cardinality at most k, unless D is the Singer difference set belonging to $PG(2, 4)$ (in which case $|M| = 6$).*

4.5   Planar difference sets

By a planar difference set we mean a difference set with parameters $(v, k, \lambda) = (n^2 + n + 1, n + 1, 1)$. Such a difference set is equivalent to a projective plane with a regular automorphism group (Singer group), see Beth, Jungnickel,

Lenz (1986). The structure of the multiplier groups of planar difference sets was studied by Ho in a sequence of papers [Ho (1993, 1993a, 1993b, 1994, 1995, submitted (a))]; to give an impression of his results, we mention the following.

Theorem 4.10 *Let* Π *be a projective plane of order* n *with Singer group* G *(not necessarily abelian) and difference set* $D \subset G$, *and let* M *be the multiplier group of* D. *Then the Sylow 2-subgroup* S *of* M *is a cyclic direct factor of* M, *and hence* M *is solvable. Moreover, the following hold.*

a) Write $n = m2^a$, *where* m *is not a square. Then* $|S| \leq 2^a$; *if* M *is abelian, then actually* $|S| = 2^a$ *and* $|M| \leq (m+1)2^a$.

b) M *fixes a line of* Π.

c) If M *has even order, then each subgroup of* G *is invariant under the unique involution in* M, *except possibly if* $n = 16$ *and* G *is nonabelian.*

d) Let H *be an abelian subgroup of* M. *If* H *has odd order, then* $|H| \leq n + 1$. *If* $|H| = n + 1$, *then* $n^2 + n + 1$ *is a prime.*

e) If M *is abelian, then either* $|M| \leq n + 1$ *or* n *is a square.*

f) If G *is abelian, then* $|M| \leq n + 1$ *except for* $n = 4$, *where* $|M| = 6$.

g) If M *is abelian and* n *is a square, then the Sylow 3-subgroup of* M *is cyclic.*

Recently, Ho (submitted (b)) obtained the following generalization of Ott's celebrated theorem, see Ott (1975).

Theorem 4.11 *A finite projective plane admitting more than one <u>abelian</u> Singer group is Desarguesian.*

Gordon (1994, submitted) provides some computational results on the prime power conjecture which states that the order of an abelian planar difference set must be prime power. He uses the known nonexistence results and a computer to show that the prime power conjecture is true for all orders $n \leq 2,000,000$ and to extend the list of integers that cannot divide the order of an abelian planar difference set; the previous version of this list can be found in Jungnickel (1992, Theorem 8.7).

An application of the Singer difference set of $\Pi = PG(2, q)$ was found by Jungnickel (1991); he used this difference set for the construction of a **antipolarity** in Π, i.e. a bijection α between the points and lines of Π satisfying $p \in \alpha(q) \Rightarrow q \notin \alpha(p)$.

4.6 The geometry of Singer and GMW difference sets

It is well-known that $-D$ is an oval of the projective plane corresponding to an abelian planar difference set D, see Jungnickel (1992). In the classical case, one can say more.

Proposition 4.12 *Let D be a Singer difference set corresponding to $PG(2,q)$, where q is odd, and let r be any integer. Then $rD := \{rd : d \in D\}$ is a conic provided that one of the following conditions holds for some integers i, j, k:*
(a) $rp^k(q^i + q^j) \equiv 1 \bmod q^2 + q + 1$;
(b) $rp^k \equiv 2 \bmod q^2 + q + 1$.

This result is due to Jackson, Quinn and Wild (1996). Similar questions have also been investigated in higher dimensions. In the context of constructing perfect ternary sequences, it is of interest for which values of r the set rD obtained from a difference set D corresponding to $\Pi = PG(d, q)$ is a quadric in Π. This has been shown to hold whenever there are integers i, j, k satisfying $rp^k(q^i + q^j) \equiv 1 \bmod (q^{d+1} - 1)/(q - 1)$; however, no necessary and sufficient conditions on r are known in general, and it is also not known for which values of r the resulting quadric is non-degenerate. We refer the reader to Høholdt, Justesen (1983), Games (1986), Jackson, Wild (1992) and Jackson, Quinn, Wild (1996).

Jackson, Wild (to appear) characterized the designs arising from the difference sets of Gordon-Mills-Welch as the $(\frac{q^n-1}{q-1}, \frac{q^{n-1}-1}{q-1}, \frac{q^{n-2}-1}{q-1})$-designs admitting $GL(m, q^t)$ as an automorphism group for appropriate m, t with $mt = n$.

An interesting connection between hyperovals in $PG(2, 2^d)$ and difference sets with Singer parameters $(v, k, \lambda) = (2^d - 1, 2^{d-1} - 1, 2^{d-2} - 1)$ was discovered by A. Maschietti (submitted). In particular, this yields a method to construct three infinite series of difference sets with Singer parameters some of which are probably non-equivalent to the known ones.

4.7   Tables

The table concerning the existence of abelian difference sets with $n \le 30$ in Jungnickel (1992) contained three open entries belonging to the parameters $(v, k, \lambda) = (90, 20, 4)$. The table is now complete, since these entries have been answered: Arasu, Sehgal (1995a) and Arasu, Davis, Jedwab, Ma, McFarland (1996) showed that in two of these cases ($\mathbf{Z}_4 \times \mathbf{Z}_8 \times \mathbf{Z}_3$ and $(\mathbf{Z}_2)^2 \times \mathbf{Z}_8 \times \mathbf{Z}_3$) no difference set can exist. The remaining case ($\mathbf{Z}_2 \times (\mathbf{Z}_4)^2 \times \mathbf{Z}_3$) was settled via construction by Arasu, Sehgal (1995).

There are some new tables available. The CRC handbook of combinatorial designs contains a table of abelian difference sets [Jungnickel, Pott (1996)] as well as of nonabelian difference sets [Smith (1996)]; these tables do not only deal with the existence question, but also provide a lot of explicit examples of difference sets.

Further tables of difference sets are Kopilovich (1989) [abelian noncyclic difference sets with $k \le 100$] and Vera Lopez, Garcia Sanchez (to appear) [abelian difference sets with $100 < k \le 150$].

Up to our knowledge, the only open cases of abelian (v, k, λ)-difference sets with $k \leq 100$ are the following.

$$(640, 72, 8), \quad \mathbf{Z}_2 \times \mathbf{Z}_4^3 \times \mathbf{Z}_5;$$
$$(640, 72, 8), \quad \mathbf{Z}_2^3 \times \mathbf{Z}_4^2 \times \mathbf{Z}_5;$$
$$(320, 88, 24), \quad \mathbf{Z}_4^3 \times \mathbf{Z}_5.$$

This is an update of the table in Jungnickel, Pott (1996). The updates are:

a) Iiams (in preparation) excluded the following cases.

$$(288, 42, 6), \quad \mathbf{Z}_4 \times \mathbf{Z}_8 \times \mathbf{Z}_3^2;$$
$$(288, 42, 6), \quad \mathbf{Z}_2^2 \times \mathbf{Z}_8 \times \mathbf{Z}_3^2;$$
$$(189, 48, 12), \quad \mathbf{Z}_3^3 \times \mathbf{Z}_7;$$
$$(176, 50, 14), \quad \mathbf{Z}_4^2 \times \mathbf{Z}_{11};$$
$$(176, 50, 14), \quad \mathbf{Z}_2^2 \times \mathbf{Z}_4 \times \mathbf{Z}_{11};$$
$$(176, 50, 14), \quad \mathbf{Z}_2^4 \times \mathbf{Z}_{11}.$$

b) The cases $(160, 54, 18)$, $\mathbf{Z}_2 \times \mathbf{Z}_{16} \times \mathbf{Z}_5$ and $\mathbf{Z}_4 \times \mathbf{Z}_8 \times \mathbf{Z}_5$ were excluded by Ma, Schmidt (1997).

c) Abelian $(320, 88, 24)$-difference sets were constructed by Davis, Jedwab (1996) in all abelian groups of order 320 and exponent not exceeding 20, except in $\mathbf{Z}_4^3 \times \mathbf{Z}_5$.

d) Arasu and Ma (in preparation) showed that no abelian $(891, 90, 9)$-difference sets exist in groups of exponent exceeding 33.

e) The case of a $(783, 69, 6)$-difference set in $\mathbf{Z}_3^3 \times \mathbf{Z}_{29}$ was excluded by Schmidt (to appear).

Acknowledgement The authors are grateful to J.A. Davis, J. Jedwab, S.L. Ma and Q. Xiang for several useful suggestions concerning this survey.

5 References

K.T. Arasu, J.A. Davis, J. Jedwab: A nonexistence result for abelian Menon difference sets using perfect binary arrays. Combinatorica, 15 (1995), 311-317.

K.T. Arasu, J.A. Davis, J. Jedwab, S.L. Ma, R.L. McFarland: Exponent bounds for a family of abelian difference sets. In: Groups, Difference Sets, and the Monster. Eds. K.T. Arasu, J.F. Dillon, K. Harada, S.K. Sehgal, R.L. Solomon. DeGruyter Verlag, Berlin/New York (1996), 129-143.

K.T. Arasu, J.A. Davis, J. Jedwab, S.K. Sehgal: New constructions of Menon difference sets, J. Combin. Theory Ser. A 64 (1993), 329-336.

K.T. Arasu, S.L. Ma: Abelian Groups Admitting McFarland Difference Sets of Order 81. In preparation.

K.T. Arasu, A. Pott: Impossibility of a certain cyclotomic equation with applications to difference sets. Des. Codes Cryptogr. 8 (1996), 23-28.

K.T. Arasu, S.K. Sehgal: Some new difference sets. J. Combin. Theory Ser. A 69 (1995), 170-172.

K.T. Arasu, S.K. Sehgal: Difference sets in abelian groups of p-rank two. Des. Codes Cryptogr. 5 (1995a), 5-12.

K.T. Arasu, Q. Xiang: Multiplier Theorems. J. Combin. Des. 3 (1995), 257-267.

E.F. Assmus, J.D. Key: Hadamard matrices and their designs: a coding theoretic approach. Trans. Amer. Math. Soc. 330 (1992), 269-293.

E.F. Assmus, J.D. Key: Designs and their Codes: Cambridge University Press, Cambridge (1992a).

E.F. Assmus, J.D. Key: Designs and codes: an update. Des. Codes Cryptogr. 9 (1996), 7-27.

T. Beth, D. Jungnickel, H. Lenz: Design Theory. Cambridge University Press, Cambridge (1986).

W.J. Broughton: A note on Table 1 of "Barker sequences and difference sets". L'Enseignement Math. 50 (1994), 105-107.

Z. Cao: Two number-theoretic conjectures and abelian difference sets with multiplier -1. Private communication.

W.K. Chan: Necessary Conditions for Menon Difference Sets. Des. Codes Cryptogr. 3 (1993), 147-154.

W.K. Chan, S.L. Ma, M.K. Siu: Non-existence of certain perfect arrays. Discrete Math. 125 (1994), 107-113.

Y.Q. Chen: On the existence of abelian Hadamard difference sets and generalized Hadamard difference sets. Submitted.

Y.Q. Chen, Q. Xiang, S.K. Sehgal: An exponent bound on skew Hadamard abelian difference sets. Des. Codes Cryptogr. 4 (1994), 313-317.

J.A. Davis: Difference sets in abelian 2-groups. J. Combin. Theory Ser. A 57 (1991), 262-286.

J.A. Davis (1992): A generalization of Kraemer's result on difference sets. J. Combin. Theory Ser. A 59, 187-192.

J.A. Davis, J.E. Iiams: Hadamard difference sets in nonabelian 2-groups with high exponent. Submitted.

J.A. Davis, J. Jedwab: A unifying construction of difference sets. Technical Report HPL-96-31, Hewlett-Packard Labs., Bristol (1996).

J.A. Davis, J. Jedwab: A summary of Hadamard difference sets. In: Groups, Difference Sets, and the Monster. Eds. K.T. Arasu, J.F. Dillon, K. Harada,

S.K. Sehgal, R.L. Solomon. DeGruyter Verlag, Berlin/New York (1996a), 145-156.

J.A. Davis, J. Jedwab: Nested Hadamard difference sets. Submitted.

J.A. Davis, J. Jedwab: Recent developments in difference sets. In preparation.

J.A. Davis, K.W. Smith: A construction of difference sets in high exponent 2-groups using representation theory. J. Algebraic Combin. 3 (1994), 137-151.

J.F. Dillon: A survey of difference sets in 2-groups. Presented at the Marshall Hall Memorial Conference, Vermont (1990).

J.F. Dillon: Difference sets in 2-groups. In: Finite Geometries and Combinatorial Designs. Contemp. Math. 111 (1990a), 65-72.

S. Eliahou, M. Kervaire: Barker sequences and difference sets. L'Enseignement Math. 38 (1992), 345-382.

S. Eliahou, M. Kervaire: Corrigendum to "Barker sequences and difference sets". L'Enseignement Math. 40 (1994), 109-111.

S. Eliahou, M. Kervaire, B. Saffari: A new restriction on the length of Golay complementary sequences. J. Combin. Theory Ser. A 55 (1990), 49-59.

M. van Eupen, V.D. Tonchev: Linear codes and the existence of a reversible Hadamard difference set in $Z_2 \times Z_2 \times Z_5^4$. Preprint.

C.T. Fan, S.L. Ma, M.K. Siu: Difference sets in dihedral groups and interlocking difference sets. Ars Combin. 20 A (1985), 99-107.

R.A. Games: The geometry of quadrics and correlation of sequences. IEEE Trans. Inform. Theory 32 (1986), 423-426.

D. Ghinelli: Regular groups on generalized quadrangles and nonabelian difference sets with multiplier -1. Geom. Dedicata 41 (1992), 165-174.

D.M. Gordon: The prime power conjecture is true for $n < 2,000,000$. Electron. J. Combin. 1, R6 (1994).

D.M. Gordon: Some restrictions on orders of abelian planar difference sets. Submitted.

C.Y. Ho: Planar Singer groups with even order multiplier groups. In: Finite Geometry and Combinatorics. Eds. F. De Clerck et al., Cambridge University Press, Cambridge (1993), 187-198.

C.Y. Ho: Projective planes with a regular collineation group and a question about powers of a prime. J. Algebra 154 (1993a), 141-151.

C.Y. Ho: Singer groups, an approach from a group of multipliers of even order. Proc. Amer. Math. Soc. 119 (1993b), 925-930.

C.Y. Ho: Subplanes of a tactical decomposition and Singer groups of a projective plane. Geom. Dedicata 53 (1994), 307-326.

C.Y. Ho: Some basic properties of planar Singer groups. Geom. Dedicata 55

(1995), 59-70.

C.Y. Ho: Arc subgroups of planar Singer groups. Submitted (a).

C.Y. Ho: Finite projective planes with abelian transitive collineation groups. Submitted (b).

T. Høholdt, J. Justesen: Ternary sequences with perfect periodic autocorrelation. IEEE Trans. Inform. Theory 29 (1983), 597-600.

J.E. Iiams: On difference sets in groups of order $4p^2$. J. Combin Theory Ser. A 72 (1995), 256-276.

J.E. Iiams: Lander's tables are complete. In Preparation.

K. Ireland, M. Rosen: A Classical Introduction to Modern Number Theory. Spring- er, Berlin/Heidelberg/New York (1990).

W.-A. Jackson, K.A.S. Quinn, P.R. Wild: Quadrics and difference sets. Ars Combin. 42 (1996), 97-106.

W.-A. Jackson, P.R. Wild: Relations between two perfect ternary sequence constructions. Des. Codes Cryptogr. 2 (1992), 325-332.

W.-A. Jackson, P.R. Wild: On GMW designs and cyclic Hadamard designs. Des. Codes Cryptogr. 10 (1997), 185-192.

J. Jedwab (1992): Generalized perfect arrays and Menon difference sets. Des. Codes Cryptogr. 2, 19-68.

J. Jedwab: Non-existence of certain perfect binary arrays. Electron. Letters 29 (1993), 99-101.

J. Jedwab, S. Lloyd: A note on the non-existence of Barker sequences. Des. Codes Cryptogr. 2 (1992), 93-97.

D. Jungnickel: An anti-polarity in $PG(2,q)$. Bull. Inst. Combin. Appl. 3 (1991), 78.

D. Jungnickel: Difference Sets. In: J.H. Dinitz and D.R. Stinson, eds., Contemporary Design Theory: A Collection of Surveys. Wiley, New York (1992), 241-324.

D. Jungnickel, A. Pott: Difference sets: abelian. In: The CRC handbook of combinatorial designs. Eds. C.J. Colbourn, J. Dinitz. CRC Press, Boca Raton (1996), 297-307.

L.E. Kopilovich: Difference sets in non-cyclic groups. Kibernetika 2 (1989), 20-23.

R.G. Kraemer: Proof of a conjecture on Hadamard 2-groups. J. Combin. Theory Ser. A (1993), 1-10.

M. Le, Q. Xiang: A result on Ma's conjecture. J. Combin. Theory Ser. A 73 (1996), 181-184.

K.H. Leung, S.L Ma, Y.L. Wong (1992): Difference sets in dihedral groups. Des. Codes Cryptogr. 1, 333-338.

R.A. Liebler: The Inversion Formula. J. Combin. Math. Combin. Comput.

13 (1993), 143-160.

R.A. Liebler, K.W. Smith: On difference sets in certain 2-groups. In: Coding Theory, Design Theory, Group Theory. Eds. D.Jungnickel, S.A. Vanstone. Wiley, New York (1993), 195-212.

C. Lin, W.D. Wallis: On the circulant Hadamard conjecture. In: Coding theory, Design Theory, Group Theory. Eds. D. Jungnickel, S.A. Vanstone. Wiley, New York (1993), 213-217.

S.L. Ma: Polynomial addition sets and symmetric difference sets. In: IMA Vol. Math. Appl. 21: Coding Theory and Design Theory, Ed. D. Ray-Chaudhuri. Springer-Verlag, New York (1990), 273-279.

S.L. Ma: McFarland's conjecture on abelian difference sets with multiplier minus one. Des. Codes Cryptogr. 1 (1991), 312-332.

S.L. Ma: A survey of partial difference sets. Des. Codes Cryptogr. 4 (1994), 221-261.

S.L. Ma: Planar functions, relative difference sets and character theory. J. Algebra, to appear.

S.L. Ma, B. Schmidt: On (p^a, p, p^a, p^{a-1})-relative difference sets. Des. Codes Cryptogr. 6 (1995), 57-71.

S.L. Ma, B. Schmidt: The structure of abelian groups containing McFarland difference sets. J. Combin. Theory Ser. A 70 (1995a), 313-322.

S.L. Ma, B. Schmidt: A sharp exponent bound for McFarland difference sets with $p = 2$. Submitted.

S.L. Ma, B. Schmidt: Difference sets corresponding to a class of symmetric designs. Des. Codes Cryptogr. 10, (1997), 223-236.

H.B. Mann, R.L. McFarland: On Hadamard difference sets. In: J.N. Srivastava et. al. (eds.), A Survey of Combinatorial Theory. North-Holland, Amsterdam (1973), 333-334.

H.B. Mann, S.K. Zaremba: On multipliers of difference sets. Illinois J. Math. 13 (1969), 378-382.

A. Maschietti: Difference sets and hyperovals. Submitted.

R.L. McFarland: A family of difference sets in non-cyclic abelian groups. J. Combin. Theory Ser. A (1973), 1-10.

R.L. McFarland (1989): Difference sets in abelian groups of order $4p^2$. Mitt. Math. Sem. Giessen 192, 1-70.

R.L. McFarland: Sub-difference sets of Hadamard difference sets. J. Combin. Theory Ser. A 54 (1990), 112-122.

D.B. Meisner: Families of Menon difference sets. Ann. Discrete Math. 52 (1992), 365-380.

D.B. Meisner: A difference set construction of Turyn adapted to semi-direct products. In: Groups Difference Sets and the Monster. Eds. K.T. Arasu,

J.F. Dillon, K. Harada, S.K. Sehgal, R.L. Solomon. DeGruyter Verlag, Berlin/New York (1996), 169-174.

D.B. Meisner: New classes of groups containing Menon difference sets. Des. Codes Cryptogr. 8 (1996a), 319-325.

P.K. Menon: On difference sets whose parameters satisfy a certain relation. Proc. Amer. Math. Soc. 13 (1962), 739-745.

M. Muzychuk: Difference sets with $n = 2p^m$. Submitted.

U. Ott: Endliche zyklische Ebenen. Math. Z. 144 (1975), 195-215.

A. Pott: On abelian difference set codes. Des. Codes Cryptogr. 2 (1992), 263-271.

A. Pott: Finite geometry and character theory. Springer, New York (1995).

A. Pott: A survey on relative difference sets. In: Groups, Difference Sets, and the Monster. Eds. K.T. Arasu, J.F. Dillon, K. Harada, S.K. Sehgal, R.L. Solomon. DeGruyter Verlag, Berlin/New York (1996), 195-232.

W. Qiu: Proving the multiplier theorem using representation theory of groups. Northeast. Math. J. 9 (1993), 169-172.

W. Qiu: The multiplier conjecture for elementary abelian groups. J. Combin. Des. 2 (1994), 117-129.

W. Qiu: On the multiplier conjecture. Acta. Math. Sinica, New Series 10 (1994), 49-58.

W. Qiu: A method of studying the multiplier conjecture and some partial solutions to it. Ars. Combin. 39 (1995), 5-23.

W. Qiu: The multiplier conjecture for the case $n = 4n_1$. J. Combin. Des. 3 (1995), 393-397.

W. Qiu: A necessary condition on the existence of abelian difference sets. Discrete Math. 137 (1995a), 383-386.

W. Qiu: Further results on the multiplier conjecture for the case $n = 2n_1$. J. Comb. Math. Comb. Comp. 20 (1996), 27-31.

W. Qiu: A character approach to the multiplier conjecture and a new result on it. Submitted (a)

W. Qiu: Further results on the multiplier conjecture for $2 = 2n_1$ and $n = 3n_1$. Submitted (b).

D.K. Ray-Chaudhuri, Q. Xiang: Constructions of partial difference sets and relative difference sets using Galois rings. Des. Codes Cryptogr. 8 (1996), 215-228.

D.K. Ray-Chaudhuri, Q. Xiang: New necessary conditions for abelian Hadamard difference sets. J. Statist. Plann. Inference, to appear (a).

B. Schmidt: Nonexistence of a $(783, 69, 6)$-difference set. To appear in Discrete Math.

B. Schmidt: There are no symmetric (v, k, λ)-designs with $v = 2p^m$. Submitted.

B. Schmidt: Decomposition groups, class groups and difference sets. In preparation.

B. Schmidt: Nonexistence results for Chen and Davis-Jedwab difference sets. Preprint.

W.C. Shiu: Difference sets in groups containing subgroups of index 2. Ars Combin. 42 (1996), 199-205.

K.W. Smith: Non-abelian Hadamard difference sets. J. Combin. Theory Ser. A 70 (1995), 144-156.

K.W. Smith: Difference sets: nonabelian. In: The CRC handbook of combinatorial designs. Eds. C.J. Colbourn, J. Dinitz. CRC Press, Boca Raton (1996), 308-312.

R.J. Turyn: The multiplier theorem for difference sets. Canad. J. Math. 16 (1964), 386-388.

R.J. Turyn: Character sums and difference sets. Pacific J. Math. 15 (1965), 319-346.

R.J. Turyn: A special class of Williamson matrices and difference sets. J. Combin. Theory Ser. A 36 (1984), 195-228.

A. Vera Lopez, M.A. Garcia Sanchez: On the existence of abelian difference sets with $100 < k \leq 150$. J. Combin. Math. Combin. Comput., to appear.

R. Wilson, Q. Xiang: Constructions of Hadamard difference sets. Submitted.

M.Y. Xia: Some infinite classes of special Williamson matrices and difference sets, J. Combin. Theory Ser. A 61 (1992), 230-242.

Q. Xiang: Some results on -1 multiplier of difference lists. Ann. Discrete Math. 52 (1994), 559-566.

Q. Xiang: Some results on multipliers and numerical multiplier groups of difference sets. Graphs Combin. 10 (1994), 293-304.

Q. Xiang: On reversible abelian Hadamard difference sets. Submitted.

Q. Xiang, Y.Q. Chen: On the size of the multiplier groups of cyclic difference sets. J. Combin. Theory Ser. A 69 (1995), 168-169

Q. Xiang, Y.Q. Chen: On Xia's Construction of Hadamard Difference Sets. Finite Fields Appl. 2 (1996), 86-95.

6 "Old" References

The following are references of papers which had been mentioned in the previous survey Jungnickel (1992), but had not appeared at that time.

K.T. Arasu, V.C. Mavron: Biplanes and Singer groups. In: Coding theory, Design Theory, Group Theory. Eds. D. Jungnickel, S.A. Vanstone. Wiley, New York (1993), 111-119.

J.A. Davis: A note on non-abelian $(64, 28, 12)$-difference sets. Ars. Combin. 32 (1991), 311-314.

S. Gao, W. Wei: On non-abelian group difference sets. Discrete Math. 112 (1993), 93-102.

D. Hachenberger: On the existence of translation nets. J. Algebra 152 (1992), 207-229.

D. Hachenberger: On a combinatorial problem in group theory. J. Combin. Theory Ser. A 64 (1993), 79-101.

J.W.P. Hirschfeld: Projective spaces of square size. Simon Stevin 65 (1991), 319-329.

C.Y. Ho: On bounds for groups of multipliers of planar difference sets. J. Algebra 148 (1992), 325-336.

J. Jedwab, C. Mitchell, F. Piper, P. Wild: Perfect binary arrays and difference sets. Discrete Math. 125 (1994), 241-254.

D. Jungnickel: On Lander's multiplier theorem for difference lists. J. Combin. Inform. System Sci. 17 (1992), 123-129.

K.H. Leung, S.L. Ma: Constructions of partial difference sets and relative difference sets on p-groups. Bull. London Math. Soc. 22 (1990), 533-539.

S. Long: A generalization of the notion of ovals to symmetric designs. In: Coding theory, Design Theory, Group Theory. Eds. D. Jungnickel, S.A. Vanstone. Wiley, New York (1993), 219-225.

A. Pott: A generalization of a construction of Lenz. Sankhyā Ser. A 54 (1992), 315-318.

A. Pott: New necessary conditions for abelian difference sets. Combinatorica 12 (1992), 89-93.

Dieter Jungnickel, Mathematisches Institut, Universität Augsburg, Universitätsstraße 15, 86135 Augsburg, Germany.
email: jungnickel@math.uni-augsburg.de

Bernhard Schmidt, Mathematisches Institut, Universität Augsburg, Universitätsstraße 15, 86135 Augsburg, Germany.
email: bernhard.schmidt@math.uni-augsburg.de

Computational results for the known biplanes of order 9

J. D. Key   *V. D. Tonchev*

Abstract

The ternary codes associated with the five known biplanes of order 9 were examined using the computer language Magma. The computations showed that each biplane is the only one to be found among the weight-11 vectors of its ternary code, and that none of the biplanes can be extended to a 3-(57,12,2) design. The residual designs of the biplanes, and designs associated with $\{12; 3\}$-arcs were also examined.

1   Introduction

There are five known biplanes of order 9, i.e. 2-(56,11,2) designs. Following the notation commenced by Denniston [8] we denote these by $B1$ to $B5$ where

1. $B1$ was found by Hall, Lane and Wales [10];

2. $B2$ was found by Mezzaroba and Salwach [14];

3. $B3$ was found by Denniston [8];

4. $B4$ was found by Denniston [8];

5. $B5$ was found by Janko and Trung [12].

The order of a biplane of block size 11 is 9, and thus only the ternary codes of such a biplane will be of interest. We computed the dimensions of the code of the design, its orthogonal (dual) code, the hull (the intersection of the code and its orthogonal), and the orthogonal of the hull, in each case, and found the minimum weight of each code. We also collected the minimum words of the code of each of these designs; in all cases 11 is the minimum weight, and in all but one case the only minimum words were the scalar multiples of the incidence vectors of the blocks of the design. In the case of $B3$ there were more weight-11 vectors, but they were not constant vectors, and even so, the supports did not hold another biplane.

We also collected all (or almost all) of the constant words of weight 12 in the orthogonal code for each of the biplanes that had sufficient of these, since this would allow us to be able to find out whether or not any of these biplanes could be extended (see the argument of why this holds in Section 2). In fact three of the biplanes ($B3$,$B4$ and $B5$) had insufficient numbers of such vectors, and the other two we were able to check and show that there are not enough with the required intersection properties. Thus we conclude that none of the biplanes can be extended to a 3-(57,12,2) design. This confirmed various reports of earlier computations for the first four biplanes that we have been unable to find documented proof of, as well as the findings of Bagchi [2], who claimed to have a proof that designs with these parameters cannot exist. This claim was subsequently invalidated by Brouwer [4] (see also Brouwer and Wilbrink [5]) due to a computational error: Lemma 4.1 in [2], which is crucial to the proof, asserts that a certain 16×16 matrix is nonsingular which leads to a unique solution of a system of equations, yielding a contradiction that establishes the lemma; the matrix however has rank 15, as was shown by Brouwer and verified again by our computations with Magma. Note that the interest in the extendability of biplanes with the given parameters stems partially from Cameron's theorem [6] (see [11, Theorem 4.2] for a statement and proof) which singles out this case as being a possible candidate of a symmetric design extending to a 3-design.

All the computations concerning the ternary codes and the collection of the vectors were made using Magma (V1) [3] on SPARCstation 5 model 70 computers running Solaris 2.4; the computations to check the supports of the weight-12 vectors in the case of $B1$ and $B2$ were made initially at Michigan Technological University, and later checked with Magma. Copies of the Magma output for the weight distributions that we found are not included due to lack of space, but may be obtained on request from the first author. In Table 1 C denotes the ternary code of the biplane, and H its hull; the last two columns show the number of minimum weight words of C and the number of $\{0, 1\}$ weight-12 words of $C^{\perp}$, respectively.

The paper is organised as follows: after a brief description of our terminology and some background results, Sections 3 deals with the non-extendability of the five biplanes; Section 4 gives the results we obtained for the 16 residual designs; Section 5 gives the results we obtained for $\{12; 3\}$-arcs.

2   Background and terminology

The notation used is generally standard and we refer the reader to Assmus and Key [1]. We recall here some of the definitions that we particularly need.

An incidence structure $\mathcal{D} = (\mathcal{P}, \mathcal{B})$ with point set $\mathcal{P}$ and block set $\mathcal{B}$ is a

| Biplane | dim(C) | \|Aut\| | Minimum weight | | | | Wt 11 | Wt 12 |
			C	$C^\perp$	H	$H^\perp$	C	$C^\perp$
$B1$	20	80640	11	8	18	7	112	2100
$B2$	22	288	11	8	12	7	112	516
$B3$	26	144	11	8	12	7	120	≤ 91
$B4$	24	64	11	8	12	7	112	148
$B5$	26	24	11	8	12	7	112	≤ 22

Table 1: The five biplanes

t-(v, k, λ) design if every block is incident with precisely k points and any set of t distinct points are together incident with precisely λ blocks. It follows (see [1, Chapter 1]) that $\mathcal{D}$ is an s-design for any $s < t$; we denote the number of blocks incident with s points by λ_s. The *order* of a t-design, where $t \geq 2$, is $n = \lambda_1 - \lambda_2$. Given a t-(v, k, λ) design $\mathcal{D} = (\mathcal{P}, \mathcal{B})$, let $x \in \mathcal{P}$; the *derived structure* at x is defined to be the incidence structure $\mathcal{D}_x = (\mathcal{P} - \{x\}, \mathcal{B}_x)$, where $\mathcal{B}_x = \{B \mid B \in \mathcal{B}, x \in B\}$, with incidence as in $\mathcal{D}$. If $t \geq 2$ then $\mathcal{D}_x$ is a $(t-1)$-$(v-1, k-1, \lambda)$ design. If $\mathcal{D}$ and $\mathcal{E}$ are designs and $\mathcal{D} = \mathcal{E}_x$ for some point x in the point-set of $\mathcal{E}$, then $\mathcal{E}$ is called an *extension* of $\mathcal{D}$.

A 2-design with $|\mathcal{P}| = |\mathcal{B}|$ is called *symmetric* , and is usually denoted simply by the parameters (v, k, λ). The symmetric designs with $\lambda = 1$ are the finite *projective planes*; symmetric designs with $\lambda = 2$ are called *biplanes*. The order of a biplane with block size k is $k - 2$. Unlike the situation for planes, there are only a small finite number of biplanes known at present, and it is completely unknown if the number of biplanes is finite or infinite. For $k = 11$ there are five biplanes known, but this list is not known to be complete. The parameters in this case satisfy the necessary divisibility conditions for the biplanes to extend to 3-$(57, 12, 2)$ designs. However, no such designs are known to exist and we have now shown computationally that the five known biplanes cannot be extended.

For a symmetric design $\mathcal{D}$, the *dual structure* is also a symmetric design, and we denote it by $\mathcal{D}^t$. In this case we can also define a *residual* design for any block B of $\mathcal{D}$: this is the incidence structure $\mathcal{D}^B = (\mathcal{P} - B, \mathcal{B}^B)$ where where $\mathcal{B}^B = \{C - B \mid C \in \mathcal{B} - \{B\}\}$. Thus the residual is obtained by deleting one block and all the points on it: for example, affine planes are obtained from projective planes by precisely this method. Thus if $\mathcal{D}$ is a symmetric 2-(v, k, λ) design, then $\mathcal{D}^B$ is a 2-$(v - k, k - \lambda, \lambda)$ design. Conversely, in the case of $\lambda = 1$ or 2, a 2-$(v - k, k - \lambda, \lambda)$ design can be uniquely embedded in a symmetric design: for $\lambda = 1$ the result is classical and can be found in any textbook on geometry (or see, for example, Hughes and Piper [11]) and for biplanes the result is due to Connor and Hall [9].

For any field F, $F^{\mathcal{P}}$ is the vector space of functions from $\mathcal{P}$ to F with basis given by the characteristic functions of the singleton subsets of $\mathcal{P}$. If $\mathcal{D} = (\mathcal{P}, \mathcal{B})$ is an incidence structure, the *code* $C_F(\mathcal{D})$ of $\mathcal{D}$ over F is the subspace of $F^{\mathcal{P}}$ spanned by the characteristic functions (incidence vectors) of the blocks of $\mathcal{D}$. If $X \subseteq \mathcal{P}$, denote the characteristic function on X by v^X: thus

$$v^X(x) = \begin{cases} 1 & \text{if } x \in X \\ 0 & \text{if } x \notin X \end{cases},$$

where $v^X(x)$ denotes the value that the function v^X takes at the point x. Then $C_F(\mathcal{D}) = \langle v^B | B \in \mathcal{B} \rangle$. If F has characteristic p then the dimension of $C_F(\mathcal{D})$ is referred to as the *p-rank* of $\mathcal{D}$. It is well known that the code of a design of order n will only be of any interest or use in characterizations of the design when the prime p divides n: see [1], for example.

The *orthogonal* code $C^\perp$ (where the orthogonal is taken with respect to the standard inner product $(u, w) = \sum_{x \in \mathcal{P}} u(x)w(x)$, for $u, w \in F^v$) is defined by $C^\perp = \{u \mid u \in F^v \text{ and } (u, w) = 0 \text{ for all } w \in C\}$.

The *hull* of a design $\mathcal{D}$ with code C over the field F is the code

$$\text{Hull}_F(\mathcal{D}) = H_F(\mathcal{D}) = C \cap C^\perp.$$

Notice that the code $\langle v^B - v^C | B, C \in \mathcal{B} \rangle$ is inside the hull when $\mathcal{D}$ is a symmetric (v, k, λ) design and the characteristic p of F divides the order $k - \lambda$, and is equal to it, with codimension 1 in the code of the design, in the case where p divides the order but not k.

Lemma 2.1 *If a 2-$(56, 11, 2)$ design extends to a 3-$(57, 12, 2)$ design, then the incidence vectors of the blocks of the 3-design that do not contain the new point must be in the orthogonal of the design's ternary code.*

Proof   Since any two blocks of a biplane meet in exactly two points, any two blocks of an extension meet in three or zero points. Since our code is ternary, the result follows.   □

Thus if we can find the $\{0, 1\}$ vectors of weight 12 in the orthogonal code $C^\perp$, we must be able to find our new blocks amongst these 12-sets. Thus we look for 210 12-sets with the correct properties. Since these codes are rather large, we needed to reduce the time taken for the computation in whatever ways possible. The complete weight enumerator of the code $C^\perp$ would indicate the number of constant weight-12 words. This could be obtained from the complete weight enumerator of the hull in the following way: denoting this polynomial by $H(x, y, z)$, the complete weight enumerator for the code C of the design is given by

$$C(x, y, z) = H(x, y, z) + H(y, z, x) + H(z, x, y),$$

since $C = \text{Hull} + \langle \jmath \rangle$, where $\jmath$ denotes the all-one vector.

Now we can use the extension of the MacWilliams relations to complete weight enumerators (see, for example, [1, Chapter 2, p. 84]) to get the complete weight enumerator of $C^{\perp}$:

$$D(x, y, z) = \frac{1}{|C|} C(x + y + z, x + ay + a^2 z, x + a^2 y + az),$$

where a is a primitive third root of unity. In fact Magma now has a function to perform the transform.

3 Non-extendability

3.1 The biplane $B1$

The biplane $B1$ was known implicitly earlier than the explicit construction of Hall, Lane and Wales, through the transitive representation, discovered by Sims, of $PSL_3(F_4)$ on 56 points, and the Gerwitz strongly-regular graph on 56 vertices: see [7, Chapter 2].

We showed that the biplane $B1$ cannot be extended, confirming an earlier claim of a proof of this quoted in [2] as being due to Hall and Baumert. We obtained the complete weight-enumerators of $\text{Hull}(C)$, and thus of all the four codes C, $C^{\perp}$, $\text{Hull}(C)$ and $\text{Hull}(C)^{\perp}$ using the formulae described above. This showed that the number of constant $\{0, 1\}$ vectors of weight-12 in $C^{\perp}$ is 2100.

Theorem 3.1 *The biplane $B1$ cannot be extended.*

Proof   The proof is by computer. The complete weight enumerator of $C^{\perp}$ (where C is the ternary code of the design) told us that there were 2100 $\{0, 1\}$ vectors in that code. Using the group to assist in completing orbits, we collected all the 2100 supports of these words and were able to show that we could not collect 210 to complete the number of blocks for the extended design. This was done by finding all the words through one of the points (450) and then forming a graph with these as vertices and with adjacency defined if the supports meet in three points. We could then find the size of a maximal clique, and this turned out to be 11, instead of 45 as would be required.   □

In fact the words of weight 12 in the orthogonal code fell into three orbits under the automorphism group, that is, two of length 420 and one of length 1260. Each of these three orbits formed the blocks of a 2-$(56, 12, \lambda)$ design, with $\lambda = 18$ and 54, respectively. The two 2-$(56,12,18)$ designs so obtained are not isomorphic but each has the automorphism group of $B1$ as its full automorphism group, as does the 2-(56,12,54) design.

The weight-distributions and the complete weight-distributions of the four codes associated with $B1$ were all obtained with Magma.

3.2 The biplane $B2$

The code of the design has dimension 22. With Magma we were able to obtain the weight distribution of each of its four associated codes, and also the complete weight-enumerators in each case, as described for $B1$. There are 516 constant $\{0,1\}$ weight-12 vectors in the code $C^{\perp}$; we were able to collect 492 of these and apply the same argument as given in the case of $B1$ to find that the maximal size of a clique through one of the points is 9, still short of the required 45, even if all the remaining 24 blocks could be used. We thus showed that we could not extract 210 to complete an extension for $B2$, and thus obtained

Theorem 3.2 *The biplane $B2$ cannot be extended.*

The weight-distributions and the complete weight-distributions of the four codes associated with $B2$ were all obtained with Magma.

3.3 The biplane $B4$

The code of the design has dimension 24. With Magma we were able to obtain the complete weight enumerator of the hull of $B4$, of dimension 23. This took approximately 13 days (1131129.444 seconds). Using the formulae given in Section 2, as before, we obtained the complete weight enumerators of the code, its orthogonal and also the hull's orthogonal. This yielded that there are only 148 $\{0,1\}$ weight-12 vectors in the code $C^{\perp}$, and thus not sufficient for the design to extend.

Theorem 3.3 *The biplane $B4$ cannot be extended.*

The weight-distributions and the complete weight-distributions of the four codes associated with $B4$ were all obtained with Magma.

3.4 The biplanes $B3$ and $B5$

The ternary hulls of these two biplanes have dimension 25; noting that the 23-dimensional hull of $B4$ required 13 days for Magma to obtain the complete weight enumerator, we deduce that the complete weight enumerator for either of these would take 130 days, which is rather too long. Expecting that the orthogonal codes would not in fact have sufficient $\{0, 1\}$ vectors of weight 12, the following method was attempted: instead of taking the span of all the blocks, a subset was chosen to span a code of lower dimension that did not contain the all-one vector. Thus if C_1 denotes this code, of dimension d, then $C_2 = C_1 + \langle \jmath \rangle$ has dimension $d + 1$ and its orthogonal code $C_2^{\perp}$ will contain the design's orthogonal code $C^{\perp}$. The complete weight enumerator of C_1 could be found in at most 13 days if $d \leq 23$, and this would yield the complete weight enumerator of $C_2^{\perp}$ quite quickly, by the method described in Section 2. In fact $d = 22$ for $B3$ yielded $C_2^{\perp}$ with only 91 $\{0, 1\}$ vectors of weight 12, and $d = 22$ for $B5$ gave only 22 of the required vectors in $C_2^{\perp}$. We may conclude that in neither case does $C^{\perp}$ have sufficient weight-12 $\{0, 1\}$ vectors. Thus we have

Theorem 3.4 *The biplanes $B3$ and $B5$ cannot be extended.*

A partial search amongst the weight-12 $\{0, 1\}$ vectors in the two orthogonal codes yielded that $B3$ had at least 42, and $B5$ at least 10, such vectors.

4   The residual designs

The five biplanes yield 16 non-isomorphic residual 2-$(45, 9, 2)$ designs (affine designs); the 16 ternary codes are also non-isomorphic. The number of different affine designs corresponds to the number of orbits on blocks:

1. $B1$ has one;

2. $B2$ has three;

3. $B3$ has three;

4. $B4$ has four;

5. $B5$ has five.

Table 2 shows, for each of the 16 affine designs, the size of the automorphism group, the dimension of the ternary code, the minimum weight of the code, the number of weight-9 vectors in the code, and the minimum weight of the orthogonal code, respectively.

The affine design for $B1$ (there is only one since $B1$ has a transitive automorphism group) has the smallest code, and thus the biggest $H^{\perp} = C + C^{\perp}$. This latter code (of dimension 36) has many (140135) constant $\{0, 1\}$ weight-9 vectors, as the complete weight enumerator showed, and it is possible that other designs with these parameters might be found amongst these vectors.

Design	$|Aut|$	$\dim(C)$	Min Wt. C	Wt-9 Wds in C	Min Wt. $C^\perp$
$B1$	1440	19	9	110	8
$B2:1$	144	21	6	110	8
$B2:2$	16	21	8	126	8
$B2:3$	8	21	8	126	8
$B3:1$	72	25	6	406	10
$B3:2$	8	25	8	494	10
$B3:3$	4	25	6	438	8
$B4:1$	16	23	6	158	8
$B4:2$	16	23	7	142	8
$B4:3$	4	23	7	186	8
$B4:4$	2	23	7	198	8
$B5:1$	3	25	8	278	8
$B5:2$	4	25	7	370	8
$B5:3$	1	25	7	318	8
$B5:4$	2	25	6	398	8
$B5:5$	4	25	6	370	8

Table 2: The 16 residual designs

5 Some maximal $\{12;3\}$-arcs

The support of a weight-12 vector in the orthogonal code of any of the five biplanes will have the property that any block will meet it in zero or three points, and thus will be a (maximal) $\{12;3\}$-arc for the biplane (in the terminology of Denniston[8] and Morgan[13]). This is clear from a quick counting argument. Conversely, any $\{12;3\}$-arc in the biplane is the support of a $\{0,1\}$ weight-12 vector in the orthogonal code. Furthermore, taking the intersections with such an arc of all the blocks that meet the arc in three points will give a 2-$(12,3,2)$ design. Blocks of these designs will meet in zero, one or two points. We have candidates here for class-3 association schemes:

Definition 1 *An **association scheme of class** d is a set X together with a partition of its subsets of size 2 into d non-empty classes $\Gamma_1, \ldots, \Gamma_d$ satisfying the two conditions:*

(i) *for $x \in X$ and $i \in \{1, \ldots, d\}$, the number $n_i(x)$ of points $y \in X$ with $\{x,y\} \in \Gamma_i$ depends on i and not on x; thus write this number as n_i;*

(ii) *for $x, y \in X$ and $\{x,y\} \in \Gamma_k$, the number $p_{i,j}^k(x,y)$ of points $z \in X$ with $\{x,z\} \in \Gamma_i$ and $\{y,z\} \in \Gamma_j$ depends only on i, j, k and not on x and y; thus write this number $p_{i,j}^k$.*

(See, for example, Cameron and van Lint [7, Chapter 17] for more on association schemes.)

We computed these designs, their automorphism groups and ternary codes for all the $\{12; 3\}$-arcs that we collected. The set of 44 blocks of any of these 2-(12,3,2) designs is a candidate for an association scheme of class 3, by taking Γ_i to be the set of all pairs of blocks that meet in i points, where $i = 0, 1, 2$. Condition (i) of the definition is satisfied (counting gives $n_1 = 16, n_2 = 24, n_3 = 3$), but not (ii) for any of the $\{12; 3\}$-arcs and thus none of these give class-3 association schemes. We found designs in $B1$ isomorphic to some in $B2$ and $B5$.

Acknowledgement The authors acknowledge support of NSF grant GER-9450080 and NSA grant MDA904-95-H-1019, respectively.

References

[1] E. F. Assmus, Jr. and J. D. Key, *Designs and their Codes*, Cambridge Tracts in Mathematics **103** (Second printing with corrections, 1993), Cambridge University Press, 1992.

[2] B. Bagchi, No extendable biplane of order 9, *J. Combin. Theory Ser. A* **49** (1988), 1–12 (Corrigendum **57** (1991), 162).

[3] W. Bosma and J. Cannon, *Handbook of Magma Functions*, Department of Mathematics, University of Sydney, 1994.

[4] A. E. Brouwer, private communication.

[5] A. E. Brouwer and H. A. Wilbrink, Block designs, *Handbook of Incidence Geometry* (ed. F. Buekenhout), pp. 349–382, Elsevier, Amsterdam, 1995.

[6] P. J. Cameron, Extending symmetric designs, *J. Combin. Theory Ser. A* **14** (1973), 215–220.

[7] P. J. Cameron and J. H. van Lint, *Designs, Graphs, Codes and their Links*, London Mathematical Society Student Texts **22**, Cambridge University Press, Cambridge, 1991.

[8] R. H. F. Denniston, On biplanes with 56 points, *Ars Combin.* **9** (1980), 167–179.

[9] M. Hall, Jr. and W.S. Connor, An embedding theorem for balanced incomplete block designs, *Canad. J. Math.* **6** (1954), 35–41.

[10] M. Hall, Jr., R. Lane, and D. Wales, Designs derived from permutation groups, *J. Combin. Theory* **8** (1970), 12–22.

[11] D. R. Hughes and F. C. Piper, *Design Theory*, Cambridge University Press, Cambridge, 1985.

[12] Z. Janko and T. Van Trung, A new biplane of order 9 with a small automorphism group, *J. Combin. Theory Ser. A* **42** (1986), 305–309.

[13] E. J. Morgan, Arcs in block designs, *Ars Combin.* **4** (1977), 3–16.

[14] C. J. Salwach and J. A. Mezzaroba, The four known biplanes with $k = 11$, *Internat. J. Math. & Math. Sci.* **2** (1979), 251–260.

J. D. Key, Department of Mathematical Sciences, Clemson University, Clemson SC 29634 , U.S.A.
e-mail: keyj@math.clemson.edu

V. D. Tonchev, Department of Mathematical Sciences,
Michigan Technological University, Houghton MI 49931-1295, U.S.A.
e-mail: tonchev@math.mtu.edu

A survey of small embeddings for partial cycle systems

C. C. Lindner

1 Introduction

A *Steiner triple system* (or more simply, a triple system) is a pair (S, T), where T is a collection of edge disjoint triangles which partition the edge set of the complete undirected graph K_n with vertex set S. The number $n = |S|$ is called the *order* of the triple system (S, T) and it has been known forever (= since 1847 [4]) that the spectrum for triple systems (= the set of all n such that a triple system of order n exists) is precisely the set of all $n \equiv 1$ or $3 \pmod 6$. It is very easy to see that if (S, T) is a triple system of order n, then $|T| = n(n-1)/6$.

Example 1.1 (triple system of order 7)

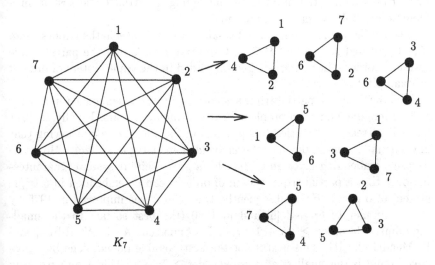

In what follows we will denote the triangle

by any cyclic shift of (a, b, c) *or* (a, c, b).

A *partial* triple system of order n is a pair (X, P), where P is a collection of edge disjoint triangles of the edge set of K_n with vertex set X. The difference between a (complete) triple system and a partial triple system is that the edge disjoint triangles belonging to a partial triple system do not necessarily include all of the edges of K_n.

Example 1.2 (partial triple system of order 6)

$$X = \{1, 2, 3, 4, 5, 6\},$$
$$P = \{(1, 2, 4), (1, 5, 6), (2, 3, 5), (3, 4, 6)\}$$

Now given a partial triple system (X, P) we can ask whether or not it is possible to decompose $E(K_n) \backslash E(P)$ into edge disjoint triangles; i.e., can a partial triple system always be *completed* to a triple system? The above example shows that this cannot always be done, since such a completion would give a Steiner triple system of order 6, which would cause Alex Rosa to froth at the mouth! Since a partial triple system cannot necessarily be completed to a triple system, the problem of embedding a partial triple system in a (complete) triple system comes to mind.

The partial triple system (X, P) is said to be *embedded* in the triple system (S, T) provided $X \subseteq S$ and $P \subseteq T$. Inspection reveals that the partial triple system of order 6 in Example 1.2 is embedded in the triple system of order 7 in Example 1.1.

Whether or not a partial triple system can always be embedded in a triple system remained an open problem for over a hundred years. Finally, in 1971 Christine Treash [13] settled this problem in the affirmative by showing that any partial triple system can be embedded in a triple system. However, Treash's embedding gives an extremely large containing system, guaranteeing only that a partial triple system of order n can be embedded in a triple system of order $< 2^{2n}$. Subsequently, this bound was improved in 1975 by C. C. Lindner to $6n + 3$ [6] and in 1980 (the best so far) to the smallest admissible order $\geq 4n + 1$ by L. D. Andersen, A. J. W. Hilton, and E. Mendelsohn [1]. This is still not the best possible bound. The best possible bound is the smallest admissible order $\geq 2n + 1$. This is an extremely difficult problem which continues to elude a solution.

Now a triangle is also a 3-cycle and so a Steiner triple system (S, T) can be described as an edge disjoint collection of 3-cycles which partitions the edge set of K_n (based on S). Since there is nothing particularly sacred about the number 3, we can certainly ask the same questions for m-cycle systems that are asked for triple systems. In particular, for $m \geq 4$, we can ask for the spectrum of m-cycle systems as well as an embedding (as small as possible) of partial m-cycle systems. We have an obvious definition here. An m-cycle system of order n is a pair (S, C), where C is an edge disjoint collection of m-cycles which partitions the edge set of K_n with vertex set S.

Example 1.3 (6-cycle system of order 9)

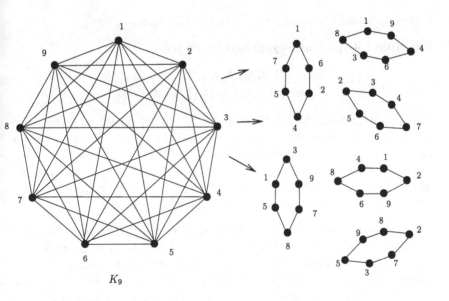

K_9

A *partial* m-cycle system of order n is a pair (X, P), where P is a collection of edge disjoint m-cycles of the edge set of K_n (which does not necessarily include all of the edges of K_n).

In what follows we will denote the m-cycle

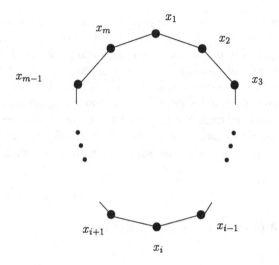

by any cyclic shift of $(x_1, x_2, x_3, \ldots, x_m)$ or $(x_1, x_m, x_{m-1}, \cdots x_2)$.

Example 1.4 (partial 6-cycle system of order 7)

$$X = \{1, 2, 3, 4, 5, 6, 7\}$$
$$P = \{(1, 6, 2, 4, 5, 7), (2, 3, 4, 7, 6, 5)\}.$$

Now clearly in Example 1.4, the edges $E(K_7) \backslash E(P)$ cannot be decomposed into 6-cycles for the simple reason that $|E(K_7) \backslash E(P)| = 9$. Therefore we can ask whether or not it is possible to *embed* (X, P) in a 6-cycle system. That is, does there exist a 6-cycle system (S, C) such that $X \subseteq S$ and $P \subseteq C$. Inspection reveals that the partial 6-cycle system (X, P) in Example 1.4 is embedded in the 6-cycle system of order 9 in Example 1.3.

In general, of course, the partial m-cycle system (X, P) is *embedded* in the m-cycle system (S, C) provided $X \subseteq S$ and $P \subseteq C$. Now Examples 1.1, 1.2, 1.3 and 1.4 illustrate the easily believable fact that, in general, a partial m-cycle system cannot necessarily be *completed* to an m-cycle system. That is to say, if (X, P) is a partial m-cycle system of order n, $E(K_n) \backslash E(P)$ cannot necessarily be decomposed into edge disjoint m-cycles. Hence the problem of finding an algorithm which will embed a partial m-cycle system in an m-cycle system. Additionally, we would like the containing m-cycle system to be as small as possible.

In [9] it is shown that a partial m-cycle system of order n can be embedded in an m-cycle system of order $m(2n+1)$ when m is ODD, and embedded in an m-cycle system of order $2nm + 1$ if m is EVEN [10]. Although the bounds are roughly the same, the techniques used for odd and even are starkly different! Odd-cycle systems are a lot harder to embed than even-cycle systems!

What follows is a reasonably elementary account of the struggle to obtain these embeddings. None of these embeddings are best possible and so there is the never ending problem of reducing the size of the containing system in general *and* in specific cases. So, for example, the result by L. D. Andersen, A. J. W. Hilton, and E. Mendelsohn [1] that a partial 3-cycle system of order n can be embedded in a 3-cycle system of order the smallest admissible order $\geq 4n + 1$ is a specific improvement of the general result that for odd m a partial m-cycle of order n can be embedded in an m-cycle system of order $m(2n + 1)$. In addition to this result for 3-cycle systems, the author is aware of only two other specific improvements for embedding partial m-cycle systems; *both* for even m. They are: a partial 4-cycle system of order n can be embedded in a 4-cycle system of order *approximately* $2n + \sqrt{2n}$ [3] and a partial 6-cycle system of order n can be embedded in a 6-cycle system of order *approximately* $3n$ [7]. We will include the constructions for these improvements for $m = 4$ and 6, but not for $m = 3$. There is an excellent reason for including the improvements for $m = 4$ and 6 and not for $m = 3$. The constructions for $m = 4$ and 6 are relatively easy to explain, while the construction for $m = 3$ is excruciatingly difficult! Since the object of this paper is an attempt at popularizing embedding theorems for partial cycle systems, the author does not wish to punish the reader with excruciating details! Rather than obscure the essence of the constructions with Professor Backwards type details, the author has chosen to illustrate the general construction for odd-cycle systems using 3-cycle and 5-cycle systems. The general construction for embedding even-cycle systems is easy enough to explain without illustrating the construction with specific cases. So we will give the general construction for even-cycle systems.

2   Embedding triple systems

As mentioned in the introduction, the serious history of embedding partial odd-cycle systems began in 1971 with Christine Treash's result that a partial triple system of order n can always be embedded in a triple system of order $< 2^{2n}$. In 1975 this was dramatically improved to $6n + 3$ by C. C. Lindner and subsequently in 1980 to the smallest admissible order $\geq 4n + 1$ by L. D. Andersen, A. J. W. Hilton, and E. Mendelsohn. Although this is the best result to date, we will content ourselves here with a description of the $6n + 3$ embedding. The principal reason being that the general embedding result for 5-cycle systems (as well as odd-cycle systems in general) is a *generalization* of the $6n + 3$ embedding and *not* the $\geq 4n + 1$ embedding. Additionally, the $\approx 4n + 1$ embedding is long and difficult and, in the author's opinion, out of place in an elementary survey article. When you come to think about it, these are two pretty good reasons!

The $6n + 3$ embedding for partial triple systems is based on a remarkable result due to Allan Cruse on embedding partial idempotent commutative quasigroups. A few preliminaries are in order.

A *partial* idempotent quasigroup is a partial quasigroup $(P, \circ)$ with the additional requirement that $x \circ x$ *is defined* for every $x \in P$ and $x \circ x = x$. In other words, the word "partial" quantifies products of the form $x \circ y$ where $x \neq y$. A partial idempotent commutative $(x^2 = x, xy = yx)$ quasigroup is a partial idempotent quasigroup $(P, \circ)$ with the additional requirement that if $x \circ y$ is defined then so is $y \circ x$ and furthermore $x \circ y = y \circ x$.

Example 2.1 (partial $x^2 = x$, $xy = yx$ quasigroup of order 4)

$$(P, \circ) =$$

$\circ$	1	2	3	4
1	1	4		2
2	4	2		1
3			3	
4	2	1		4

Now we can ask the same questions for partial $x^2 = x$, $xy = yx$ quasigroups that were asked for partial triple systems. Namely, can a partial $x^2 = x$, $xy = yx$ quasigroup be "completed" to an $x^2 = x$, $xy = yx$ quasigroup? And if not, can it be embedded in an $x^2 = x$, $xy = yx$ quasigroup? The partial quasigroup $(P, \circ_1)$ is *embedded* in the quasigroup $(Q, \circ_2)$ if and only if $x \circ_1 y = x \circ_2 y$ for all x, y for which $x \circ_1 y$ is defined. In other words, when considered as latin squares the upper left hand corner of $(Q, \circ_2)$ agrees with $(P, \circ_1)$. Example 2.1 shows that the answer to the first question is NO, since we cannot define $3 \circ 4 = 4 \circ 3$ without violating the cancellation law. Since a partial $x^2 = x$, $xy = yx$ quasigroup cannot necessarily be completed to an $x^2 = x$, $xy = yx$ quasigroup the problem of embedding becomes paramount.

Example 2.2 The partial $x^2 = x$, $xy = yx$ quasigroup $(P, \circ)$ of order 4 in Example 2.1 is embedded in the $x^2 = x$, $xy = yx$ quasigroup $(Q, \circ)$ of order 9.

∘	1	2	3	4
1	1	4		2
2	4	2		1
3			3	
4	2	1		4

$(P, \circ)$

∘	1	2	3	4	5	6	7	8	9
1	1	4	7	2	9	8	3	6	5
2	4	2	9	1	8	7	6	5	3
3	7	9	3	8	6	5	1	4	2
4	2	1	8	4	7	9	5	3	6
5	9	8	6	7	5	3	4	2	1
6	8	7	5	9	3	6	2	1	4
7	3	6	1	5	4	2	7	9	8
8	6	5	4	3	2	1	9	8	7
9	5	3	2	6	1	4	8	7	9

$(Q, \circ)$

In 1974 Allan Cruse obtained the best possible bound for embedding partial $x^2 = x, xy = yx$ quasigroups.

Theorem 2.3 (Allan Cruse [2]). *A partial $x^2 = x, xy = yx$ quasigroup of order n can be embedded in an $x^2 = x, xy = yx$ quasigroup of order t for every ODD $t \geq 2n + 1$.*

Cruse's Theorem is the best possible result in that it is always possible to construct a partial $x^2 = x, xy = yx$ quasigroup of order n which *cannot* be embedded in an $x^2 = x, xy = yx$ quasigroup of order $< 2n + 1$, for every $n \geq 4$.

Lemma 2.4 (Folk Lemma) *Let (X, P) be a (partial) triple system of order n and define a partial groupoid $(X, \circ)$ as follows: (i) $x \circ x = x$ for all $x \in X$, and (ii) if $x \neq y$, $x \circ y$ and $y \circ x$ are defined and $x \circ y = y \circ x = z$ if and only if $\{x, y, z\} \in P$. Then $(X, \circ)$ is a (partial) $x^2 = x, xy = yx$ quasigroup.*

Example 2.5 *em Let (X, P) be the partial triple system defined by $X = \{1, 2, 3, 4, 5, 6\}$ and $P = \{(1, 2, 4), (1, 5, 6), (2, 3, 5), (3, 4, 6)\}$ (Example 1.1). Then $(X, \circ)$ is given by the accompanying table.*

$\circ$	1	2	3	4	5	6
1	1	4		2	6	5
2	4	2	5	1	3	
3		5	3	6	2	4
4	2	1	6	4		3
5	6	3	2		5	1
6	5		4	3	1	6

$(X, \circ) =$ partial $x^2 = x$,
$xy = yx$ quasigroup

The $6n+3$ embedding. Let (X, P) be a partial triple system of order n and $(X, \circ)$ the partial $x^2 = x$, $xy = yx$ quasigroup constructed from (X, P). By Allan Cruse's Theorem we can embed $(X, \circ)$ in an $x^2 = x$, $xy = yx$ quasigroup $(Q, \circ)$ of order $2n+1$. Let $S = Q \times \{1, 2, 3\}$ and define a collection of triangles T as follows:

(1) $((x, 1), (x, 2), (x, 3)) \in T$, all $x \in Q$.

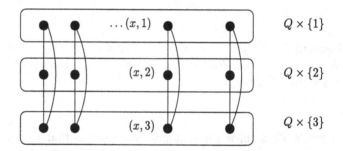

$Q \times \{1\}$

$Q \times \{2\}$

$Q \times \{3\}$

(2) For each triangle $t \in P$, take a fixed representation (x, y, z) and define a collection of 9 triangles as follows: let $(\{1, 2, 3\}, \otimes)$ be the idempotent quasigroup given by

$\otimes$	1	2	3
1	1	3	2
2	3	2	1
3	2	1	3

and for each ordered pair $(i, j) \in \{1, 2, 3\}$ (i and j not necessarily distinct) place the triangle $((x, i), (y, j), (z, i \otimes j))$ in T.

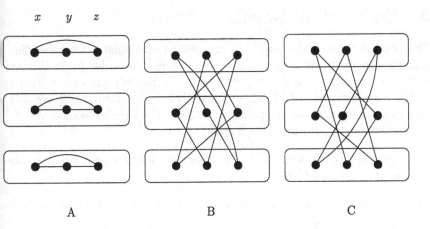

$x \quad y \quad z$

A   B   C

(3) If $x \neq y$ and the edge $\{x, y\}$ does not belong to a triangle in P, place
    the three triangles $((x, 1), (y, 1), (x \circ y, 2))$, $((x, 2), (y, 2), (x \circ y, 3))$, and
    $((x, 3), (y, 3), (x \circ y, 1))$ in T.

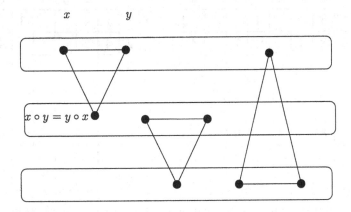

$x \qquad y$

$x \circ y = y \circ x$

It is straightforward to see that (S, T) is a triple system of order $6n + 3$
(just count the number of triangles and show that each edge is in one of the
triangles described above. The triangles in (2)A guarantee that three disjoint
copies of (X, P) are contained in (S, T).

We have the following theorem.

Theorem 2.6 (C. C. Lindner [6]) *A partial triple system of order n can
be embedded in a triple system of order $6n + 3$.*

3   Embedding 5-cycle systems

The obvious thing to do here is to attempt to extrapolate the embedding construction for triple systems to 5-cycle systems. In order to do this we need to figure out a reasonable way to define a (partial) quasigroup from a partial 5-cycle system. If (S, C) is a (partial) 5-cycle system, there are only two ways to define a binary operation "∘" from the 5-cycles belonging to C with any hope of $(S, ∘)$ being a quasigroup:

(1) $a ∘ a = a$, for all $a \in S$, and if $a \neq b$, $a ∘ b = c$ and $b ∘ a = e$ if and only if $(a, b, c, d, e) \in C$, OR

(2) $a ∘ a = a$, for all $a \in S$, and if $a \neq b$, $a ∘ b = b ∘ a = d$ if and only if $(a, b, c, d, e) \in C$.

Example 3.1 Let (X, P) be the partial 5-cycle system given by $X = \{1, 2, 3, 4, 5, 6, 7, 8\}$ and $P = \{(1, 2, 3, 4, 5), (2, 6, 7, 4, 8)\}$. Let $∘_1$ be the binary operation defined by (1) and $∘_2$ the binary operation defined by (2).

$∘_1$	1	2	3	4	5	6	7	8
1	1	3			4			
2	5	2	4			7		4
3		1	3	5				
4			2	4	1		6	2
5	2			3	5			
6		8				6	4	
7				8		2	7	
8		6		7				8

$∘_2$	1	2	3	4	5	6	7	8
1	1	4			3			
2	4	2	5			4		7
3		5	3	1				
4			1	4	2		2	6
5	3			2	5			
6		4				6	8	
7				2		8	7	
8		7		6				8

The above example shows that, unlike the case for (partial) triple systems, it is not always possible to define a (partial) quasigroup from a (partial) 5-cycle system. So an obvious extrapolation of the $6n + 3$ embedding for triple systems applied to 5-cycle systems will not work.

The reason for the trouble in Example 3.1 is because the vertices 2 and 4 are joined by a path of length 2 in two different 5-cycles of P. This forces such things as $2 ∘_1 3 = 2 ∘_1 8 = 4$ and $4 ∘_2 5 = 4 ∘_2 7 = 2$, which guarantees that neither operation gives a quasigroup. A bit of reflection reveals, however, that if (S, C) is a (partial) 5-cycle system with the additional property that each pair of vertices is joined by (at most) *exactly* one path of length two belonging to a 5-cycle of C then both $∘_1$ and $∘_2$ produce (partial) quasigroups.

Such a (partial) 5-cycle system is said to be 2-*perfect*. Since there is no guarantee that a given (partial) 5-cycle system is 2-perfect we will need to use something other than a (partial) quasigroup if we are going to try to embed 5-cycle systems with a construction similar to the embedding of Steiner triple systems. The following is just what the doctor ordered.

A partial groupoid $(X, \circ)$ with the following properties will be called a partial *embedding groupoid*:

(1) $x \circ x = x$, for all $x \in X$ (idempotent),

(2) $x \circ y$ is defined if and only if $y \circ x$ is defined (but $x \circ y$ and $y \circ x$ are *not necessarily equal*),

(3) $(X, \circ)$ is ROW latin $= (a \circ x = a \circ y$ implies $x = y)$, and

(4) each $x \in X$ occurs as a product an ODD number of times.

Example 3.2 A partial embedding groupoid of order 8.

∘	1	2	3	4	5	6	7	8
1	1	2			5			
2	1	2	3			6		8
3		2	3	4				
4			3	4	5		7	8
5	1			4	5			
6		2				6	7	
7				4		6	7	
8		2		4				8

The following generalization of Cruse's Theorem is proved in [8].

Theorem 3.3 (C. C. Lindner and C. A. Rodger [8]). *A partial embedding groupoid of order n can always be embedded in an idempotent groupoid of order $2n + 1$ which is row latin and such that the partial groupoid consisting of the main diagonal plus all products not defined by the embedding groupoid is a partial idempotent commutative quasigroup (of order $2n + 1$).*

Example 3.4 The groupoid of order 17 given below contains the partial embedding groupoid of order 8 in Example 3.2 and has all of the properties stated in Theorem 3.3.

∘	1	2	3	4	5	6	7	8	9	10	11	12	13	14	15	16	17
1	1	2	15	10	5	16	11	14	3	4	17	7	8	9	13	12	6
2	1	2	3	9	16	6	14	8	12	5	7	10	17	11	4	15	13
3	15	2	3	4	14	13	9	12	5	6	1	8	16	10	7	17	11
4	10	9	3	4	5	17	7	8	6	11	2	1	12	13	16	14	15
5	1	16	14	4	5	11	12	10	7	17	15	2	6	8	3	13	9
6	16	2	13	17	11	6	7	9	8	14	4	15	10	3	1	4	12
7	11	14	9	4	12	6	7	17	15	13	5	16	3	1	10	2	8
8	14	2	12	4	10	9	17	8	1	15	13	11	5	16	6	7	3
9	3	12	5	6	7	8	15	1	9	16	10	13	2	17	14	11	4
10	4	5	6	11	17	14	13	15	16	10	13	3	9	7	2	8	1
11	17	7	1	2	15	4	5	13	10	12	11	9	14	6	8	3	16
12	7	10	8	1	2	15	16	11	13	3	9	12	4	5	17	6	4
13	8	17	16	12	6	10	3	5	2	9	14	4	13	15	11	1	7
14	9	11	10	13	8	3	1	16	17	7	6	5	15	14	12	4	2
15	13	4	7	16	3	1	10	6	14	2	8	17	11	12	15	9	5
16	12	15	17	14	13	5	2	7	11	8	3	6	1	4	9	16	10
17	6	13	11	15	9	12	8	3	4	1	16	4	7	2	5	10	17

It turns out that Theorem 3.3 is exactly what is needed to obtain a $10n + 5$ embedding á la the $6n + 3$ embedding for triple systems. Here goes!

The $10n + 5$ **Embedding.** Let (X, P) be a partial 5-cycle system of order n and define a partial groupoid $(X, \circ)$ as follows: (i) $x \circ x = x$ for all $x \in X$, and (ii) if $x \neq y, x \circ y = y$ and $y \circ x = x$ if and only if $\{x, y\}$ belongs to a 5-cycle of P. It is straightforward to see that $(X, \circ)$ is a partial embedding groupoid. (For example, if $X = \{1, 2, 3, 4, 5, 6, 7, 8\}$, and $P = \{(1, 2, 3, 4, 5), (2, 6, 7, 4, 8)\}$, then $(X, \circ)$ is the partial embedding groupoid given in Example 3.2.) By Theorem 3.3 we can embed $(X, \circ)$ in a groupoid $(Q, \circ)$ of order $2n + 1$ having the properties guaranteed by Theorem 3.3. Let $S = Q \times \{1, 2, 3, 4, 5\}$ and define a collection of 5-cycles C as follows:

(1) $((x, 1), (x, 2), (x, 3), (x, 4), (x, 5))$ and $((x, 1), (x, 3), (x, 5), (x, 2), (x, 4)) \in$
     C for all $x \in Q$.

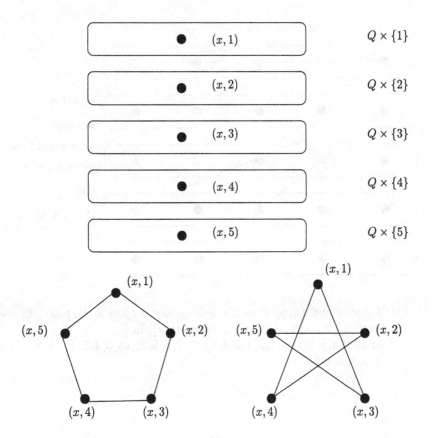

(2) For each 5-cycle $c \in P$ take a fixed representation $(x_1, x_2, x_3, x_4, x_5)$ of c and define a collection of size 25 of 5-cycles as follows: Let $(\{1, 2, 3, 4, 5\}, \otimes)$ be the idempotent quasigroup given by

$\otimes$	1	2	3	4	5
1	1	3	5	2	4
2	5	2	4	1	3
3	4	1	3	5	2
4	3	5	2	4	1
5	2	4	1	3	5

(any idempotent quasigroup will do) and for each ordered pair $(i, j) \in \{1, 2, 3, 4, 5\}$ (i and j not necessarily distinct) place the 5-cycle $((x_1, i), (x_2, j), (x_3, i \otimes j), (x_4, j), (x_5, i \otimes j))$ in C.

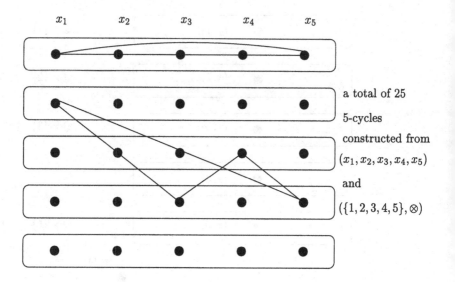

a total of 25
5-cycles
constructed from
$(x_1, x_2, x_3, x_4, x_5)$
and
$(\{1, 2, 3, 4, 5\}, \otimes)$

(3) If $x \neq y$ and $\{x, y\}$ does not belong to a 5-cycle in P, place the five
5-cycles $((x, i), (y, i), (x, j), (x \circ y, k), (y, j))$ in C,
for all $(i, i, j, k, j) \in \{(1, 1, 2, 4, 2), (2, 2, 3, 5, 3), (3, 3, 4, 1, 4), (4, 4, 5, 2, 5),$
$(5, 5, 1, 3, 1)\}$.

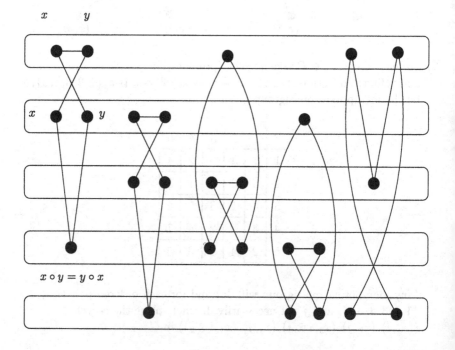

It is IMPORTANT to note here that if $\{x, y\}$ does not belong to a 5-cycle in P, then $x \circ y = y \circ x$ is computed in a partial idempotent commutative quasigroup. Although a bit more tedious than the $6n + 3$ construction, it is straightforward to show that (S, C) is a 5-cycle system of order $10n + 5$ (count the number of 5-cycles and show that each edge is in one of the 5-cycles described above). The 5-cycles in (2) guarantee that five disjoint copies of (X, P) are contained in (S, C).

We have the following theorem.

Theorem 3.5 (C. C. Lindner and C. A. Rodger [9]) *A partial 5-cycle system of order n can be embedded in a 5-cycle system of order $10n + 5$.*

4 Embedding odd-cycle systems

We have shown that a partial 3-cycle system of order n can be embedded in a 3-cycle system of order $3(2n + 1)$, and that a partial 5-cycle system can be embedded in a 5-cycle system of order $5(2n+1)$. An obvious generalization of these embeddings to partial odd-cycle systems in general gives the following theorem.

Theorem 4.1 (C. C. Lindner and C. A. Rodger [9]) *If m is ODD, a partial m-cycle system of order n can be embedded in an m-cycle system of order $m(2n + 1)$.*

The embedding is obtained by modifying the $10n + 5$ embedding for 5-cycles as follows: Define an embedding groupoid and embed it exactly as in the $10n + 5$ embedding. Let $S = Q \times \{1, 2, 3, \ldots, m\}$; in (1) use an m-cycle system of order m (this is just a Hamiltonian decomposition of K_m); in (2) use an idempotent quasigroup of order m (any idempotent quasigroup will do); and finally in (3) for each edge not in one of the m-cycles in the partial m-cycle system define m-cycles (m of them) á la the "5-cycles" in (3). This requires a bit of trickery which can be found in [9]. Since the object of this survey is to popularize the subject of embedding partial cycle-systems, trickery has no place in this paper. The interested reader can pore over the details in [9].

We now turn our attention to the much easier problem of embedding even-cycle systems.

5 Embedding even-cycle systems

Embedding partial even-cycle systems is a good deal easier than embedding odd-cycle systems, the primary reason being that quasigroups/groupoids are not involved in any way. In fact, even-cycle systems are so much easier than odd-cycle systems that we will give the *general construction* for embedding partial even-cycle systems. No need to sneak up on the construction as we did for odd-cycle systems. In particular, we will show that if m is even, a partial m-cycle system of order n can be embedded in an m-cycle system of order $2mn + 1$. In order to do this we will need the following two results.

Theorem 5.1 (A. Kotzig [5] and A. Rosa [11]) *There exists an m-cycle system of order $2m + 1$ for every even m.*

Theorem 5.2 (D. Sotteau [12]) *The necessary and sufficient conditions to decompose the complete bipartite graph $K_{m,n}$ into $2k$-cycles are: (i) m and n are even, (ii) $m \geq k$, $n \geq k$, and (iii) $2k \mid mn$.*

The $2mn + 1$ embedding. Let m be *even* and (X, P) a partial m-cycle system of order n. Let $S = \{\infty\} \cup (X \times \{1, 2, 3, \ldots, 2m\})$ and define a collection of m-cycles as follows:

(1) For each $x \in X$, define an m-cycle system on $\{\infty\} \cup (X \times \{1, 2, 3, \ldots, 2m\})$ and place these m-cycles in C.

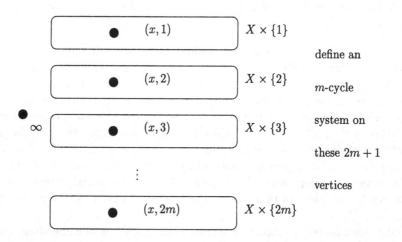

(2) For each m-cycle $c \in P$ take a fixed representation $(x_1, x_2, x_3, \ldots, x_m)$ of c and define a collection of $4m^2$ m-cycles as follows: Let $(\{1, 2, 3, \ldots, 2m\}, \otimes)$ be any idempotent quasigroup and for each ordered pair $(i, j) \in \{1, 2, 3, \ldots, 2m\}$ place the m-cycle
$$((x_1, i), (x_2, j), (x_3, i \otimes j), (x_4, j), (x_5, i \otimes j), (x_6, j), \ldots,$$
$$(x_{2m-1}, i \otimes j), (x_{2m}, j)) \text{ in } C.$$

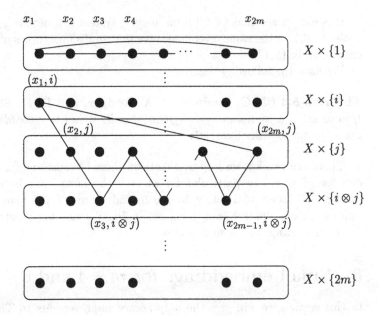

(3) If $a \neq b$ and the edge $\{a, b\}$ does not belong to an m-cycle in P, use Sotteau's Theorem to decompose $K_{2m,2m}$ (with parts $\{a\} \times \{1, 2, 3, \ldots, 2m\}$ and $\{b\} \times \{1, 2, 3, \ldots, 2m\}$) into m-cycles and place these m-cycles in C.

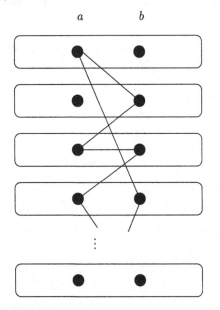

Use Sotteau's Theorem
to decompose $K_{2m,2m}$ into
$4m$ m-cycles

It is easy to see that (S, C) is an m-cycle system of order $2mn + 1$ and the cycles in (2) on the same level guarantee that m disjoint copies of (X, P) are embedded in (S, C).

We have the following theorem.

Theorem 5.3 (C. C. Lindner, C. A. Rodger, and D. R. Stinson [10]) *If m is even, a partial m-cycle system of order n can be embedded in an m-cycle system of order $2mn + 1$.*

There are two known improvements on these bounds; one for $m = 4$ and one for $m = 6$. In particular for $m = 4$ the bound has been improved from $8n + 1$ to approximately $2n + \sqrt{2n}$ and for $m = 6$ the bound has been improved from $12n + 1$ to approximately $3n$. We give the constructions for these embeddings in the next section.

6 Small embeddings for $m = 4$ and 6

In this section we will give the only known improvements to Theorem 5.3, first for $m = 4$ and then for $m = 6$.

The $2n + \sqrt{2n}$ embedding (for partial 4-cycle systems). Let (Q, P) be a partial 4-cycle system of order n. Let x be the smallest *odd* positive integer such that $\binom{x}{2} \geq n$. Let X be a set of size x and $Y \supseteq Q$ a set of size $\binom{x}{2}$. Let $S = (Y \times \{1, 2\}) \cup X$ and define a collection of 4-cycles C of the edge set of K_{x^2} with vertex set S as follows:

(1) For each 4-cycle in P take a fixed representation (a, b, c, d) and place the four 4-cycles $((a, 1), (b, 1), (c, 1), (d, 1))$, $((a, 2), (b, 2), (c, 2), (d, 2))$, $((a, 1), (b, 2), (c, 1), (d, 2))$, and $((a, 2), (b, 1), (c, 2), (d, 1))$ in C.

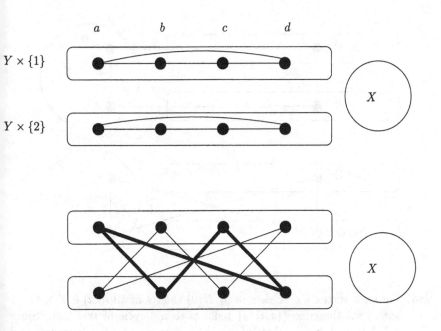

(2) If $a \neq b$ and the edge $\{a, b\}$ does not belong to a 4-cycle in P, place the
4-cycle $((a,1),(b,1),(a,2),(b,2))$ in C.

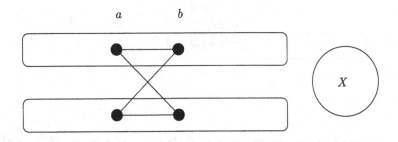

(3) Let α be any $1-1$ mapping from the edge set of K_x with vertex set
X onto the set Y. For each $a \in Y$, place *exactly one* of the 4-cycles
$((a,1),(a,2),f,e)$ or $((a,1),(a,2),e,f)$ in C, where $\{e,f\}\alpha = a$.

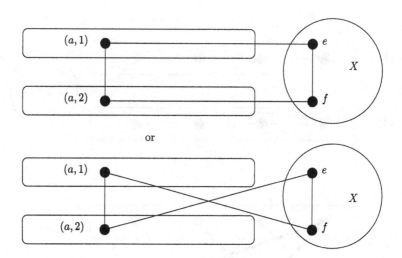

(4) For each vertex $v \in X$ denote by $N(v)$ the set of all $(a,i) \in Y \times \{1,2\}$ such that the edge $\{(a,i),v\}$ belongs to a 4-cycle of type (3). Since v belongs to $x-1$ edges of K_x, $|N(v)| = x - 1$. Furthermore, and this is *important*, the collection $\pi = \{N(v) \mid v \in X\}$ is a partition of $Y \times \{1,2\}$. By Sotteau's Theorem 5.2, $K_{N(v),X\setminus\{v\}}$ can be partitioned into 4-cycles. For each $v \in X$ place these $(x-1)^2/4$ 4-cycles in C.

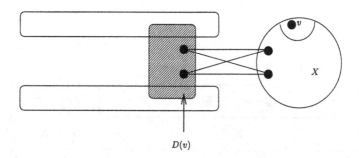

$$D(v)$$

It is an easy exercise to see that (S,C) is a 4-cycle system of order $2\binom{x}{2} + x = x^2$ and the cycles of type (1) guarantee that two disjoint copies of (Q,P) are contained in (S,C).

Theorem 6.1 (J. D. Horton, C. C. Lindner, and C. A. Rodger [3])
A partial 4-cycle system of order n can be embedded in a 4-cycle system of order x^2, where x is the smallest odd integer such that $\binom{x}{2} \geq n$.

Remark. If x is the smallest odd integer such that $\binom{x}{2} \geq n$, then $x \approx \sqrt{2n}$. Hence $x^2 = 2\binom{x}{2} + x \approx 2n + \sqrt{2n}$. Without going into details, the best possible embedding is approximately $n + \sqrt{n}$.

The $3n$ embedding (for partial 6-cycle systems). Let (Q, P) be a partial 6-cycle system of order n. Let $12k \geq n$ with k as small as possible and let $Y \supseteq Q$ be a set of size $12k$. Let Z be a set of size 9, $S = (Y \times \{1, 2, 3\}) \cup Z$, and define a collection of 6-cycles H of the edge set of K_{36k+9} with vertex set S as follows:

(1) For each 6-cycle in P take a fixed representation (a, b, c, d, e, f) and place the six 6-cycles $((a, 1), (b, 1), (c, 1), (d, 1), (e, 1), (f, 1))$, $((a, 1), (b, 2), (c, 1), (d, 2), (e, 1), (f, 2))$, $((a, 2), (b, 1), (c, 2), (d, 1), (e, 2), (f, 1))$, $((a, 2), (b, 3), (c, 2), (d, 3), (e, 2), (f, 3))$, $((a, 3), (b, 2), (c, 3), (d, 2), (e, 3), (f, 2))$, and $((a, 3), (b, 3), (c, 3), (d, 3), (e, 3), (f, 3))$ in C.

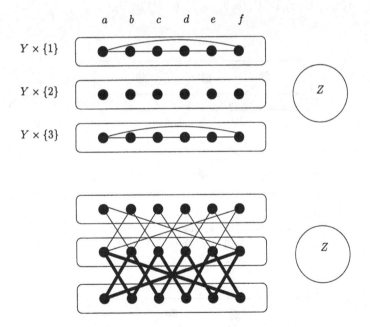

(2) Define a 6-cycle system on $(Y \times \{2\}) \cup Z$ and place these 6-cycles in H. (The spectrum for 6-cycle systems is the set of all $n \equiv 1$ or 9 (mod 12).)

(3) For each edge $\{a, b\}$ *not* in a 6-cycle of P, place the 6-cycle $((a, 1), (b, 1), (a, 2), (b, 3), (a, 3), (b, 2))$ in H.

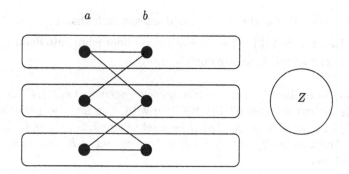

Let (Y, G, B) be a group divisible design of order $|Y| = 12k$, with groups G of size 2 and blocks B of size 3. (Delete a point from a Steiner triple system of order $12k + 1$.)

(3) For each block $\{a, b, c\} \in B$ place the 6-cycle
$((a, 1), (b, 3), (c, 1), (a, 3), (b, 1), (c, 3))$ in H.

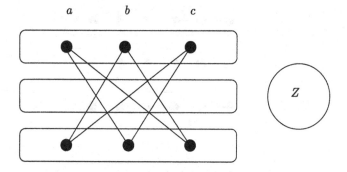

(4) For each group $\{d, e\} \in G$, place the 6-cycle
$((d, 1), (d, 2), (d, 3), (e, 1), (e, 2), (e, 3))$ in H.

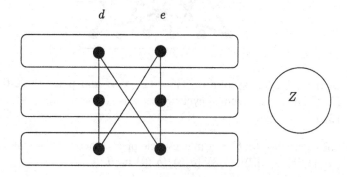

(5) Let ∞_1 and ∞_2 be any two points in Z, and for each group $\{d, e\}$ place
the 6-cycle
$(\infty_1, (e, 1), (e, 3), \infty_2, (d, 3), (d, 1))$ in H.

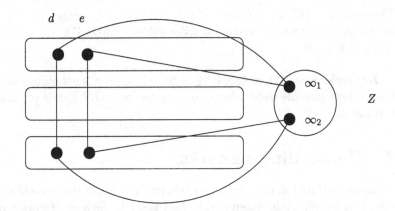

(6) Partition the complete bipartite graph with parts $Y \times \{1\}$ and $Z \backslash \{\infty_1\}$
into 6-cycles and place these 6-cycles in H. (Sotteau's Theorem 5.2.)

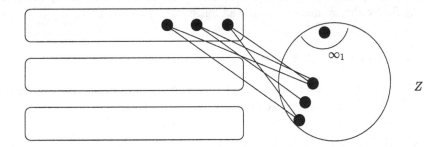

(7) Partition the complete bipartite graph with parts $Y \times \{3\}$ and $Z \backslash \{\infty_1\}$
into 6-cycles and place these 6-cycles in H. (Sotteau's Theorem 5.2.)

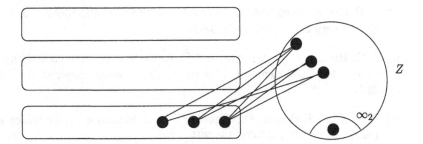

It is routine to see that (S, C) is a 6-cycle system of order $36k + 9$ and the cycles of type (1) guarantee that two disjoint copies of (Q, P) are contained in (S, C).

Theorem 6.2 (C. C. Lindner [7]) *A partial 6-cycle system of order n can be embedded in a 6-cycle system of order $36k + 9$, where $12k \geq n$ and k is as small as possible.*

Remark. Since $12k \approx n$, $36k + 9 \approx 3n$. The author is not quite sure what the smallest possible embedding is for 6-cycles, but is absolutely certain that it is not $36k + 9$!

7   Concluding remarks

As mentioned frequently throughout this paper, *none* of the embeddings obtained for partial cycle systems is the best possible. So a lot of work remains to be done. However, quite recently, the author and E. J. Billington have generalized Theorem 6.2 to reduce the size of the containing cycle system when $m = 6k$ from $2mn + 1$ to *approximately* $3kn$. This result is still being written up and so will appear elsewhere. This paper is long enough as it is. As Porky Pig would say: That's all folks!

Acknowledgement

The author wishes to thank Elizabeth Billington and Anne Street for many helpful comments during the preparation of this paper.

References

[1] L. D. Andersen, A. J. W. Hilton and E. Mendelsohn, Embedding partial Steiner triple systems, *J. London Math. Soc.* 41 (1980) 554-576.

[2] A. B. Cruse, On embedding incomplete symmetric latin squares, *J. Combin. Theory Ser. A* 16 (1974) 18-27.

[3] J. D. Horton, C. C. Lindner, and C. A. Rodger, A small embedding for partial 4-cycle systems, *J. Combin. Math. Combin. Comput.* 5 (1989), 23-26.

[4] Rev. T. P. Kirkman, On a problem in combinations, *Cambridge and Dublin Math. J.* 2 (1847), 191-204.

[5] A. Kotzig, On the decomposition of complete graphs into $4k$-gons, *Mat.-Fyz. Časopis Sloven. Akad. Vied.* 15 (1965), 229-233.

[6] C. C. Lindner, A partial Steiner triple system of order n can be embedded in a Steiner triple system of order $6n + 3$, *J. Combin. Theory Ser. A* 18 (1975) 349-351.

[7] C. C. Lindner, A small embedding for partial hexagon systems, *Australas. J. Combin.*, to appear.

[8] C. C. Lindner and C. A. Rodger, Generalized embedding theorems for partial latin squares, *Bull. Inst. Combin. Appl.* 5 (1992), 81-99.

[9] C. C. Lindner and C. A. Rodger, A partial $m = (2k + 1)$-cycle system of order n can be embedded in an m-cycle system of order $(2n + 1)m$, *Discrete Math.* 117 (1993), 151-159.

[10] C. C. Lindner, C. A. Rodger, and D. R. Stinson, Embedding cycle systems of even length, *J. Combin. Math. Combin. Comput.* 3 (1988), 65-69.

[11] A. Rosa, On cyclic decompositions of the complete graph into $(4m + 2)$-gons, *Mat.-Fyz. Časopis Sloven. Akad. Vied.* 16 (1966), 349-353.

[12] D. Sotteau, Decompositions of $K_{m,n}(K^*_{m,n})$ into cycles (circuits) of length $2k$, *J. Combin. Theory Ser. B* 30 (1981), 75-81.

[13] C. Treash, The completion of finite incomplete Steiner triple systems with applications to loop theory, *J. Combin. Theory Ser. A* 10 (1971), 259-265.

C. C. Lindner, Department of Discrete and Statistical Sciences, 120 Mathematics Annex, Auburn University, Auburn, Alabama 36849-5307, U.S.A.
email: lindncc@mail.auburn.edu

Rosa triple systems

Alan C.H. Ling   Charles J. Colbourn

Abstract

A Rosa triple system is a triple system of order congruent to 2 modulo 3 whose chromatic index is minimum, having the largest possible number of maximum parallel classes in such a block colouring. The existence of Rosa triple systems is settled completely. Together with known results on Kirkman, Hanani, and almost resolvable twofold triple systems, the existence of Rosa triple systems is used to settle completely the existence of triple systems with minimum chromatic index.

1   Introduction

A *balanced incomplete block design* (BIBD) is a pair $(V, \mathcal{B})$ where V is a v-set and $\mathcal{B}$ is a collection of b k-subsets of V (*blocks*) such that each element of V is contained in exactly r blocks and any 2-subset of V is contained in exactly λ blocks. The numbers v, b, r, k, λ are *parameters* of the BIBD. Since $bk = vr$ and $\lambda(v - 1) = r(k - 1)$, the notation BIBD$(v, k, \lambda)$ is used for a BIBD on v points, block size k and index λ. A *triple system* on v points and *index* λ (TS(v, λ)) is a BIBD with $k = 3$.

When $v \equiv 0 \pmod{k}$, a *parallel class* is a set of $\frac{v}{k}$ mutually disjoint blocks. A design $\mathcal{D} = (V, \mathcal{B})$ is *resolvable* if its blocks can be partitioned into parallel classes.

Let K and G be sets of nonnegative integers and let λ be a positive integer. A *group divisible design of index λ and order v* $((K, \lambda)$-GDD$)$ is a triple $(V, \mathcal{G}, \mathcal{B})$, where V is a finite set of cardinality v, $\mathcal{G}$ is a partition of V into parts (*groups*) whose sizes lie in G, and $\mathcal{B}$ is a family of subsets (*blocks*) of V which satisfy the properties:

(1) If $B \in \mathcal{B}$, then $|B| \in K$.

(2) Every pair of distinct elements of V occurs in exactly λ blocks or one group, but not both.

(3) $|G| > 1$.

The *group-type (type)* of the GDD is the multiset $\{|G| : G \in \mathcal{G}\}$. An "exponential" notation describes group-type: $g_1^{u_1} \cdots g_s^{u_s}$ denotes u_i groups of size g_i for $1 \le i \le s$. If $K = \{k\}$, then the (K, λ)-GDD is a (k, λ)-GDD. If $\lambda = 1$, the GDD is a K-GDD. Furthermore, a $(\{k\}, 1)$-GDD is a k-GDD.

Let k be a positive integer and let v and α be positive integers. Let V be a set of cardinality v. Then a (k, α)-*partial parallel class* is a collection $\mathcal{C}$ of blocks such that every element of V occurs in either exactly α or exactly zero blocks of $\mathcal{C}$. The set of elements of V not occurring in the partial parallel class is the *complement* of the class.

Let k, α, and λ be positive integers. A $(\lambda, \alpha; k)$-*frame* is a triple $(V, \mathcal{G}, \mathcal{P})$ where V is a set of cardinality v, $\mathcal{G}$ is a partition of V into parts (*groups*), and $\mathcal{P}$ is a collection of (k, α)-partial parallel classes of V which satisfies the conditions:

(1) The complement of each (k, α)-partial parallel class P of $\mathcal{P}$ is a group $G \in \mathcal{G}$;

(2) Each unordered pair $\{x, y\}$ of V which does not lie in some group $G \in \mathcal{G}$ lies in precisely λ blocks of $\mathcal{P}$;

(3) No unordered pair $\{x, y\}$ of elements of V which lies in some group G of $\mathcal{G}$ also lies in a block of $\mathcal{P}$.

These partial parallel classes are called *frame parallel classes*. The *type* of the $(\lambda, \alpha; k)$-frame is the multiset $T = [|G| : G \in \mathcal{G}]$. If $\mathcal{G}$ contains a_i groups of size g_i for $i = 1, 2, \ldots, s$, then the exponential notation $g_1^{a_1} g_2^{a_2} \ldots g_s^{a_s}$ is also used. By convention, factors of the type 0^a can be included in the exponential form of the type to accommodate null groups. A (k, λ)-*frame* of type T is a $(\lambda, 1; k)$-frame of type T.

In a BIBD(v, k, λ), a set of m blocks form a *partial parallel class* if any two distinct blocks are disjoint. Evidently, $m \le \lfloor \frac{v}{k} \rfloor$. A partial parallel class is a *maximum partial parallel class* if $m = \lfloor \frac{v}{k} \rfloor$. Let $d = b \bmod m$. If the blocks of a BIBD(v, k, λ) can be partitioned into $\lfloor \frac{b}{m} \rfloor$ maximum parallel classes, and (when $d > 0$) one partial parallel class with d blocks, the BIBD is called *minci*.

The definition is a generalization of many type of designs. A resolvable BIBD is a minci BIBD. An NRB(v, k) [3] is a minci BIBD$(v, k, k - 1)$. Kirkman triple systems are minci BIBD$(6x + 3, 3, 1)$. Hanani triple systems are minci BIBD$(6x + 1, 3, 1)$ [8].

A *Rosa triple system* RTS(v, λ) is a minci TS(v, λ) when $v \equiv 2 \pmod 3$. Rosa [7] posed the question of the existence of such systems in general, because of the role they play in block colouring problems.

For a BIBD(v, k, λ) $\mathcal{D} = (V, \mathcal{B})$, a *block-colouring* of $\mathcal{D}$ is a mapping $\psi : \mathcal{B} \to C$ such that if $\psi(B) = \psi(B')$ for $B, B' \in \mathcal{B}, B \ne B'$, then B is disjoint from B'. If the set of colours C has size q, then ψ is a *q-block-colouring*. For each $c \in C$, the set $\psi^{-1}(c)$ is a *block-colour class*. The *chromatic index* of $\mathcal{D}$, $\psi'(\mathcal{D})$, is the smallest q such that there exists a *q-block-colouring* of $\mathcal{D}$. Since any block-colour class can contain at most $\lfloor \frac{v}{k} \rfloor$ blocks, $\psi'(\mathcal{D}) \geq |\mathcal{B}|/\lfloor v/k \rfloor$. A minci BIBD gives the minimum chromatic index among all designs with the same parameters.

2   Rosa Triple Systems when $v \equiv 5 \pmod 6$

In this section, the existence of Rosa triple systems is settled when $v \equiv 5 \pmod 6$. The number of blocks in a TS$(6x+5,3)$ is $(9x+9)(2x+1)+1$. So, if an RTS$(6x+5,3)$ exists, the number of blocks not in a maximum parallel class is precisely one.

Let $V = \{0,1,2,\ldots,6x+4\}$ and $H = \{6x, 6x+1, \ldots, 6x+4\}$. A RTS$(6x+5)-RTS(5)$ is defined to be a collection of blocks of size three so that every pair of points in V, but not both in H, appears in exactly three blocks. Each pair of points in H appears in no block. In addition, the blocks admit a partition so that there are nine sets of $2x$ blocks, and each set of $2x$ blocks partitions $V \setminus H$. The other blocks partition into maximum parallel classes on V.

Lemma 2.1 *There exists an* RTS$(5,3)$.

Proof All 3-subsets of $\{1,2,3,4,5\}$ give a TS$(5,3)$, which is an RTS as each maximum parallel class only contains one block. $\Box$

Lemma 2.2 *If there exists an* RTS$(6x+5) -$ RTS(5), *then there exists an* RTS$(6x+5,3)$.

Proof In the RTS$(6x+5)-$RTS(5), there are exactly nine partial parallel classes missing the hole. Associate a different block of RTS$(5,3)$ defined on H with each partial parallel class. This gives nine more maximum parallel classes; only one block is left over. $\Box$

Lemma 2.3 *There exists an* RTS$(17) -$ RTS(5).

Proof Parallel classes are listed, with the hole on points $\{12,13,14,15,16\}$:

$$\{0,4,12\},\{1,10,15\},\{2,8,16\},\{5,9,13\},\{7,11,14\}$$
$$\{0,1,12\},\{2,3,13\},\{4,6,15\},\{5,10,14\},\{8,9,16\}$$
$$\{0,6,13\},\{1,10,15\},\{3,11,16\},\{4,9,14\},\{5,7,12\}$$
$$\{0,9,15\},\{1,5,12\},\{2,10,14\},\{3,4,16\},\{8,11,13\}$$
$$\{1,9,16\},\{2,8,12\},\{3,11,14\},\{4,7,13\},\{5,6,15\}$$

$$\{0,8,14\},\{1,3,15\},\{2,9,13\},\{6,7,16\},\{10,11,12\}$$
$$\{0,8,16\},\{1,7,14\},\{3,11,15\},\{4,9,13\},\{6,10,12\}$$
$$\{0,6,14\},\{2,4,16\},\{3,9,12\},\{5,11,15\},\{7,10,13\}$$
$$\{0,3,11\},\{1,11,14\},\{2,5,16\},\{4,7,15\},\{8,10,13\}$$
$$\{0,5,16\},\{1,2,14\},\{3,4,15\},\{6,11,13\},\{7,9,12\}$$
$$\{0,11,15\},\{2,6,13\},\{4,8,12\},\{5,9,16\},\{7,10,14\}$$
$$\{1,3,13\},\{2,9,15\},\{4,5,14\},\{6,8,12\},\{10,11,16\}$$
$$\{0,9,14\},\{1,2,12\},\{3,10,16\},\{5,7,13\},\{6,8,15\}$$
$$\{1,11,13\},\{2,6,14\},\{3,5,12\},\{7,10,16\},\{8,9,15\}$$
$$\{0,11,16\},\{1,3,13\},\{4,9,14\},\{6,10,12\},\{7,8,15\}$$
$$\{0,7,15\},\{2,11,12\},\{3,8,14\},\{4,6,16\},\{5,10,13\}$$
$$\{0,4,13\},\{1,6,16\},\{2,10,15\},\{5,8,14\},\{7,9,12\}$$
$$\{0,8,13\},\{1,7,16\},\{2,5,15\},\{3,6,14\},\{4,11,12\}$$
$$\{0,1,2\},\{3,4,5\},\{6,7,8\},\{9,10,11\}$$
$$\{0,1,7\},\{2,3,6\},\{4,5,10\},\{8,9,11\}$$
$$\{0,2,4\},\{1,8,11\},\{3,9,10\},\{5,6,7\}$$
$$\{0,2,10\},\{1,5,8\},\{3,6,9\},\{4,7,11\}$$
$$\{0,3,5\},\{1,6,9\},\{2,7,11\},\{4,8,10\}$$
$$\{0,3,7\},\{1,4,10\},\{2,5,8\},\{6,9,11\}$$
$$\{0,5,11\},\{1,4,6\},\{2,7,9\},\{3,8,10\}$$
$$\{0,6,10\},\{1,5,9\},\{2,4,11\},\{3,7,8\}$$
$$\{0,9,10\},\{1,4,8\},\{2,3,7\},\{5,6,11\}$$

□

Lemma 2.4 *There exists an* RTS(v) − RTS(5) *for* $v = 23, 35,$ *and* 47.

Proof Write $v = 6x + 5$. Let the points be $\mathbb{Z}_{6x} \cup \{X, \overline{X}, Y, \overline{Y}, \infty\}$. Define two permutations on this point set:

$$\alpha = (0, 1, \ldots, 6x - 1)(X, \overline{X})(Y, \overline{Y})(\infty)$$

$$\beta = (0, 1, \ldots, 6x - 1)(X)(\overline{X})(Y)(\overline{Y})(\infty)$$

Now define three sets of blocks as follows. First, there are three *special* blocks, each consisting of three elements of $\mathbb{Z}_{6x}$ that are distinct modulo 3. Next there is a *half base class* of $x + 1$ triples. This is a set of blocks containing $\{\infty, 0, 3x\}$, and a set of other triples all of whose elements are distinct and nonzero modulo $3x$; of these, one triple contains X and one contains Y, while none contains $\overline{X}$ or $\overline{Y}$. Finally define a *whole base class* of $2x + 1$ triples, of which $X, \overline{X}, Y, \overline{Y}, \infty$ are each contained in one and all other elements are distinct modulo $6x$. The primary requirement is that, calculating differences between every pair of elements from $\mathbb{Z}_{6x}$ among the special triples, and the triples of the half and the whole base classes, one finds every nonzero element of $\mathbb{Z}_{6x}$ as a difference precisely three times.

Forming the orbits of the special triples and whole base class under β, and the triples of the half base class under α, yields a partial triple system in which every pair is covered three times, except those on $\{X, \overline{X}, Y, \overline{Y}, \infty\}$, which are uncovered. To show that the system is an RTS(v)$-$RTS(5), proceed as follows.

For each special triple B, and for $j \in \{0, 1, 2\}$, the blocks $\{\beta^{3i+j}(B) : 0 \leq i \leq 2x\}$ form a parallel class (B has all three elements distinct modulo 3), and hence each special triple yields three holey parallel classes (missing the hole $\{X, \overline{X}, Y, \overline{Y}, \infty\}$). Now for the half parallel class $\mathcal{H}$, since x is odd, form the set of blocks $\mathcal{H} \cup (\alpha^{3x}(\mathcal{H}))$, suppressing the duplication of $\{\infty, 0, 3x\}$. Since all elements in the half parallel class are distinct modulo $3x$, the result is a whole parallel class $\mathcal{P}$. Now $3x$ parallel classes are obtained as $\{\alpha^i(\mathcal{P}) : 0 \leq i < 3x\}$. Finally for the whole parallel class $\mathcal{F}$, form the $6x$ parallel classes $\{\beta^i(\mathcal{F}) : 0 \leq i < 6x\}$.

It remains only to produce the special triples, half base classes, and whole base classes for the required cases. These are given in Table 1. $\square$

Order	Part	
23	Special	$\{0,1,2\}$, $\{0,1,8\}$, $\{0,4,8\}$
	Half	$\{1,7,13\}$, $\{X,2,5\}$, $\{Y,3,8\}$, $\{\infty,0,9\}$
	Whole	$\{3,12,14\}$, $\{8,11,13\}$, $\{X,1,4\}$, $\{\overline{X},5,9\}$, $\{Y,10,15\}$, $\{\overline{Y},2,9\}$, $\{\infty,17,7\}$
35	Special	$\{0,1,2\}$, $\{0,1,8\}$, $\{0,4,8\}$
	Half	$\{6,12,20\}$, $\{9,18,28\}$, $\{11,23,25\}$, $\{X,1,4\}$, $\{Y,2,7\}$, $\{\infty,0,15\}$
	Whole	$\{1,10,20\}$, $\{2,11,17\}$, $\{7,19,21\}$, $\{15,18,28\}$, $\{12,24,29\}$, $\{8,14,25\}$, $\{X,0,3\}$, $\{\overline{X},5,9\}$, $\{Y,22,27\}$, $\{\overline{Y},16,23\}$, $\{\infty,6,13\}$
47	Special	$\{0,1,2\}$, $\{0,1,8\}$, $\{0,4,8\}$
	Half	$\{3,14,31\}$, $\{5,17,30\}$, $\{6,19,34\}$, $\{20,29,39\}$, $\{12,15,32\}$ $\{X,1,4\}$, $\{Y,2,7\}$, $\{\infty,0,21\}$
	Whole	$\{0,2,26\}$, $\{35,3,21\}$, $\{7,17,37\}$, $\{8,10,32\}$, $\{9,14,20\}$, $\{33,39,6\}$, $\{41,5,28\}$, $\{15,24,36\}$, $\{12,31,38\}$, $\{29,40,13\}$ $\{X,1,4\}$, $\{\overline{X},30,34\}$, $\{Y,22,27\}$, $\{\overline{Y},16,23\}$, $\{\infty,11,19\}$

Table 1: RTS(v)$-$RTS(5) for Small Orders v

Lemma 2.5 *There is an* RTS(11, 3).

Proof Use the ingredients in Table 2 as follows. Over $\mathbb{Z}_{10}$, the special block $B = \{0, 1, 2\}$ produces three maximum parallel classes $\{B + j, B + 3 +$

$j, B + 6 + j$}, for $j = 0, 1, 2$, with one block $(B + 9)$ remaining. For the half class, add 5 to each block and form a maximum parallel class by suppressing the duplication of the block containing ∞. Develop modulo 10 to get five maximum parallel classes. Finally develop the whole class modulo 10 to get ten maximum parallel classes. □

Order	Part	
11	Special	{0,1,2}
	Half	{∞,0,5}, {1,3,7}
	Whole	{∞,0,1}, {2,5,8}, {7,9,4}

Table 2: RTS(11, 3)

Lemma 2.6 *There exists an* RTS(29, 3).

Proof There is a cyclic TS(29,3) with base blocks {0,1,2}, {0,4,8}, {0,5,10}, {0,6,12}, {0,10,20}, {0,1,22}, {2,9,17}, {3,5,18}, {4,24,27}, {7,16,19}, {8,11,25}, {10,14,21}, {12,23,28}, {13,15,26}. The last nine starter blocks form a base parallel class; develop modulo 29 to produce 29 parallel classes. Each of the five remaining base blocks has the form {0, a, 2a}. It is easy to see that, whenever *any two* blocks are removed from the 29 blocks in the orbit of {0,1,2}, the remaining 27 blocks partition into three parallel classes. Multiplying such a partition by a establishes the same statement for the orbit of {0, a, 2a} (since a and 29 are relatively prime). Hence it remains only to choose two blocks from the orbit of each of the first five base blocks, and from these ten blocks to form a parallel class leaving one block over. This extra parallel class is {19,20,21}, {26,27,28}, {0,4,25}, {9,13,17}, {5,10,15}, {6,11,16}, {1,18,24}, {2,8,14}, {3,12,22}; the extra block is {0,10,20}. □

Lemma 2.7 *There exists an* RTS(41, 3).

Proof The proof is similar to that of Lemma 2.6. Form a base parallel class with starter blocks {0,1,21}, {2,6,9}, {3,5,8}, {4,26,28}, {7,22,38}, {10,15,18}, {11,17,33}, {12,29,35}, {13,19,27}, {14,25,32}, {16,23,31}, {20,24,36}, {30,34,39} and develop modulo 41 to form 41 parallel classes. The starter blocks {{0, a, 2a} : $a = 1, 9, 10, 11, 12, 13, 14$} complete a TS(41,3), so it suffices to partition the corresponding blocks into parallel classes with one block left over. Choose the single parallel class {30,31,32}, {34,35,36}, {4,13,22}, {7,16,39}, {0,10,20}, {1,11,21}, {6,17,28}, {18,29,40}, {2,14,26}, {3,15,27}, {12,25,38}, {5,19,33}, {9,23,37}. Leave {0,13,26} as the extra block. Then the remaining $7 \cdot 39$ blocks partition into 21 parallel classes (as in the previous lemma).   □

Recursive constructions are now introduced to settle the spectrum of Rosa triple systems when $v \equiv 5 \pmod 6$.

Lemma 2.8 *Suppose there exists a $(3,3)$-frame of type $g_1 g_2 \ldots g_n$ where the size of each group is a multiple of six. If, for each $i = 1, 2, \ldots, n-1$, there exists an $RTS(g_i + 5) - RTS(5)$ and there exists an $RTS(g_n + 5, 3)$, then there is an $RTS(g_1 + g_2 + \ldots + g_n + 5, 3)$.*

Proof The number of frame parallel classes missing a group of size g_i is precisely $3g_i/2$, which is also the number of maximum parallel classes in a $RTS(g_i + 5) - RTS(5)$. Associate each frame parallel class with a maximum parallel class to obtain a maximum parallel class in the $TS(v, 3)$ where $v = g_1 + g_2 + \ldots + g_n + 5$, for all groups where $i = 1, 2, \ldots, n-1$. Finally, there are nine holey parallel classes on each group; together with the frame parallel classes missing the last group, there is a set of $3g_n/2 + 9$ parallel classes missing the set of $g_n + 5$ points. Simple counting shows that the number of maximum parallel classes in a $RTS(g_n + 5, 3)$ is precisely $3g_n/2 + 9$. Append each maximum parallel class in $RTS(g_n + 5, 3)$ to a frame parallel class. All but one block in the $RTS(g_n + 5, 3)$ are exhausted in this way; hence it is an $RTS(v, 3)$.   □

Lemma 2.9 *If $v \equiv 5 \pmod{12}$, then there is an $RTS(v, 3)$.*

Proof The cases when $v = 5, 17, 29, 41$ are dealt with in Lemmas 2.1, 2.3, 2.6 and 2.7. There exists a $(3,3)$-frame of type 12^r for all $r \geq 4$ [4]. Apply Lemma 2.8 to obtain the result; $RTS(17) - RTS(5)$ exists by Lemma 2.3.   □

Lemma 2.10 *Suppose there exists a 4-GDD of type $g_1 g_2 \ldots g_n$. Then there exists a $(3,1)$-frame of type $(2g_1)(2g_2) \ldots (2g_n)$ and hence a $(3,3)$-frame of type $(2g_1)(2g_2) \ldots (2g_n)$.*

Proof This is just a fundamental frame construction, giving weight 2 to each point, using a $(3,1)$-frame of type 2^4. For more details, see [3].   □

Lemma 2.11 *There exist 4-GDDs of type $6^4 3^1$, $6^4 9^1$, $6^6 3^1$, 9^5, $9^5 6^1$, $6^5 12^1 15^1$, $9^5 18^1$ and $6^8 21^1$.*

Proof A 4-GDD of type 9^5 is contained in [2]. In addition, 4-GDDs of type $9^5 18^1$ and $6^8 21^1$ are obtained by completing a resolvable 3-GDD of type 9^5 and 6^8 [5]. For the remaining designs, see [6].   □

Lemma 2.12 *If $v \equiv 11 \pmod{12}$, then there exists an $RTS(v, 3)$.*

Proof If $v = 11, 23, 35, 47$, apply Lemma 2.5 and Lemma 2.4. For $59 \leq v \leq 143$, use the 4-GDDs in Lemma 2.11 in Lemma 2.10 to obtain a $(3,3)$-frame, and apply Lemma 2.8 to yield the interval. Finally, take a $TD(6, n)$ and truncate a group to obtain a $\{5, 6, n\}$-GDD of type $5^n x^1$ where $0 \leq x \leq n-1$. One can also truncate a group of size five to obtain a $\{4, 5, 6, n-1, n\}$-GDD of type $5^{n-1} x^1$. In these GDDs, give weight six to each point and apply fundamental construction for frames [4] to obtain a $(3,3)$-frame of type $(30)^n (6x)^1$ or $(30)^{n-1} (6x)^1$. Use this frame together with Lemma 2.8 and Lemma 2.4 to obtain a RTS$(30n + 6x + 5, 3)$ or RTS$(30(n-1) + 6x + 5, 3)$. This gives an RTS$(v, 3)$ for all $v \geq 143$ and $v \equiv 11 \pmod{12}$. $\square$

3   Rosa Triple Systems when $v \equiv 2 \pmod 6$

In this section, the existence of Rosa triple systems of order v when $v \equiv 2 \pmod 6$ is settled. An RTS$(6x+2, 6)$ has $(18x+9)(2x)+2$ blocks; hence the number of blocks not contained in a maximum parallel class is precisely two.

The following holey structure is needed. Let $V = Z_6 \cup \{\infty_1, \infty_2\}$. Let $B = \{0, 1, 2\}$; then $B + i, B + 3 + i$ for $i = 0, 1, 2$ gives three parallel classes missing the points ∞_1 and ∞_2. Also, developing $\{\infty_1, 0, 2\}$ and $\{\infty_2, 1, 4\}$ over Z_6 gives six maximum parallel classes on V. The three starter blocks give a TS$(8, 2)$–TS$(2, 2)$ (a triple system on eight points with a hole of size two and index 2). Take three copies to obtain a set of blocks on eight points with 27 maximum parallel classes, nine of which miss the hole. Denote this configuration by RTS(8)–RTS(2).

Lemma 3.1 *There is an* RTS$(v, 6)$ *for* $v = 8$, *14, and 20.*

Proof Employ the ingredients in Table 3 as follows. In each case, the specified triples are developed under the action of Z_{v-1}. For each special triple B, three maximum parallel classes are formed as $\{\{B + 3i + j : 0 \leq i < (v-2)/3\} : j = 0, 1, 2\}$. For each of the first, second, and third base classes, apply the action of Z_{v-1} to produce $v - 1$ maximum parallel classes from each. $\square$

Lemma 3.2 *If* $v \equiv 2 \pmod 6$, *there exists an* RTS$(v, 6)$.

Proof There exists a $(3, 6)$-frame of type 6^r for all $r \geq 4$ [3]. An RTS$(6r + 2, 6)$ is constructed on the $6r$ points in the frame and two extra points. The blocks of RTS(8)–RTS(2) are placed on each group with the two extra points aligned on ∞_1 and ∞_2. The number of frame parallel classes is 18 which the the same as the number of maximum parallel classes in the RTS(8)–RTS(2). Associate each frame parallel class with a maximum parallel class to obtain

Order	Part	
8	Special	{0,1,2}, {0,1,2}
	First	{∞,0,2}, {4,5,1}
	Second	{∞,0,3}, {1,2,6}
	Third	{∞,0,3}, {2,4,6}
14	Special	{0,1,2}, {0,1,2}
	First	{∞,0,2}, {3,7,8}, {9,10,5}, {4,6,11}
	Second	{∞,0,3}, {2,4,11}, {6,8,1}, {9,12,5}
	Third	{∞,0,3}, {6,9,1}, {8,12,2}, {11,4,7}
20	Special	{0,1,2}, {0,1,2}
	First	{∞,0,2}, {3,4,15}, {5,6,12}, {7,9,14}, {8,10,17}, {16,1,11}
	Second	{∞,0,3}, {2,6,8}, {17,1,12}, {7,10,15}, {5,11,14}, {9,13,18}
	Third	{∞,0,3}, {4,10,1}, {8,12,2}, {5,9,16}, {7,14,18}, {17,6,11}

Table 3: RTS($v, 6$) for Some Small Orders v

a maximum parallel class on the $6r + 2$ points, for $r - 1$ groups. In each group, nine parallel classes remain on the six points in the group. Together with the frame parallel classes, this yields 27 holey parallel classes missing the remaining eight points. Associate each frame parallel class with a maximum parallel class in a RTS(8,6), to obtain 27 further maximum parallel classes on $6r + 2$ points leaving two blocks unused. This is an RTS($6r + 2, 6$). This construction, together with Lemma 3.1, yields the result.   □

4 Triple Systems with Minimum Chromatic Index

Now the main result can be proved; see [7] for the existence results on resolvable, almost resolvable, and Hanani triple systems used in the proof.

Theorem 4.1 *There is a triple system* TS(v, λ) *that admits a block-colouring in* $\lceil \frac{\lambda v(v-1)/6}{\lfloor v/3 \rfloor} \rceil$ *colours, in which at most one colour class is not a maximum parallel class, whenever* $\lambda \equiv 0 \pmod{\gcd(v - 2, 6)}$ *and* $v \geq 3$, *except when* $v = 6$ *and* $\lambda \equiv 2 \pmod 4$, $(v, \lambda) = (13, 1)$, *or* $v = 7$ *and* $\lambda \in \{1, 3, 5\}$.

Proof When $v \equiv 3 \pmod 6$, simply take λ copies of a Kirkman triple system. When $v \equiv 0 \pmod 6$ and $v \neq 6$, take $\lambda/2$ copies of a resolvable TS($v, 2$). There is a resolvable TS(6, 4). However, a simple counting argument shows that if a $TS(6, \lambda)$ is resolvable, then $\lambda \equiv 0 \pmod 4$. When $v \equiv 4 \pmod 6$, take $\lambda/2$ copies of an almost resolvable TS($v, 2$). When $v \equiv 1$

$\{0,3,7\}$ $\{1,10,11\}$ $\{2,5,6\}$ $\{4,9,12\}$ $\{0,2,8\}$ $\{1,9,12\}$ $\{4,6,11\}$ $\{5,7,10\}$
$\{0,3,6\}$ $\{1,2,11\}$ $\{4,7,9\}$ $\{8,10,12\}$ $\{0,4,11\}$ $\{1,5,12\}$ $\{2,6,7\}$ $\{3,8,10\}$
$\{0,7,9\}$ $\{1,10,12\}$ $\{2,3,11\}$ $\{4,6,8\}$ $\{1,5,11\}$ $\{2,10,12\}$ $\{3,4,9\}$ $\{6,7,8\}$
$\{0,9,10\}$ $\{1,2,7\}$ $\{4,5,12\}$ $\{6,8,11\}$ $\{0,8,12\}$ $\{1,4,7\}$ $\{2,3,6\}$ $\{5,9,11\}$
$\{0,1,5\}$ $\{3,7,11\}$ $\{4,6,12\}$ $\{8,9,10\}$ $\{0,3,5\}$ $\{1,4,8\}$ $\{2,9,12\}$ $\{7,10,11\}$
$\{0,1,6\}$ $\{2,8,9\}$ $\{3,5,12\}$ $\{4,10,11\}$ $\{0,4,8\}$ $\{2,9,11\}$ $\{3,7,12\}$ $\{5,6,10\}$
$\{0,6,12\}$ $\{1,3,9\}$ $\{2,5,8\}$ $\{4,7,10\}$ $\{0,9,10\}$ $\{1,3,4\}$ $\{5,8,11\}$ $\{6,7,12\}$
$\{0,5,10\}$ $\{1,3,8\}$ $\{2,7,12\}$ $\{6,9,11\}$ $\{0,11,12\}$ $\{1,7,8\}$ $\{2,3,10\}$ $\{5,6,9\}$
$\{0,7,11\}$ $\{1,6,10\}$ $\{2,4,5\}$ $\{3,8,9\}$ $\{0,2,4\}$ $\{3,6,10\}$ $\{5,7,9\}$ $\{8,11,12\}$
$\{1,6,9\}$ $\{2,4,10\}$ $\{3,11,12\}$ $\{5,7,8\}$ $\{0,1,2\}$ $\{3,4,5\}$

Table 4: A minci TS$(13,3)$

(mod 6) and $v \notin \{7,13\}$, take λ copies of a Hanani triple system of order v; in alternate copies, place the short parallel class of $\frac{v-1}{6}$ triples on two disjoint sets of $\frac{v-1}{2}$ points; then in pairs of copies, adjoin the two short parallel classes to form one maximum parallel class. The lack of Hanani triple systems of orders 7 and 13 poses a problem for higher indices. There are almost resolvable TS$(v,2)$s for $v \in \{7,13\}$, so take $\lambda/2$ copies whenever λ is even. There is a TS$(13,3)$ which has chromatic index 20, and has all but one colour class being a maximum parallel class; see Table 4. If $\lambda \geq 3$ is odd, the union of a minci TS$(13,3)$ and a minci TS$(13,\lambda-3)$ provides the required system. For TS$(7,\lambda)$, there are 1, 10, and 109 nonisomorphic systems for $\lambda = 1,3,5$, respectively. A complete backtrack verified that none is minci. However, a minci TS$(7,7)$ does exist, on $\{0,1,2,3,a,b,c\}$ with parallel classes as follows:

$\{0,2,a\},\{1,3,b\}$ $\{0,2,a\},\{1,3,b\}$ $\{0,2,b\},\{1,3,a\}$ $\{0,2,b\},\{1,3,c\}$
$\{0,2,c\},\{1,3,a\}$ $\{0,1,a\},\{2,3,c\}$ $\{0,1,c\},\{2,3,a\}$ $\{0,3,b\},\{1,2,c\}$
$\{0,3,c\},\{1,2,b\}$ $\{0,1,3\},\{a,b,c\}$ $\{0,2,3\},\{a,b,c\}$ $\{1,2,3\},\{a,b,c\}$
$\{\{1,2,a\},\{3,b,c\}\}$ under the permutation $(0\ 1\ 2\ 3)(a\ b\ c)$

with short parallel class $\{\{0,1,2\}\}$. By unions, this gives a minci TS$(7,\lambda)$ for all odd $\lambda \geq 7$.

Now if $v \equiv 5$ (mod 6), form an RTS$(v,3)$ in which the short parallel class contains just the triple $\{0,1,2\}$. Form $\lambda/3$ copies by adding $3i$ to each block of the RTS for $0 \leq i < \lambda/3$. Then the extra blocks in up to $(v-2)/3$ consecutive copies form a partial parallel class, so these are adjoined in runs of $(v-2)/3$, with the remaining extra blocks (if any) forming a short parallel class. When $v \equiv 2$ (mod 6), proceed similarly, assuming the short parallel class is $\{\{0,1,2\},\{3,4,5\}\}$, and replace 3 by 6 in the argument given. $\square$

References

[1] T. Beth, D. Jungnickel and H. Lenz, *Design Theory*, Cambridge University Press, Cambridge, England, (1986).

[2] A.E. Brouwer, A. Schrijver and H. Hanani, Group divisible designs with block size 4, *Discrete Math.* 20 (1977), 1-10.

[3] S. Furino, J.X. Yin and Y. Miao, *Frames and Resolvable Designs*, CRC Press, Boca Raton FL, 1996.

[4] R.C. Mullin and H-D.O.F. Gronau, PBDs, Frames, and Resolvability in: *CRC Handbook of Combinatorial Designs* (C.J. Colbourn, J.H. Dinitz; eds.), CRC Press, Boca Raton FL, 1996, pp. 224-226.

[5] R.S. Rees, Two new direct product-type constructions for resolvable group-divisible designs, *J. Combin. Des.* 1 (1993) 15-26.

[6] R.S. Rees and D.R. Stinson, On the existence of incomplete designs of block size four having one hole, *Utilitas Math.* 35 (1989), 119-152.

[7] A. Rosa and C.J. Colbourn, "Colorings of block designs", *Contemporary Design Theory* (J.H. Dinitz and D.R. Stinson, eds.), John Wiley & Sons, New York, 1992, pp. 401–430.

[8] S.A. Vanstone, D.R. Stinson, P.J. Schellenberg, A. Rosa, R. Rees, C.J. Colbourn, M.J. Carter and J.A. Carter, Hanani triple systems, *Israel J. Math.* 83 (1993) 305-319.

Alan C.H. Ling, Department of Combinatorics and Optimization, University of Waterloo, Waterloo, Ontario, Canada N2L 3G1.
e-mail: chaling@barrow.uwaterloo.ca

Charles J. Colbourn, Department of Combinatorics and Optimization, University of Waterloo, Waterloo, Ontario, Canada N2L 3G1.
e-mail: cjcolbou@math.uwaterloo.ca

Searching for spreads and packings

Rudolf Mathon

Abstract

New algorithms are presented for finding spreads and packings of sets with applications to combinatorial designs and finite geometries. An efficient deterministic method for spread enumeration is used to settle several existence problems for t-designs and partial geometries. Randomized algorithms based on tabu search are employed to construct new Steiner 5-designs and large sets of combinatorial designs. In particular, partitions are found of the 4-subsets of a 16-set into 91 disjoint affine planes of order 4.

1 Introduction

In combinatorial applications it is often required to find a specified substructure in a given structure or to decompose a structure into prescribed substructures. For example, in a graph we may want to find one or all cliques of a particular size, or partition its vertices into a number of independent sets; in a combinatorial design we may want to find a parallel class of disjoint blocks, or resolve the set of all blocks into non-intersecting parallel classes. Problems of this type arise in many different areas such as operations research, scheduling and transportation, network analysis, design of experiments, coding theory and cryptography.

From an algorithmic point of view it is well known that subconfigurations and decompositions are hard to find. They belong to the class of NP-complete problems for which no efficient algorithms are known. Several common algorithmic approaches are used to search for combinatorial configurations. These techniques are usually divided into two broad classes depending on whether they search systematically for all possible solutions or attempt to construct only one solution if it exists.

Among exhaustive algorithms backtracking plays a prominent role [6, 14]. A backtrack algorithm attempts to construct a solution by recursively building up feasible partial solutions, one step at a time. At every step all possible extensions are considered in some orderly fashion after which one backtracks.

161

Backtrack algorithms produce all solutions and hence can be used to establish non-existence results. A search tree is usually pruned by rejecting partial solutions which do not have the correct structure or are isomorphic to ones generated before. In general, the more we know about a solution the more it is possible to prune the search tree. Unfortunately, both lookahead and isomorph rejection are often too expensive to be carried out at all levels and so only restricted forms of them are feasible. Another feature of backtracking for combinatorial structures is the potential for producing a large number of solutions. To reduce space, a "solution" graph can be constructed in which the solutions are represented implicitly as weighted paths. This can be used to count and store solutions without having to generate them explicitly [16].

Among non-exhaustive techniques randomized search has been successfully used to approximate solutions of large and difficult combinatorial optimization problems. Randomized algorithms such as hill-climbing, simulated annealing and tabu search have also been applied to construct a variety of combinatorial configurations. In a randomized algorithm approximate solutions are subjected to small changes which move them through the solution space. At each point of the space a cost function is defined which measures the "goodness" of a solution and achieves a global minimum at all points representing the desired configuration. The goal of an algorithm is to find a solution of minimum cost by applying random downhill or sideways moves. Different strategies are used to avoid getting trapped in a local minimum. A hill-climbing algorithm is restarted from a different random initial state whenever it gets stuck. In simulated annealing occasional uphill moves are allowed with a decreasing probability as the search goes on. A tabu search avoids local minima and cycling by maintaining a list of prohibited moves. Hill-climbing and simulated annealing have been successfully applied to a number of problems concerning triple systems and latin squares [6, 7].

The rest of the paper is organized as follows. In Section 2 we describe in detail a backtrack algorithm for spread enumeration and a tabu search for both spreads and packings. The remaining sections describe various computational experiments and highlight the main results obtained using the new algorithms. In Section 3 a tabu search is applied to construct Steiner 5-designs invariant under $\mathrm{PSL}_2(p)$ for several prime orders p yielding the first examples of Steiner systems $S(5, 6, 168)$. In Section 4 we use spread enumeration to search for partial geometries in a number of classical pseudo-geometric strongly regular graphs. In the final section we exhibit four interesting decompositions of configurations into large sets. Among them is a partition of the 4-subsets of a 16-set into 91 disjoint affine planes of order 4 which is one of the largest known decompositions found by a computer.

2 Algorithms

In order to describe the new algorithms we need to formalize the concept of a spread and packing. Let $\Sigma = (\mathcal{P}, \mathcal{L}, \mathrm{I})$ be an incidence structure consisting of a set $\mathcal{P}$ of *points*, a collection $\mathcal{L}$ of subsets of $\mathcal{P}$ called *lines* and an incidence relation I of containment. A subset $\mathcal{S} \subset \mathcal{L}$ is called a *spread* if the lines in $\mathcal{S}$ partition $\mathcal{P}$, i.e., if any point in $\mathcal{P}$ is incident with exactly one line of $\mathcal{S}$. A collection $\mathcal{R}$ of spreads in Σ is called a *packing* if it partitions the set of lines $\mathcal{L}$, i.e., if any line in $\mathcal{L}$ belongs to exactly one spread of $\mathcal{R}$.

Our first algorithm is an exhaustive search for spreads in a structure $\Sigma = (\mathcal{P}, \mathcal{L}, \mathrm{I})$. A spread $\mathcal{S}$ is constructed one line at a time. If $\mathcal{S}_i$ is a partial spread of i disjoint lines and $\mathcal{P}_i \in \mathcal{P}$ are points not covered by $\mathcal{S}_i$ a new line l must lie in the set $\mathcal{L}_i \in \mathcal{L}$ of lines disjoint from all members of $\mathcal{S}_i$. For $\mathcal{S}_i$ to have a completion, every point in $\mathcal{P}_i$ must be incident with some line in $\mathcal{L}_i$. This is a strong condition which allows an early detection of bad partial spreads (a lookahead) and motivates the following powerful heuristic for line selection. Choose a point $p \in \mathcal{P}_i$ which is incident with a minimum number (say t) of lines in $\mathcal{L}_i$ and choose the next line from this set. If $t = 0$ then a backtrack step is necessary, if $t = 1$ the unique line through p is forced; in general, the fewest number of lines is being examined. This can be formulated succinctly in terms of a recursive procedure.

Algorithm "spreads" Spread enumeration in $\Sigma = (\mathcal{P}, \mathcal{L}, \mathrm{I})$ by backtracking

```
procedure Spread(i, Pᵢ, Lᵢ, Sᵢ);
begin
   if Pᵢ = ∅
   then record Sᵢ as a solution;
   else begin
      Find a point p ∈ Pᵢ incident with the fewest lines Lᵢ(p) in Lᵢ;
      for each l ∈ Lᵢ(p) do
      begin
         Find the set P(l) of points in Pᵢ incident with l;
         Find the set L(l) of lines in Lᵢ intersecting l;
         Pᵢ₊₁ := Pᵢ \ P(l);
         Lᵢ₊₁ := Lᵢ \ L(l);
         Sᵢ₊₁ := Sᵢ ∪ {l};
         Spread(i + 1, Pᵢ₊₁, Lᵢ₊₁, Sᵢ₊₁);
      end
   end
end

P₀ := P;
```

$\mathcal{L}_0 := \mathcal{L};$
$\mathcal{S}_0 := \emptyset;$
$\text{Spread}(0, \mathcal{P}_0, \mathcal{L}_0, \mathcal{S}_0);$

An actual implementation of this algorithm requires good data structures to facilitate fast and efficient computations in, and updating of the various point and line sets.

Our next algorithm is a tabu search for spreads in $\Sigma = (\mathcal{P}, \mathcal{L}, \mathrm{I})$. A spread $\mathcal{S}$ is assembled line by line according to a randomized procedure based on local optimality. If $\mathcal{P}_F$ are the points not covered by $\mathcal{S}$ the next line is chosen from the lines $\mathcal{L}_F = \mathcal{L} \setminus \mathcal{S}$ in two steps. First, a point $p \in \mathcal{P}_F$ is selected at random; next, a random line is chosen from the most promising lines $\mathcal{L}(p)$ incident with p. A cost function $c(\mathcal{S}) = |\mathcal{P}_F|$ measures the incompleteness of $\mathcal{S}$. For a line $l \in \mathcal{L}_F$ we can calculate the cost change $\Delta c(l)$ if l is added to $\mathcal{S}$ and the lines $\mathcal{S}(l) \in \mathcal{S}$ conflicting with l are deleted. Then $\Delta c(l) = |\mathcal{P}_S(l)| - |\mathcal{P}(l)|$, where $\mathcal{P}(l)$ and $\mathcal{P}_S(l)$ are the respective point sets of l and $\mathcal{S}(l)$. A line $l \in \mathcal{L}(p)$ is *optimal* if $\Delta c(l)$ is minimum in $\mathcal{L}(p)$. To prevent cycling a list is defined which contains T lines most recently removed from $\mathcal{S}$. A line appearing more than H number of times in this list is *tabu* and can not be used by the algorithm to extend $\mathcal{S}$. The parameters T and H are called tabu length and tabu height, respectively. The number of moves N and trials M as well as T and H are selected by the user.

Algorithm "spreadr" A tabu search for a spread in $\Sigma = (\mathcal{P}, \mathcal{L}, \mathrm{I})$
  repeat
    $\mathcal{P}_F := \mathcal{P};$
    $\mathcal{S} := \emptyset;$
    $k := 0;$
    repeat N **times**
    begin
        Choose a random point $p \in \mathcal{P}_F$;
        Find the set $\mathcal{L}(p) \subset \mathcal{L}_F$ of optimal lines incident with p;
        Choose a random non-tabu line $l \in \mathcal{L}_{(p)}$;
        Find the set $\mathcal{S}(l)$ of lines in $\mathcal{S}$ intersecting l;
        Find the points $\mathcal{P}(l)$ of l and $\mathcal{P}_S(l)$ of $\mathcal{S}(l)$;
        $\mathcal{P}_F := \mathcal{P}_F \setminus \mathcal{P}_S(l) \cup \mathcal{P}(l)$;
        $\mathcal{S} := \mathcal{S} \setminus \mathcal{S}(l) \cup \{l\}$;
        Push the lines $\mathcal{S}(l)$ into the tabu list;
        if $\mathcal{P}_F = \emptyset$ **then** exit loop;
    end
    $k := k + 1;$
  until $k \geq M$ **or** $\mathcal{P}_F = \emptyset$

As before, efficient data structures and techniques are needed to perform a fast move. The tabu strategy can be enhanced by using aspiration and diversification [8, 9]. An aspiration condition is equivalent to a tabu override in case that a prohibited move improves the overall minimum of the cost function. To achieve diversification we perform from time to time a specified number of random moves ignoring optimality and the tabu list.

Our final algorithm is a tabu search for packings in $\Sigma = (\mathcal{P}, \mathcal{L}, \mathrm{I})$. At every stage of the search we have a partition $\mathcal{R}$ of the line set $\mathcal{L}$ consisting of r cells $\mathcal{L}_i$ which approximates a packing. A cost function $c(\mathcal{R})$ measures how much the cells of $\mathcal{R}$ differ from a spread. If $c_i = \sum_{k,l \in \mathcal{L}_i,\ k \neq l} |k \cap l|$ is the sum of the number of points in all pairwise line intersections in $\mathcal{L}_i$ then $c(\mathcal{R}) = c_1 + \cdots + c_r$. The change in cost if a line l is moved from cell i to j is

$$\Delta c(l; i, j) = \sum_{k \in \mathcal{L}_j} |k \cap l| - \sum_{k \in \mathcal{L}_i,\ k \neq l} |k \cap l|.$$

A move is called *optimal* in $\mathcal{R}$ if it minimizes $\Delta c(l; i, j)$ over all i, j and non-tabu lines $l \in \mathcal{L}_i$. The algorithm selects a random optimal move and replaces the oldest entry in the tabu list by l. To find the set of optimum moves a list of $\Delta c(l; i, j)$ is maintained for all lines in $\mathcal{L}$ and all cells of $\mathcal{R}$ together with a list of optimal moves between any two cells. Using appropriate data structures one can update these lists in $O(sb)$ and search them in $O(r^2)$, where b is the number of lines in $\mathcal{L}$ and s measures the average cell size.

Algorithm "packingr" A tabu search for a packing in $\Sigma = (\mathcal{P}, \mathcal{L}, \mathrm{I})$
    repeat
        $k := 0$;
        Generate a random partition $\mathcal{R}$ of $\mathcal{L}$ into r cells;
        Compute the cost $c(\mathcal{R})$;
        repeat N **times**
        begin
            Find cells $\mathcal{L}_i$, $\mathcal{L}_j$ and a non-tabu line $l \in \mathcal{L}_i$ minimizing $\Delta c(l; i, j)$;
            Update $\mathcal{R}$ by moving l from $\mathcal{L}_i$ to $\mathcal{L}_j$;
            $c(\mathcal{R}) := c(\mathcal{R}) + \Delta c(l; i, j)$;
            Push the line l into the tabu list;
            if $c(\mathcal{R}) = 0$ **then** exit loop;
        end
        $k := k + 1$;
    until $k \geq M$ **or** $c(\mathcal{R}) = 0$;

As before, aspiration and diversification should be applied to improve the behavior of the algorithm. It has been found that if a nonlinear cost function

$$c(\mathcal{R}) = \sum_{i=1}^{r} d_i, \quad d_i = f\Big(\sum_{k,l \in \mathcal{L}_i, k \neq l} |k \cap l| \Big)$$

is used, such as $f(0) = 0$, $f(x) = x + E$, for $x > 0$ and some $E > 0$, the algorithm will tend to preserve conflict-free cells (spreads). Note that $E = 0$ corresponds to the original cost function. For packings with a large number of spreads a good choice of $E > 0$ can often dramatically speed up the search.

3   Steiner 5-designs

A Steiner t-design $S(t, k, v)$ is a v-set V together with a family $\mathcal{B}$ of k-subsets of V such that every t-subset of V is contained in exactly one member of $\mathcal{B}$. If $\Sigma = (\mathcal{P}, \mathcal{C}, \mathrm{I})$ is the incidence structure with points $\mathcal{P}$ the $\binom{v}{t}$ t-subsets, lines $\mathcal{L}$ the $\binom{v}{k}$ k-subsets of V and I the containment relation then a Steiner t-design S corresponds to a spread (or parallel class) of lines. If S is invariant under the action of some group $\mathcal{G}$ then it can be viewed as a spread in a smaller incidence structure in which the points and lines are representatives of the corresponding $\mathcal{G}$-orbits of $\mathcal{P}$ and $\mathcal{L}$.

All known Steiner 5-designs are of order $q + 1$ where $q \equiv 3 \pmod 4$ is a prime power and have an automorphism group containing $\mathrm{PSL}_2(q)$ [3, 10]. It is therefore natural to assume that the group $\mathrm{PSL}_2(q)$ acts on the sets $\mathcal{P}$ and $\mathcal{L}$. We have investigated Steiner 5-designs with parameters $S(5, 6, p+1)$, where p is a prime congruent to 11, 23, 47 $\pmod{60}$. It is well-known that the group $\mathrm{PSL}_2(p)$ is 2-homogeneous on the $(p+1)$-set $V = \mathrm{GF}(p) \cup \infty$. For $p \in \{23, 47, 71, 83, 107, 131, 167\}$ $\mathrm{PSL}_2(p)$ is generated by (1) $x \to x + 1 \pmod p$, (2) $x \to \alpha x \pmod p$, where $\alpha = 2$ for $p \in \{23, 47, 71, 167\}$ and $\alpha = 3$ $p \in \{83, 107, 131\}$ and (3) $x \to -\frac{1}{x}$, where $-\frac{1}{0} = \infty$ and $-\frac{1}{\infty} = 0$.

p	# of 5-set orbits	# of short 6-set orbits	# of full 6-set orbits	# of solutions	Algorithm
47	33	86	144	459	spreads
71	79	203	682	4204 (s)	spreads
83	108	279	1200	38717 (s)	spreads
107	182	466	2776	≥ 1000 (s)	spreadr
131	276	704	5314	≥ 100 (s)	spreadr
167	451	1145	11459	≥ 4 (s)	spreadr

Table 3.1: Steiner 5-designs invariant under $\mathrm{PSL}_2(p)$

Let $\mathcal{P}$ be the set of all 5-subset orbits and denote by $\mathcal{L}$ collection of 6-subset orbits in terms of the 5-orbits they contain. A spread S of lines from $\mathcal{L}$ is equivalent to a Steiner system $S(5, 6, p + 1)$ invariant under $\mathrm{PSL}_2(p)$. In addition to full orbits there are short orbits of 5-sets and 6-sets. For $p = 71$, 131 there is a unique fifth 5-orbit and for all p there are half, third and sixth orbits of 6-subsets. The total number of 5-orbits, short and full 6-orbits for different primes p are listed in Table 3.1 An easy counting argument shows that asymptotically there are $O(p^2)$ 5-orbits and $O(p^3)$ 6-orbits. It has been observed by Denniston [5], who found the first examples of $S(5, 6, p+1)$ for $p = 23$, 47, 84, that many solutions are composed of short 6-orbits only. This has been confirmed later in [10], where all solutions invariant under $\mathrm{PSL}_2(p)$ were enumerated for $p = 47$ and all short 6-orbit solutions were found for $p = 71$, 83. Short orbit solutions for $p = 107$, 131, 167 in Table 3.1 have been found using the tabu search **spreadr**, however for the first two orders **spreads** also yields solutions.

p	Max. # of moves/trial	Tabu length	Tabu height	# of solutions	Average # of moves/trial	CPU time in seconds
47	200	5	1	1000	23	2.4
71	3000	15	2	1000	322	53
83	10000	15	2	1000	1291	239
107	100000	25	3	702	39216	11464

Table 3.2: Performance statistics for 1000 trials of the algorithm *spreadr*

As we have seen in the previous section a tabu search algorithm depends on several parameters which are set by the user. These include the maximum number of moves per trial, the length of the tabu list and the number of times a move can appear in the list before it is tabu (tabu height). Optimal values for these parameters are listed in Table 3.2 together with the number of solutions, the average number of moves and total CPU times on a Sun SPARC2 for the orders $p = 47$, 71, 83, 107 and 1000 trials. The number of moves required to find a solution for a fixed p follows a typical distribution pattern. For example, for $p = 83$ and 1000 trials the number of solutions generated in n moves, $1000i < n \leq 1000(i + 1)$, $i = 0, \cdots, 9$ was 555, 248, 102, 49, 25, 15, 5, 1, 0, 0, respectively, and no solution required more than 10000 moves. The number of moves required to find a solution grows rapidly as p increases. It took more than 10 hours of CPU time to find a solution for $p = 167$. Omitting the common elements ∞ 0 1, the 160 base blocks of an $S(5, 6, 168)$ are:

2	3	85	2	5	113	2	7	162	2	8	25	2	9	16	2	10	131
2	11	20	2	12	92	2	13	24	2	14	78	2	18	60	2	19	150
2	21	40	2	22	94	2	28	69	2	31	75	2	34	72	2	36	107
2	43	164	2	45	106	2	47	70	2	56	86	2	59	118	3	4	83
3	8	41	3	10	30	3	11	95	3	12	68	3	13	39	3	14	135
3	15	81	3	17	147	3	18	134	3	19	132	3	20	109	3	21	51
3	22	115	3	23	123	3	24	117	3	25	156	3	27	116	3	28	131
3	31	90	3	33	69	3	37	44	3	40	45	3	43	54	3	48	110
3	49	89	3	52	73	3	57	162	3	61	153	3	62	118	3	63	99
3	64	107	3	65	100	3	67	87	3	70	76	3	77	146	3	79	165
3	94	128	3	98	119	4	5	20	4	13	45	4	14	131	4	15	59
4	17	46	4	18	98	4	19	31	4	21	95	4	23	60	4	24	130
4	25	115	4	26	104	4	27	37	4	29	148	4	30	125	4	32	97
4	34	92	4	39	73	4	40	81	4	41	48	4	43	138	4	47	124
4	49	91	4	50	63	4	52	82	4	55	152	4	62	157	4	64	99
4	68	113	4	71	143	4	77	135	4	79	149	4	87	100	4	136	142
5	6	148	5	11	34	5	12	60	5	16	108	5	19	146	5	21	48
5	24	87	5	37	141	5	39	131	5	41	49	5	44	74	5	66	163
6	7	138	6	13	75	6	14	33	6	15	65	6	17	142	6	20	64
6	22	154	6	23	104	6	25	157	6	35	131	6	39	145	6	43	139
6	52	100	6	66	78	6	108	134	6	147	159	7	11	123	7	12	128
7	15	110	7	20	71	7	22	54	7	23	107	7	41	95	7	50	78
7	60	66	7	73	159	7	77	102	7	82	119	7	98	149	7	115	133
8	10	58	8	13	124	8	20	141	8	34	44	8	55	131	8	63	77
9	23	40	9	39	159	9	47	117	9	53	155	9	59	107	9	89	157
9	90	121	9	96	131	10	81	103	11	15	149	12	30	86	13	40	142
13	79	152	15	31	119	18	34	58	33	34	81						

To obtain 3 more non-isomorphic solutions replace either 2 13 24 by 2 13 21 or 33 34 81 by 33 34 120 or make both changes, respectively. It is interesting to note that an S(5,6,168) has 175036708 blocks and that $|\text{PSL}_2(167)| = 168 \cdot 167 \cdot 83 = 2328648$.

Using current techniques a computer search is feasible for orders up to 200. To find Steiner systems $S(5, 6, v)$ for larger orders v, which no doubt exist in profusion, several strategies are possible. Ideally perhaps, an infinite family can be discovered by analyzing the structure of the known systems. Additional assumptions on the solutions would restrict the search space and extend the range of orders for which randomized techniques can be successfully applied. Finally, combined approaches based on tabu search and local backtrack may be used to find Steiner t-designs for $t > 5$ and for larger block sizes.

4   Partial geometries

A *partial geometry* $pg(s, t, \alpha)$ is an incidence structure $\mathcal{S} = (\mathcal{P}, \mathcal{L}, \mathrm{I})$ with $\mathcal{P}$ and $\mathcal{L}$ the sets of points and lines respectively, and with a symmetric incidence relation satisfying the following axioms [4].

1. Each point is incident with $t + 1$ ($t \geq 1$) lines and two distinct points are incident with at most one line.

2. Each line is incident with $s + 1$ ($s \geq 1$) points and two distinct lines are incident with at most one point.

3. If a point x and a line L are not incident, then there are α ($\alpha \geq 1$) points which are collinear with x and incident with L.

A partial geometry $pg(s, t, \alpha)$ has $v = |\mathcal{P}| = (s + 1)(st + \alpha)/\alpha$ points and $b = |\mathcal{L}| = (t + 1)(st + \alpha)/\alpha$ lines. The point graph of a $pg(s, t, \alpha)$ is strongly regular [1] with parameters $srg(v, k, \lambda, \mu)$, where $k = s(t + 1)$ is the number of points collinear with a given point, $\lambda = s + t\alpha - t - 1$ and $\mu = \alpha(t + 1)$ is the number of neighbors common to two collinear and non-collinear points, respectively. A strongly regular graph is called *pseudo-geometric* if it has the same parameters as the point graph of a partial geometry, and *geometric* if it is the point graph of a $pg(s, t, \alpha)$. To show that a pseudo-geometric graph G is geometric it is sufficient to identify maximum cliques with lines and to find a set $\mathcal{S}$ of mutually edge-disjoint cliques which covers all edges of G. Let $\mathcal{P}$ be the edges of G and denote by $\mathcal{L}$ the set of all cliques (lines) in terms of their edges. Then a partial geometry $\mathcal{S}$ is a spread of lines in $\mathcal{L}$.

No.	Parameters of $srg(v, k, \lambda, \mu)$	Parameters of $pg(s, t, \alpha)$	# of maximum cliques (orbits)	Geometry	Comments
1	(117,80,52,60)	(8, 9, 6)	17641 (6)	no	AG(3,3)
2	(125,52,15,26)	(4, 12, 2)	1025 (2)	no	$\mathcal{H}(5)$
3	(130,81,48,54)	(9, 8, 6)	8424 (2)	no	PG(3,3)
4	(275,112,30,56)	(4, 27, 2)	15400 (1)	?	McLaughlin
5	(280,117,44,52)	(9, 12, 4)	2464 (3)	no	from $\binom{9}{3}$
6	(280,144,68,80)	(9, 15, 5)	13104 (2)	no	$J_2/3.\mathrm{PGL}_9$
7	(343,150,53,75)	(6, 24, 3)	12789 (4)	no	$\mathcal{H}(7)$
8	(495,238,109,119)	(14, 16, 7)	25245 (2)	no	$\binom{12}{4}$
9	(729,328,127,164)	(8, 40, 4)	76545 (4)	yes	$\mathcal{H}(9)$

Table 4.1: Some pseudo-geometric strongly regular graphs

A number of interesting pseudo-geometric graphs has been examined utilizing the algorithm **spreads**. Among the examples listed in Table 4.1 are several well-known "classical" graphs with large automorphism groups. The graph symmetries have been used to speed up our computer experiments. We end this section with some comments on the graphs and results reported in Table 4.1.

- The graphs No.1 and 3 are complements of the line graphs of AG(3,3) and PG(3,3), respectively. Here maximum cliques correspond to spreads of lines in the corresponding 3-space. Hence, the 2 orbits of cliques in No.3 are related to the 2 translation planes of order 9.

- The graphs No.2, 7 and 9 are defined for odd prime powers q in terms of a hermitian parabola in AG$(2, q^2)$, that is, the hermitian curve in PG$(2, q^2)$ with one point deleted. The vertices of $\mathcal{H}(q)$ are pairs (x, y) satisfying $x^{q+1} + y + y^q = 0$; two vertices (x_1, y_1), (x_2, y_2) are adjacent if $x_1 x_2^q + y_1 + y_2^q$ is a square or non-square in GF(q^2) according as $q \equiv 3$ *or* 1 (*mod* 4). The automorphism group of $\mathcal{H}(q)$ is a subgroup of the classical group PΓU$(3, q^2)$ acting transitively on the points (x, y) of $\mathcal{H}(q)$. It is interesting to note that $\mathcal{H}(9)$ contains many new non-isomorphic geometries $pg(8, 40, 4)$ and is the first member of an infinite family of hermitian graphs $\mathcal{H}(3^{2n})$ which are geometric [15].

- The graph No.5 corresponds to a rank 5 action of S_9 on the 280 1-factors (3 disjoint triples) of the 3-uniform 9-hypergraph $\mathcal{K}_9^3$ with valencies 1, 27, 36, 54 and 162. The graph No.8 is generated by a rank 5 action of S_{12} on the 495 4-subsets of a 12-set, two vertices being adjacent if the corresponding subsets have 0 or 2 elements in common. It is interesting to note that this graph has a much larger automorphism group of order $2^{21} \, 3^6 \, 5^2 \, 7 \cdot 11 \cdot 17$ but is not geometric.

- The graph No.6 is generated by a rank 4 action of the group J$_2$/3.PGL$_9$ on 280 points with one of the valencies corresponding to a generalized quadrangle GQ(9,3). Finally, No.4 is the unique Mc Laughlin graph. It has 15400 edges and the same number of 5-cliques from which a partial geometry with 1540 lines is to be formed. A complete search is not feasible, however we could show, that a geometry does not exist if an automorphism of order 11, 7 or 5 is assumed. On the other hand up to 1120 lines have been found which are consistent with the axioms of a partial geometry.

5   Large sets

The algorithms **spreads** and **packingr** described in Section 2 have been used to find a number of new sets of configurations. All our examples can be viewed as block decompositions C of a design B into a number of other designs D. If B is a complete design consisting of all possible blocks then the constituent D's are said to form a *large set*. In order to apply our algorithms the D's must be interpreted as spreads S and C as a packing R of lines L. In most cases it will assumed that R is invariant under the action of some group G in which case the points and lines are representatives of the corresponding G-orbits.

- *A large set of 105 mutually disjoint 1-factorizations of K_{10}*

  Let K_{10} be the complete graph on 10 vertices. Then P is formed by the $\binom{10}{2}=45$ edges of K_{10} and L consists of the $9 \cdot 7 \cdot 5 \cdot 3 = 945$ 1-factors formed by 5 disjoint edges. A 1-factorization of K_{10} is a partition of its edges into 9 1-factors. These correspond to spreads S in L. We searched for packings invariant under a Frobenius group of order 21 with generators $\alpha = (0\ 1\ 2\ 3\ 4\ 5\ 6)$ and $\beta = (1\ 2\ 4)(3\ 6\ 5)(7\ 8\ 9)$. This group partitions the 945 lines of L into 45 orbits of length 21. A modified version of **spreads** has been employed to find 5 1-factorizations with each factor belonging to a different orbit.

01 23 45 67 89	01 23 46 57 89	01 24 35 67 89
02 16 35 49 78	02 17 39 45 68	02 19 34 57 68
03 15 28 47 69	03 16 24 59 78	03 16 25 49 78
04 18 26 39 57	04 12 36 58 79	04 12 37 58 69
05 19 27 38 46	05 14 28 37 69	05 18 29 36 47
06 17 29 34 58	06 15 29 38 47	06 13 27 48 59
07 13 24 59 68	07 18 26 35 49	07 15 28 39 46
08 14 25 36 79	08 19 25 34 67	08 14 23 56 79
09 12 37 48 56	09 13 27 48 56	09 17 26 38 45

01 24 36 57 89	01 24 37 59 68
02 19 35 48 67	02 18 34 57 69
03 12 46 59 78	03 17 29 48 56
04 16 23 58 79	04 13 26 58 79
05 13 29 47 68	05 16 23 49 78
06 18 27 39 45	06 19 25 38 47
07 15 26 38 49	07 12 36 45 89
08 14 25 37 69	08 15 27 39 46
09 17 28 34 56	09 14 28 35 67

To obtain all 105 factorizations expand the 5 base factors using α and β. The full automorphism group of the resulting large set has order 63, the 5 base 1-factorizations have groups of order 6,3,6,6,3, respectively. This is the first example of a large set for K_{10}.

A related problem of interest is an edge-decomposition of the complement of the line graph of K_{10} into 63 edge disjoint 1-factors. With factors as lines a spread can be shown to form a partial geometry $pg(4, 6, 3)$. The two non-isomorphic solutions are easily found using **spreads** [12].

- *10 mutually orthogonal 1-factorizations of $\mathcal{K}_9^3$*

Let $\mathcal{K}_9^3$ be the 3-uniform 9-hypergraph. Then $\mathcal{P}$ is formed by the $\binom{9}{3}$=84 edges (triples) of $\mathcal{K}_9^3$ and $\mathcal{L}$ consists of the $\binom{8}{3}\binom{5}{3}$=280 1-factors formed by 3 disjoint edges. A 1-factorization of $\mathcal{K}_9^3$ is a partition of its edges into 28 1-factors which which can be interpreted as spreads $\mathcal{S}$ in $\mathcal{L}$. Two factorizations are said to be *orthogonal* if any two factors belonging to different factorizations have at most one edge in common. Since 1-factors of $\mathcal{K}_9^3$ contain 3 edges we see that any two factorizations $\mathcal{S}$ of a packing $\mathcal{R}$ must be mutually orthogonal. Factorizations of $\mathcal{K}_9^3$ have been studied in [13]. We employed a modified version of **spreads** to search for packings invariant under the Frobenius group of order 21 with generators $\alpha = (0\ 1\ 2\ 3\ 4\ 5\ 6)$ and $\beta = (1\ 2\ 4)(3\ 6\ 5)$. This group partitions the 280 blocks of $\mathcal{L}$ into 13 orbits of length 21 and one short orbit of length 7. The following base factors represent one of the 12 non-isomorphic solutions.

012 345 678	013 268 457	015 236 478	016 248 357
017 234 568	018 257 346	035 127 468	037 168 245
038 157 246	078 124 356		

012 346 578	014 257 368	015 237 468	018 246 357

Use β on the first 10 base factors and α on the remaining 4 to get two base factorizations. Expand the first one via α and the second via β to obtain 7 and 3 factorizations of the solution, respectively. Several random packings with trivial automorphism groups have been found using a tabu search implemented in **packingr**. Interpreting the 1-factorizations as resolutions of the complete design 2-(9,3,7) this is the first known example realizing the upper bound on the number of mutually orthogonal resolutions.

- *12 mutually orthogonal resolutions of the 280 1-factors of $\mathcal{K}_9^3$ into $AG(2, 3)$'s*

As in the previous example, we start with the edges (triples) and 1-factors (3 disjoint triples) of $\mathcal{K}_9^3$. Then $\mathcal{P}$ is formed by the 280 1-factors and $\mathcal{L}$ consists of quadruples of 1-factors with edges forming an affine plane of order 3. There are 840 lines in $\mathcal{L}$, or equivalently, 840 distinct affine planes in $\mathcal{K}_9^3$. A spread $\mathcal{S}$ in $\mathcal{L}$ corresponds to a resolution of the 280 1-factors into 70 affine planes. In addition to finding spreads in $\mathcal{L}$ we are interested in packings of mutually orthogonal spreads. To find such spreads and packings we will assume an automorphism $\alpha = (2\ 4\ 7\ 3\ 5)$ of order 5 yielding 56 point orbits and 168 line orbits. Employing **spreads** we found the following solution.

012 345 678	037 146 258	048 157 236	056 138 247
012 346 578	037 148 256	045 167 238	068 135 247
012 347 568	036 157 248	045 138 267	078 146 235
012 348 567	035 168 247	046 137 258	078 145 236
012 356 478	034 157 268	058 146 237	067 138 245
012 357 468	038 147 256	045 136 278	067 158 234
012 358 467	037 156 248	045 178 236	068 134 257
012 367 458	034 168 257	056 147 238	078 135 246
012 368 457	037 158 246	048 167 235	056 134 278
012 378 456	036 148 257	047 135 268	058 167 234
016 234 578	028 137 456	035 148 267	047 125 368
016 235 478	028 145 367	034 127 568	057 138 246
018 234 567	027 135 468	036 147 258	045 126 378
018 235 467	024 136 578	037 145 268	056 127 348

012 345 678	037 158 246	048 136 257	056 147 238
012 346 578	037 156 248	045 138 267	068 147 235
012 347 568	035 148 267	046 157 238	078 136 245
012 348 567	036 147 258	045 168 237	078 135 246
012 356 478	038 146 257	045 137 268	067 158 234
012 357 468	036 158 247	045 167 238	078 134 256
012 358 467	034 178 256	057 136 248	068 145 237
012 367 458	034 157 268	056 138 247	078 146 235
012 368 457	037 146 258	048 135 267	056 178 234
012 378 456	035 168 247	048 157 236	067 134 258
016 234 578	027 138 456	035 147 268	048 125 367
016 235 478	024 158 367	038 127 456	057 134 268
018 234 567	026 147 358	037 125 468	045 136 278
018 235 467	024 157 368	037 126 458	056 134 278

First, use α to expand each set of 14 base $AG(2,3)$'s into a base resolution. Then use the group S_3 generated by $(0\ 1)$ and $(0\ 1\ 6)$ to obtain a maximum set of 12 orthogonal resolutions.

- *A large set of 91 mutually disjoint affine planes $AG(2,4)$*

Let $X = \{0, \ldots, 15\}$ be a set with 16 elements. Then $\mathcal{P}$ are the $\binom{16}{2} = 120$ pairs in X and $\mathcal{L}$ are the $\binom{16}{4} = 1820$ quadruples. Note that every line in $\mathcal{L}$ is incident with 6 points of $\mathcal{P}$. A spread in $\mathcal{L}$ corresponds to the 20 lines of a Steiner 2-(16,4,1) design (affine plane of order 4). To obtain a packing of spreads in $\mathcal{L}$ we assume the action of a group $\mathcal{H}_8$ of order 56 with a doubly transitive action on each of the subsets $\{0,1,2,3,4,5,6,7\}$, $\{8,9,10,11,12,13,14,15\}$ of X. $\mathcal{H}_8$ is generated by $\alpha = (0\ 1\ 5\ 6\ 3\ 4\ 2)(8\ 9\ 13\ 14\ 11\ 12\ 10)$ and $\beta = (0\ 1)(2\ 3)(4\ 5)(6\ 7)(8\ 9)(10\ 11)(12\ 13)\ (14\ 15)$ and partitions the lines $\mathcal{L}$ into 30 orbits of length 56 and 20 orbits of length 7. With the help of **spreads** we found all 1391 non-equivalent spreads with blocks belonging to different 56-orbits (so called transversal spreads). For each spread we use α to regroup the blocks of the unused orbits (i.e. 10 orbits of size 56 and 20 orbits of size 7) into 100 orbits of size 7. An exhaustive search utilizing **spreads** yielded a unique set of 5 disjoint transversal spreads hitting each 7-orbit in one block. This solution has $\mathcal{H}_8$ as its full automorphism group and can be generated from the following 3 base designs.

0	1	2	8	3	5	7	12	4	11	13	15	6	9	10	14
0	3	4	10	1	11	12	14	2	6	7	15	5	8	9	13
0	5	6	11	1	4	7	9	2	10	12	13	3	8	14	15
0	9	12	15	1	3	6	13	2	4	5	14	7	8	10	11
0	7	13	14	1	5	10	15	2	3	9	11	4	6	8	12

0	1	2	4	3	5	6	7	8	9	11	13	10	12	14	15
0	3	9	12	1	5	13	15	2	6	8	10	4	7	11	14
0	5	10	11	1	3	8	14	2	7	12	13	4	6	9	15
0	6	13	14	1	7	9	10	2	3	11	15	4	5	8	12
0	7	8	15	1	6	11	12	2	5	9	14	3	4	10	13

0	1	2	3	4	5	6	7	8	11	13	14	9	10	12	15
0	4	11	12	1	5	9	14	2	6	8	15	3	7	10	13
0	5	13	15	1	4	8	10	2	7	12	14	3	6	9	11
0	6	10	14	1	7	11	15	2	4	9	13	3	5	8	12
0	7	8	9	1	6	12	13	2	5	10	11	3	4	14	15

Using α and β we obtain orbits of disjoint $AG(2,4)$ of length 56, 28 and 7 respectively. We note that the last set of 7 affine planes forms the unique affine Steiner quadruple system SQS(16) which contains exactly 56 distinct resolutions into 7 affine planes. Replacing the 7-orbit of $AG(2,4)$ in our solution by resolutions of the affine quadruple system yields several non-isomorphic large sets.

Of the 4 new decompositions exhibited in this section the last is the most interesting one. Although existence is settled for large sets of Steiner triple systems, very little is known for Steiner 2-designs with blocks of size $k \geq 4$. Apart from our set of 91 disjoint $AG(2,4)$ only one other large set is known [11], namely the 55 disjoint projective planes of order 3 found by Chouinard in 1983 [2].

Acknowledgement This research was supported by NSERC grant OGP0008651.

References

[1] Brouwer, A. E. and van Lint, J. H., Strongly regular graphs and partial geometries, *Enumeration and Design*, D. M. Jackson and S. A. Vanstone (eds.), Academic Press (1984).

[2] Chouinard II, L. G., Partitions of the 4-subsets of a 13-set into disjoint projective planes, *Discrete Math.* **45** (1983), 297-300.

[3] Colbourn, C. J., Mathon, R., Steiner Systems, *The CRC Handbook of Combinatorial Designs*, C.J. Colbourn, J.H. Dinitz (eds.), CRC Press, Boca Raton, 1996, 66-75.

[4] De Clerck, F. and Van Maldeghem, H., Some classes of rank 2 geometries, *Handbook of Incidence Geometry: buildings and foundations*, F. Buekenhout (ed.), Elsevier, Amsterdam and New York, 1995, 433-475.

[5] Denniston, R. H. F., Some new 5-designs, *Bull. London Math. Soc.* **8** (1976), 263-267.

[6] Gibbons, P. B., Computational methods in design theory, *The CRC Handbook of Combinatorial Designs*, C.J. Colbourn, J.H. Dinitz (eds.), CRC Press, Boca Raton, 1996, 718-740.

[7] Gibbons, P. B., Mathon, R., The use of hill-climbing to construct orthogonal Steiner triple systems, *J. Combin. Designs* **1**, (1993), 27-50.

[8] Glover, F., Laguna, M., Tabu search, *Modern heuristic techniques for combinatorial problems*, C. Reeves (ed.), Halsted Press, New York, 1993.

[9] Glover, F., Taillard, E., de Werra, D, A user's guide to tabu search, *Annals of Operations Research* **41** (1993), 3-28.

[10] Grannell, M. J., Griggs, T. S., Mathon, R., Some Steiner 5-designs with 108 and 132 points, *J. Combin. Designs* **1**, (1993), 213-238.

[11] Kramer, E. S., Magliveras, S. S., Stinson, D. R., Some small large sets of t-designs, *Australas. J. Combin.* **3** (1991), 191-205.

[12] Mathon, R., The partial geometries $pg(5, 7, 3)$, *Congressus Numerantium* **31**, Utilitas Math. Publ. Inc., Winnipeg (1981), 129-139.

[13] Mathon, R., and Rosa, A., A census of 1-factorizations of K_9^3 : Solutions with a group of order > 4, *Ars Combin.*, **16** (1983), 129-147.

[14] Mathon, R., Computational methods in design theory, *Surveys in Combinatorics, 1991*, London Math. Soc. Lecture Note Series **166**, A. D. Keedwell (ed.), Cambridge University Press, Cambridge, 100-117.

[15] Mathon, R., A new family of partial geometries, to appear.

[16] Schmalz, B., The t-designs with prescribed automorphism group, new simple 6-designs, *J. Combin. Designs* **1** (1993), 125-170.

Rudolf Mathon, Department of Computer Science, University of Toronto, Toronto, Canada M5S 3G.
email: combin@cs.toronto.edu

A note on Buekenhout-Metz unitals

Klaus Metsch

Abstract

Buekenhout has given a construction of unitals in $PG(2, q^2)$ using the André representation of $PG(2, q^2)$ in the space $PG(4, q)$. Metz has shown that this construction produces hermitian and non-hermitian unitals. In this note, we give a geometric criterion in $PG(4, q)$ to decide whether the unital in $PG(2, q^2)$ is hermitian or not.

1 Introduction

Using the method of André [1] and Bruck-Bose [5] to model $PG(2, q^2)$ inside $PG(4, q)$, Buekenhout [6] has given two constructions for unitals in $PG(2, q^2)$, called the hyperbolic and parabolic method. While Metz [9] showed that the parabolic method can give non-hermitian unitals, Barwick [3] showed that the hyperbolic method always produces hermitian unitals. Analytic conditions have been given to decide when the parabolic method gives hermitian unitals, see for instance [2, 7, 8]. In this note a geometric criterion is given that tells if a unital, constructed by the parabolic method, is hermitian.

The model of $PG(2, q^2)$ in $PG(4, q)$ is constructed using a regular spread S of a hyperplane H of $PG(4, q)$ (see section 2). A Buekenhout-Metz unital in this model is related to an orthogonal elliptic cone C in $PG(4, q)$ given by a quadratic form $f(x) = 0$ with the property that C meets H in a line of S. If one considers $PG(4, q)$ as a subgeometry of $PG(4, q^2)$, then the quadratic form $f(x) = 0$ determines an orthogonal hyperbolic cone in $PG(4, q^2)$. Furthermore, there exist precisely two lines s and s' in $PG(4, q^2)$ that miss $PG(4, q)$ but meet every line of S (see [4]). The lines s and s' are conjugate under the Baer involution of $PG(4, q^2)$ with respect to $PG(4, q)$ and we call them the *generator lines* of S; each of them determines S. In this note, we prove the following theorem.

Theorem 1.1 *Let U be a Buekenhout-Metz unital in $PG(2, q^2)$, defined in $PG(4, q)$ by use of a regular spread S in a hyperplane and an elliptic cone given by a quadratic form $f(x) = 0$. If s is a generator line of S in $PG(4, q^2)$, then U is hermitian if and only if s lies on the hyperbolic cone of $PG(4, q^2)$ corresponding to f.*

177

2   The proof

Let S be a regular spread of a hyperplane H of $PG(4, q)$. The structure $A(S)$ whose points are the points of $PG(4, q) \setminus H$, whose lines are the planes of $PG(4, q)$ that meet H in a line of S, and with incidence induced from $PG(4, q)$ is an affine plane of order q^2. For the projective closure $P(S)$ of $A(S)$, we may take the lines of S as the points at infinity and the spread S as the line at infinity with the obvious extension of the incidence relation. Since S is regular, $P(S)$ is isomorphic to $PG(2, q^2)$, see [1, 5].

We consider $PG(4, q)$ as an F_q-vector space with basis $\{v_1, \ldots, v_5\}$ and represent points $\langle \sum x_i v_i \rangle$ also by their homogeneous coordinates $x = (x_1, \ldots, x_5)$. A quadratic form $f(x)$ with a one dimensional radical $P := \langle v \rangle$ determines an *orthogonal cone* with *vertex* P consisting of all points x with $f(x) = 0$. The lines on P that are contained in C form an elliptic or hyperbolic quadric in the quotient space on P and the cone is then called *elliptic* or *hyperbolic*.

Consider an orthogonal elliptic cone C in $PG(4, q)$ such that $H \cap C$ is a line of S. The line $H \cap C$ together with the points of $C \setminus H$ can be seen as a set U of $q^3 + 1$ points of $P(S)$. Since every line of $P(S)$ meets U in one or $q + 1$ points, U is a unital. The unitals of $P(S)$ defined in this way are called *Buekenhout-Metz unitals*.

Consider $PG(4, q)$ as a subgeometry of $PG(4, q^2)$ and let s be a line of $PG(4, q^2)$ that is a generator line of the spread S. We shall show that U is hermitian if and only if s lies on the hyperbolic cone of $PG(4, q^2)$ defined by $f(x) = 0$. To see this, we investigate the form f.

Let $p \in F_q[x]$ be an irreducible polynomial of the form $p(x) = x^2 - x - e$, and let $\mu \in F_{q^2}$ be a root of p. Then $1 - \mu$ is the second root, the elements of F_{q^2} have the form $a + b\mu$ with $a, b \in F_q$, and the isomorphism $z \to \bar{z} := z^q$ of F_{q^2} is given by $(a + b\mu)^q = a + b - b\mu$.

Since the regular spreads of H are projectively equivalent, we may assume that the generator line s of S is $s := \langle ev_2 + \mu v_3, ev_4 + \mu v_5 \rangle$, so that the lines of S are $l_\infty := \langle v_4, v_5 \rangle$ and $l_{a,b} := \langle v_2 + av_4 + bv_5, v_3 + bev_4 + (a+b)v_5 \rangle$ with $a, b \in F_q$. If one represents points of $PG(2, q^2)$ by homogeneous coordinates (x_1, x_2, x_3), then a collineation g of $P(S)$ onto $PG(2, q^2)$ is given by $g(l_\infty) := (0, 0, 1)$, $g(l_{a,b}) := (0, 1, a + b\mu)$ and $g(1, a, b, c, d) = (1, a + b\mu, c + d\mu)$ for $a, b, c, d \in F_q$. To see this, check that for any two points P_1, P_2 of $PG(4, q) \setminus H$, the points $g(P_1), g(P_2)$, and $g(l)$ are collinear in $PG(2, q^2)$, where l is the line of S that meets the line on P_1 and P_2.

The only lines containined in C are the lines on the vertex Z of C, so Z is on the line $C \cap H$. Since the collineation group of $PG(4, q)$ that fixes S is transitive on the points of H, we may assume that $Z = \langle v_5 \rangle$ is the vertex of C. Then $Z \in l_\infty$ and thus $C \cap H = l_\infty$.

The cone C is uniquely determined by its vertex Z and the elliptic quadric

$E := H' \cap C$ of the hyperplane H' given by $x_5 = 0$. In H', the line $l := \langle v_2, v_3 \rangle$ is an external line of E and $\langle l, v_1 \rangle$ is one of the two tangents planes of E on l. Since the elations with axes H are transitive on the points of $PG(4, q) \setminus H$, we may assume that the second tangent plane of E on l touches E in $\langle v_1 \rangle$. Since $\langle v_5 \rangle$ is the vertex of C, the quadratic form defining C has thus the form

$$f(x_1, \ldots, x_5) = a_{14} x_1 x_4 + a_{22} x_2^2 + a_{23} x_2 x_3 + a_{33} x_3^2$$

where $a_{14} \neq 0$ and $a_{22} x^2 + a_{23} x + a_{33}$ is an irreducible polynomial of $F_q[x]$. We remark that we may choose $a_{14} = 1$. Also, since the line $\langle v_1, v_4 - v_2 \rangle$ meets E in a second point, we may assume that $\langle v_1 - v_2 + v_4 \rangle$ is on E. This implies that $a_{22} = a_{14} = 1$, so that the Buekenhout Metz unital depends on only two parameters. But this was known before.

Over F_{q^2}, the polynomial $a_{22} x^2 + a_{23} x + a_{33}$ splits into two linear factors, so that the form $f(x) = 0$ defines a hyperbolic cone of $PG(4, q^2)$. The generator $s = \langle ev_2 + \mu v_3, ev_4 + \mu v_5 \rangle$ of the spread S lies on this hyperbolic cone if and only if $f(0, e, \mu, 0, 0) = 0$, that is $a_{22} e^2 + a_{23} e \mu + a_{33} \mu^2 = 0$. Since $\mu^2 = \mu + e$, this is equivalent to $a_{22} e + a_{33} = 0 = a_{23} e + a_{33}$, and hence to $h = 0$ where

$$h := a_{22} e + a_{33} + (a_{22} - a_{23}) \mu.$$

Therefore we have to show that $g(U)$ is hermitian if and only if $h = 0$.

The unique point of $g(U)$ on the line $\langle w_2, w_3 \rangle$ is the point $g(l_\infty) = (0, 0, 1)$. To determine the points of $g(U)$ that are not on this line, consider the polynomial $f \in F_{q^2}[x, y]$ given by

$$\hat{f}(y, z) := \bar{h} y^2 + h y^{2q} + h_{22} y^{q+1} + h_{13} z^q + \bar{h}_{13} z$$

with

$$h_{22} = 2 a_{22} e + a_{23} - 2 a_{33} \quad \text{and} \quad h_{13} = a_{14}(2e + \mu).$$

Then, using $\bar{\mu} = 1 - \mu$ and $(r + s\mu)^q = r + s(1 - \mu)$ for $r, s \in F_q$, an easy calculation shows that $\hat{f}(r + s\mu, t + u\mu) = (4e + 1) f(1, r, s, t, u)$ (note that $4e + 1 \neq 0$, since $x^2 - x - e$ is irreducible in $F_q[x]$). Thus $g(U)$ consists of the point $(0, 0, 1)$ and the points $(1, a, b)$ with $\hat{f}(1, a, b) = 0$. Hence, if $h = 0$, then $g(U)$ is the hermitian unital related to the hermitian form $h(x_1, x_2, x_3) = h_{22} x_2^{q+1} + h_{13} x_1 x_3^q + \bar{h}_{13} x_1^q x_3$.

Conversely, suppose that $g(U)$ is hermitian. We have to show that $h = 0$. Since $g(U)$ is hermitian, it can be described by a hermitian form $l(x_1, x_2, x_3) = \sum_{i,j} l_{ij} x_i x_j^q$ with $\bar{l}_{ij} = l_{ji}$. Since the lines given by $x_1 = 0$ and $x_3 = 0$ are tangents of $g(U)$ at $(0, 0, 1)$ and $(1, 0, 0)$, we have $l(x_1, x_2, x_3) = l_{22} x_2^{q+1} + l_{13} x_1 x_3^q + \bar{l}_{13} x_1^q x_3$.

Since $l_{13} x_1 x_3^q + \bar{l}_{13} x_1^q x_3 = 0$ and $h_{13} x_1 x_3^q + \bar{h}_{13} x_1^q x_3 = 0$ define the same set of $q + 1$ points on the line $x_2 = 0$, we have $l_{13}/h_{13} \in F_q$ and we can therefore assume that $l_{13} = h_{13}$. For every point $(1, a, b)$ on $g(U)$, we have $\hat{f}(1, a, b) = l(1, a, b) = 0$ and hence

$$0 = \hat{f}(1, a, b) - l(1, a, b) = \bar{h}a^2 + ha^{2q} + (h_{22} - l_{22})a^{q+1}.$$

Since the tangent line of $g(U)$ on $(0, 0, 1)$ is given by $x_1 = 0$, every line with equation $ax_1 = x_2$ is a secant on $(0, 0, 1)$ and meets $g(U)$ in $q + 1$ points. Hence for all $a \in F_{q^2}$, there exists some $b \in F_{q^2}$ with $(1, a, b) \in g(U)$, so that $\bar{h}a + ha^{2q-1} + (h_{22} - l_{22})a^q = 0$ for all $a \in F_{q^2}$. Since $2q - 1 < q^2$, we obtain $h = 0$ (and $h_{22} = l_{22}$).

Remark. For $q = 2$, the only irreducible polynomial of degree two in $F_q[x]$ is $x^2 + x + 1$, so that $a_{22} = a_{23} = a_{33} = e = 1$ and therefore always $h = 0$. In fact, every unital of $PG(2, 4)$ is hermitian.

References

[1] J. André, Über nicht-Desarguesche Ebenen mit transitiver Translations-gruppe, *Math. Z.* **60** (1954), 156-186.

[2] R.D. Baker and G. L. Ebert, On Buekenhout-Metz unitals of odd order, *J. Combin. Theory Ser. A* **60** (1992), 67-84.

[3] S.G. Barwick. A characterization of the classical unital, *Geom. Dedicata* **52** (1994), 175-180.

[4] R.H. Bruck, Construction problems finite projective planes, *Conference on Combinatorial Mathematics and its Applications*, University of North Carolina, 1967, University of North Carolina Press, Chapel Hill, 1969.

[5] R.H. Bruck and R.C. Bose, The construction of translation planes from projective spaces, *J. Algebra* **1** (1964), 85-102.

[6] F. Buekenhout, Existence of unitals in finite translation planes of order q^2 with kernel of order q, *Geom. Dedicata* **5** (1976), 189-194.

[7] G. L. Ebert, On Buekenhout-Metz unitals of even order, *European J. Combin.* **13** (1992), 109-117.

[8] J. W. P. Hirschfeld, *Finite Projective Spaces of Three Dimensions*, Oxford University Press, Oxford, 1985.

[9] R. Metz, On a class of unitals, *Geom. Dedicata* **8** (1979), 125-126.

Klaus Metsch, Mathematisches Institut, Arndtstr. 2, D-35392 Giessen, Germany.
e-mail: klaus.metsch@math.uni-giessen.de

Elation generalized quadrangles of order (q^2, q)

Christine M. O'Keefe   *Tim Penttila*

Abstract

We survey some recent classification theorems for elation generalized quadrangles of order (q^2, q), q even, with particular emphasis on those involving subquadrangles of order q.

1 Introduction

A *generalized quadrangle* (GQ) of *order* (s, t) is an incidence structure $(\mathcal{P}, \mathcal{B}, I)$ in which $\mathcal{P}$ and $\mathcal{B}$ are disjoint (non-empty) sets of *points* and *lines* and for which I is a symmetric point-line incidence relation satisfying the following axioms:

(i) Each point is incident with $1 + t$ lines ($t \geq 1$) and two distinct points are incident with at most one line.

(ii) Each line is incident with $1 + s$ points ($s \geq 1$) and two distinct lines are incident with at most one point.

(iii) If x is a point and L is a line not incident with x then there is a unique pair $(y, M) \in \mathcal{P} \times \mathcal{B}$ such that $x \, I \, M \, I \, y \, I \, L$.

If $s = t$ then the GQ has order s. For an introduction to GQ, see [34].

The *dual* of the incidence structure $(\mathcal{P}, \mathcal{B}, I)$ is the incidence structure $(\mathcal{B}, \mathcal{P}, I)$. It is immediate that the dual of a GQ S of order (s, t) is a GQ $S^\wedge$ of order (t, s).

Higman (see [34, 1.2.3]) showed that in a GQ of order (s, t) where $s, t > 1$ we have $t \leq s^2$ (and, dually, $s \leq t^2$). In this paper we propose to survey some results on GQ of order (q^2, q), that is, those which have equality in Higman's bound, and therefore represent the extremal configurations. We will see below that these GQ are linked with ovoids in $\mathrm{PG}(3, q)$, and therefore with inversive planes, in two different ways. The first is via Tits' construction of a GQ of order (q, q^2) from any ovoid in $\mathrm{PG}(3, q)$, and the second occurs because ovoids in $\mathrm{PG}(3, q)$ arise in certain classical subquadrangles of GQ of order (q, q^2). The classification of ovoids in $\mathrm{PG}(3, q)$, q even, is one of the most important outstanding problems in finite geometry, and there is some hope that these links might provide further insight, or at least transmit information in both directions. Further, to GQ of order (q^2, q) constructed from a q-clan there correspond ovals, flocks and translation planes. Apart from contributing to the study of the associated GQ, these structures are important in their own right.

A fundamental concept in this area is that of an elation generalized quadrangle. Let S be a GQ of order (s, t) and let p be a point of S. An *elation about* p is either the identity collineation or a collineation of S which fixes each line incident with p and fixes no point not collinear with p. If S admits a group G of elations about p acting regularly on the points not collinear with p then S is an *elation generalized quadrangle (EGQ)* with *elation group* G and *base point p*.

The well-known connections between the various geometrical objects mentioned above mean that it should be profitable to study flocks, ovoids and GQ together. In this paper we attempt to draw together results of this nature, and also re-interpret existing results within the wider framework.

2   The known examples

As far as the authors are aware, every known GQ of order (q^2, q) is of the form either $T_3(\Omega)^\wedge$ for some ovoid Ω (Section 2.2) or $GQ(\mathcal{C})$ for some q-clan $\mathcal{C}$ (Section 2.3) or is a Roman GQ (Payne [31]). Further, each of these examples can be constructed via a 4-gonal family in an appropriate group (Section 2.3). It is interesting to note that the Roman GQ are indirectly associated with q-clans since they are the translation duals of the duals of the GQ which arise from the Ganley q-clans.

We remark that there is a unique GQ of order $(2, 4)$ [40, 41, 42, 12] and a unique GQ of order $(3, 9)$, see [34].

2.1   The classical GQ $Q^-(5, q)^\wedge$ and $H(3, q^2)$

Let Q be a non-singular elliptic quadric in $PG(5, q)$. The points and lines of Q form a GQ $Q^-(5, q)$ of order (q, q^2) (see [34, 3.1.1], [6, 9.3.2]). Let H be a non-singular Hermitian variety in $PG(3, q^2)$. The points and lines of H form a GQ $H(3, q^2)$ of order (q^2, q) (see [34, 3.1.1], [6, 9.3.2]). These two GQ are among the five so-called *classical* GQ.

Theorem 2.1 [34, 3.2.3], [6, 10.3.2] The dual of $Q^-(5, q)$ is isomorphic to $H(3, q^2)$.

2.2   Tits' GQ $T_3(\Omega)^\wedge$

Tits (see [10], [34, 3.1.2] or [6, 9.4.2]) has shown that from an ovoid Ω in $PG(3, q)$ one can construct a GQ $T_3(\Omega)$ of order (q, q^2).

Let Ω be an ovoid in $PG(3, q)$ and let $PG(3, q) = \Sigma_\infty$ be embedded as a hyperplane in $PG(4, q)$. Define an incidence structure with *points* as (i) the points of $PG(4, q) \setminus \Sigma_\infty$, (ii) the hyperplanes X of $PG(4, q)$ for which $|X \cap \Omega| = 1$ and (iii) one new symbol (∞). The *lines* are (a) the lines of $PG(4, q)$ which are not contained in Σ_∞ and which meet Ω (necessarily in a unique point) and (b) the points of Ω. *Incidence* is defined as follows: a point of type (i) is incident only with the lines of type (a) which contain it, a point of type (ii) is incident with all lines of type (a) contained in it and with the unique line of type (b) on it and the point of type (iii) is incident with no line of type (a) and all lines of type (b).

Theorem 2.2 [34, 3.2.4] Let Ω be an ovoid in $PG(3, q)$. Then $T_3(\Omega)$ is isomorphic to $Q^-(5, q)$ if and only if Ω is an elliptic quadric.

We remark that for q odd, every ovoid is an elliptic quadric (Barlotti [3] and Panella [28]), so every GQ $T_3(\Omega)$ is classical. Thus these quadrangles are only new in the case q even.

Since there are two infinite families of ovoids of $PG(2, q)$ known, namely the elliptic quadrics for all q and the Tits ovoids for $q = 2^h$, $h \geq 3$ odd, there is one infinite family of non-classical GQ $T_3(\Omega)$ known. These examples $T_3(\Omega)$ and their duals $T_3(\Omega)^\wedge$ are EGQ.

2.3 $GQ(\mathcal{C})$

First we recall Kantor's [15] construction of an EGQ with base point (∞) from a 4-gonal family, see [34, 8.2]. Let G be a group of order $s^2 t$ for some integers s, t. Let $\mathcal{F} = \{S_i : i = 0, \dots, t\}$ be a family of $t + 1$ subgroups of G, each of order s, such that for each $i = 0, \dots, t$ there is a subgroup S_i^* of G of order st containing S_i and with the following properties:
(K1) $S_i S_j \cap S_k = \{1\}$ for distinct i, j, k and
(K2) $S_i^* \cap S_j = \{1\}$ for distinct i, j.
Such a family $\mathcal{F}$ of subgroups is called a *4-gonal family for G*. It is straightforward to verify that the following incidence structure is an elation generalized quadrangle (with base point (∞)) of order (s, t) : *points:* (i) elements $g \in G$, (ii) cosets $S_i^* g$ for $g \in G$ and $i = 0, \dots, t$ and (iii) a symbol (∞), and *lines:* (a) cosets $S_i g$ for $g \in G$ and $i = 0, \dots, t$ and (b) symbols $[S_i]$ for $i = 0, \dots, t$. A point g of type (i) is incident with each line $S_i g$ for $i = 0, \dots, t$. A point $S_i^* g$ of type (ii) is incident with the line $[S_i]$ and with each line $S_i h$ contained in $S_i^* g$. The point (∞) is incident with each line $[S_i]$ for $i = 0, \dots, t$. There are no further incidences.

Let $\mathcal{C} = \{A_t : t \in GF(q)\}$ be a collection of 2×2 matrices with entries from $GF(q)$. Following Payne [31], we call $\mathcal{C}$ a *q-clan* if $A_s - A_t$ is anisotropic (that is, the equation $(x,y)(A_s - A_t)(x,y)^T = 0$ has only the trivial solution $(x,y) = (0,0)$) for all $s,t \in \mathrm{GF}(q)$ with $s \neq t$.

Let $\mathcal{G} = \{(\alpha, c, \beta) : \alpha, \beta \in \mathrm{GF}(q)^2, c \in \mathrm{GF}(q)\}$, with multiplication defined as:

$$(\alpha, c, \beta)(\alpha', c', \beta') = (\alpha + \alpha', c + c' + \beta(\alpha')^T, \beta + \beta').$$

Let $\mathcal{C}$ be a q-clan, for each $t \in \mathrm{GF}(q)$ let $K_t = A_t + A_t^T$ and define the following subgroups of $\mathcal{G}$:

$$
\begin{aligned}
A(\infty) &= \{(0,0,\beta) : \beta \in \mathrm{GF}(q)^2\} \text{ and} \\
A(t) &= \{(\alpha, \alpha A_t \alpha^T, \alpha K_t) : \alpha \in \mathrm{GF}(q)^2\}, \quad t \in \mathrm{GF}(q).
\end{aligned}
$$

For each $t \in \mathrm{GF}(q) \cup \{\infty\}$ we define $A^*(t) = A(t)Z$ where $Z = \{(0,c,0) : c \in \mathrm{GF}(q)\}$ is the centre of $\mathcal{G}$. Then $\mathcal{F} = \{A(t) : t \in \mathrm{GF}(q) \cup \{\infty\}\}$ is a 4-gonal family for $\mathcal{G}$ (q odd: Kantor [16], q even: Payne [29, 30], see [34, 10.4]).

We have therefore outlined a process by which a q-clan $\mathcal{C} = \{A_t : t \in \mathrm{GF}(q)\}$ gives rise to a 4-gonal family for the group $\mathcal{G}$ above, and hence to an EGQ $GQ(\mathcal{C})$ of order (q^2, q) and with base point (∞).

Further, if $\mathcal{C} = \{A_t : t \in \mathrm{GF}(q)\}$ is a collection of upper triangular matrices

$$A_t = \begin{pmatrix} a_t & c_t \\ 0 & b_t \end{pmatrix}$$

then Thas [43] has shown that $\mathcal{C}$ is a q-clan if and only if the planes $a_t x_0 + b_t x_1 + c_t x_2 + x_3 = 0$ are a flock of the quadratic cone $x_0 x_1 = x_2^2$ in $\mathrm{PG}(3, q)$. Thas [6, 7.10.5] lists the flocks known at the time of printing; the corresponding q-clans and GQ are later discussed by Thas [6, 9.4.4]. To this list we add, in the case of q even, the Subiaco flocks, q-clans and GQ [7] and several cyclic q-clans and their corresponding GQ [32]. In the case of q odd, nineteen new flocks were discovered by Penttila and Royle [36].

For q odd, Knarr [17] has provided a geometric construction for $GQ(\mathcal{C})$ using the associated BLT-set (Bader, Lunardon and Thas, [2]). This construction is also discussed in [6, 10.4.4].

Theorem 2.3 [34, 6] Let $\mathcal{C}$ be a q-clan. Then $GQ(\mathcal{C})$ is isomorphic to $H(3, q^2)$ if and only if $\mathcal{C}$ corresponds to the linear flock.

3　Characterisations

There are many interesting characterizations of non-classical GQ, most of which are of a combinatorial nature, see [34, 5.3.4] or [6, 10.8.6]. Other characterisations involve Property (G) [44] or Veblen's axiom [45]. In this section we consider characterizations which concentrate on subquadrangles.

3.1   Characterisations of $T_3(\Omega)$, q even

If q is even, then the classifications of ovoids in $PG(3,q)$ for small values of q have immediate corollaries in the present context.

Theorem 3.1 [4, 11, 35, 20, 21, 27] If $q = 4$ or 16 then $T_3(\Omega)$ is isomorphic to $Q^-(5,q)$. If $q = 8$ or 32 then either $T_3(\Omega)$ is isomorphic to $Q^-(5,q)$ or Ω is a Tits ovoid.

The next theorem displays a connection between the classification of ovoids in $PG(3,q)$ and the question of whether there exist GQ of order (q^2, q) which are not EGQ. The resolution of either of these problems would be a major advance.

Theorem 3.2 Let q be even and let Ω be an ovoid in $PG(3,q)$. $T_3(\Omega)^\wedge$ is an EGQ if and only if Ω is either an elliptic quadric or a Tits ovoid.

Proof   Suppose first that $T_3(\Omega)^\wedge$, the dual of the GQ $T_3(\Omega)$ as constructed in Section 2.2, is an EGQ. If Ω is an elliptic quadric then $T_3(\Omega)^\wedge$ is isomorphic to $Q^-(5,q)^\wedge$, which is an EGQ with base point p for each of its points p. Otherwise, $\mathrm{Aut}T_3(\Omega)$ fixes the point (∞). Since the group of elations is a subgroup of $\mathrm{Aut}T_3(\Omega)$ and in $T_3(\Omega)^\wedge$ fixes only the base point and the lines on the base point, it follows that the base point of $T_3(\Omega)^\wedge$ must be incident with (∞), that is, is a point of the ovoid Ω.

Thus there exists a point $p \in \Omega$ and a group G of collineations of $T_3(\Omega)$ fixing ∞ and each hyperplane of $PG(4,q) \setminus \Sigma_\infty$ which meets Σ_∞ in the tangent plane π to Ω at p. Now $\mathrm{Aut}T_3(\Omega)_{(\infty)} = A\Gamma L(4,q)_\Omega$ [26]; thus $G \leq \mathrm{Aut}T_3(\Omega)_{(\infty)} \leq A\Gamma L(4,q)_\Omega$. Since G fixes each hyperplane on π, it follows that $G \leq AGL(4,q)_\Omega$. Further, G acts regularly on the q^5 lines of $PG(4,q) \setminus \Sigma_\infty$ which meet Σ_∞ in a point of $\Omega \setminus \{p\}$. Thus $|G| = q^5$ and G is transitive on $\Omega \setminus \{p\}$. Now since G fixes π, G acts on the tangent lines to Ω on p. There is an odd number $(q+1)$ of such tangent lines and since $|G| = q^5$ is even G must fix a tangent line L to Ω at p. It follows that G acts transitively on the pencil of Ω with carrier L. Let π' be a plane of this pencil and consider $H = G_{\pi'}$. Then $|H| = q^4$. Now H fixes π and π', so acts on the $q - 1$ further planes on L in Σ_∞. As there is an odd number of these and $|H|$ is even, H fixes a further plane on L and hence H fixes every plane of the pencil with carrier L (since H acting on the planes of Σ_∞ on L is a subgroup of $PGL(1,q)$).

Let $I(\Omega)$ denote the inversive plane with *points* the points of Ω and *circles* the plane sections of Ω. A *translation about* p of an inversive plane is an automorphism which fixes a point p and every circle of a pencil with carrier p. We have shown that H is a group of translations of $I(\Omega)$ of order q^4. Glynn [13] has shown that this implies that Ω is either an elliptic quadric or a Tits ovoid.

Conversely, suppose that Ω is either an elliptic quadric or a Tits ovoid. In the first case, $T_3(\Omega)^\wedge$ is isomorphic to $Q^-(5,q)^\wedge$, which is an EGQ with any of its points as base point. In the second case, we show that $T_3(\Omega)^\wedge$ is an EGQ with base point a point of the ovoid Ω by displaying an appropriate elation group G, as follows. Let G be the subgroup of the Sylow 2-subgroup of $\mathrm{AGL}(4,q)_\Omega$ which fixes a point $p \in \Omega$ and fixes each hyperplane of $\mathrm{PG}(4,q) \setminus \Sigma_\infty$ which meets Σ_∞ in the tangent plane to Ω at p (this exists, by the first part of the proof). So G is a subgroup of $\mathrm{Aut}T_3(\Omega)_{(\infty)}$. By [18, Theorem 21.8], G is transitive on the points of $\Omega \setminus \{p\}$. Further, G acts regularly on the lines of $\mathrm{PG}(4,q) \setminus \Sigma_\infty$ meeting Σ_∞ in a point of $\Omega \setminus \{p\}$.

$\square$

Further Remarks: (1) We have shown that $T_3(\Omega)^\wedge$ is an EGQ if and only if the associated inversive plane $\mathrm{I}(\Omega)$ has Hering type greater than I which is if and only if Ω is either an elliptic quadric or a Tits ovoid. Inherent in the proof is the identification of a special pencil of the ovoid Ω; the carrier of this pencil is an axis of each translation oval in the pencil.

(2) A very similar method of proof, combined with a characterisation of translation ovals [22], can be used to give a new proof that $T_2(O)^\wedge$ is a TGQ if and only if O is a translation oval, which is if and only if $T_2(O)$ is self-dual.

Now we turn to characterisations in terms of subquadrangles. By an analogous construction to that given in Section 2.2, an oval O in $\mathrm{PG}(2,q)$ gives rise to a GQ $T_2(O)$ of order q (Tits, see [10] or [34, 3.1.2]). If the oval O is a secant plane section of the ovoid Ω, then we obtain a GQ $T_2(O)$ of order q embedded as a subquadrangle in $T_3(\Omega)$. Characterisations of ovoids in $\mathrm{PG}(3,q)$, q even, by the nature of their plane sections can be reinterpreted as characterisations of $T_3(\Omega)$ in terms of certain subquadrangles. (For a survey of such characterisations of ovoids, see [19].)

There are various characterizations of the elliptic quadrics as ovoids admitting a certain collection of plane sections which are conics, see [4, 13, 38, 39]. Since $T_2(O)$ is isomorphic to the classical GQ $Q(4,q)$ (formed by the points and lines of a non-singular quadric in $\mathrm{PG}(4,q)$) if and only if O is a conic ([34, 3.2.2]), we obtain corresponding characterizations of $Q^-(5,q)$ in terms of the existence of a certain configuration of classical subquadrangles. However, these characterizations are probably weaker than the known combinatorial characterisations of $Q^-(5,q)$ (see [34]) and Theorem 4.1 below.

On the other hand, the existing characterizations of the known ovoids as ovoids admitting a certain collection of translation oval plane sections have interesting and possibly useful interpretations in the GQ. First, if q is even then $T_2(O)$ is self-dual if and only if O is a translation oval [34, Chapter 12]. Since each plane section O of a known ovoid Ω is a translation oval, it follows that each subGQ $T_2(O)$ of $T_3(\Omega)$, for Ω a known ovoid, is self-dual. Theorem 3.4 provides a converse, and relies on the following lemma:

Lemma 3.3 [26] Let S be a subquadrangle of $T_3(\Omega)$ which contains a coregular point. Then S is $T_2(O)$ for some plane section O of Ω.

Theorem 3.4 Let Ω be an ovoid in $PG(3,q)$, q even. If each subquadrangle of $T_3(\Omega)$ of order q and containing a coregular point is self-dual then Ω is either an elliptic quadric or a Tits ovoid.

Proof By Lemma 3.3, each subquadrangle of order q and containing a coregular point is of the form $T_2(O)$ for some plane section O of Ω. Conversely, each subquadrangle of the form $T_2(O)$ for some plane section O of Ω has a coregular point. Thus each plane section O of Ω is a translation oval, and the result follows by Theorem 5.3 of Penttila and Praeger [35].   □

   In fact Penttila and Praeger's result requires only that the ovals in a pencil of plane sections of Ω be translation ovals, where the carrier of the pencil is an axis of (at least) one of the ovals. O'Keefe and Penttila [23] removed the extra assumption on the axis, thus:

Theorem 3.5 [35, 23] Let Ω be an ovoid in $PG(3,q)$, q even. If each oval O in a pencil of Ω gives rise to a self-dual GQ $T_2(O)$ then Ω is either an elliptic quadric or a Tits ovoid.

   There are stronger characterisations of the known ovoids in $PG(3,q)$, see [14, 24], thus the same result can be proved under weaker hypotheses on the subquadrangles.

3.2   Characterizations of $GQ(\mathcal{C})$, q even

Let q be even and let $GQ(\mathcal{C})$ be the EGQ arising from a q-clan $\mathcal{C}$ as in Section 2.3. The classification of flocks (and hence q-clans with each matrix upper triangular) for small values of q has an immediate consequence for the classification of the $GQ(\mathcal{C})$:

Theorem 3.6 (1) (Thas [43]) If $q = 2$ or 4 then every $GQ(\mathcal{C})$ is $Q^-(5,q)^\wedge$.
(2) (De Clerck, Gevaert and Thas [8]) If $q = 8$ then every $GQ(\mathcal{C})$ is either $Q^-(5,q)^\wedge$ or is the dual of the GQ constructed by Kantor [15] from the generalized hexagon $H(q)$ of order q associated with the group $G_2(q)$.
(3) (De Clerck and Herssens [9]) If $q = 16$ then every $GQ(\mathcal{C})$ is either $Q^-(5,q)^\wedge$ or is a non-classical GQ now known to be one of the family of Subiaco GQ constructed by Cherowitzo, Penttila, Pinneri and Royle [7].

   Results of Payne and Maneri [33] and Payne [30] guarantee that $GQ(\mathcal{C})$ admits $q + 1$ subquadrangles $T_2(O_i)$, for ovals $O_1, \ldots, O_{q+1}$, each containing the base point ∞ of $GQ(\mathcal{C})$ and any given point not collinear with the base point (usually taken as (0,0,0)). These ovals are the ovals of the *herd* $H(\mathcal{C})$ associated with $\mathcal{C}$ ([7], see [25]). Thus recent characterizations of herds can be restated as characterizations of $GQ(\mathcal{C})$ in terms of the nature of certain of its subquadrangles. The first result in this direction is due to O'Keefe and Penttila:

Lemma 3.7 [26] Let q be even and let S be a subquadrangle of $GQ(\mathcal{C})$ for some q-clan $\mathcal{C}$. If S contains the point (∞) then S is $T_2(O_i)$ for some oval $O_i \in H(\mathcal{C})$.

Theorem 3.8 Let $GQ(\mathcal{C})$ be an EGQ, q even. If $GQ(\mathcal{C})$ admits a self-dual subquadrangle of order q containing the point (∞) then $GQ(\mathcal{C})$ is either $Q^-(5,q)^\wedge$ or is the dual of the GQ constructed by Kantor [15] from the generalized hexagon $H(q)$ of order q associated with the group $G_2(q)$.

Proof By Lemma 3.7, the subquadrangle is of the form $T_2(O_i)$ for some oval $O_i \in H(\mathcal{C})$. Since $T_2(O_i)$ is self-dual, then O_i is a translation oval and the result follows from [25]. □

As a corollary, we obtain a classification of the $T_2(O)$, O a translation oval, which can occur as subGQ of a $GQ(\mathcal{C})$, under the further assumption that $O \in H(\mathcal{C})$. First recall that if we write $\mathcal{D}(\alpha) = \{(1,t,t^\alpha) : t \in \mathrm{GF}(q)\} \cup \{(0,0,1)\}$ for $\alpha \in \mathrm{AutGF}(q)$, then every translation oval is projectively equivalent to some set $\mathcal{D}(\alpha)$. Suppose there exists a q-clan $\mathcal{C}$ such that there is a translation oval $O \in H(\mathcal{C})$ with $T_2(O)$ a subGQ of $GQ(\mathcal{C})$. By Theorem 3.8, $GQ(\mathcal{C})$ is either $Q^-(5,q)^\wedge$ or is the dual of the GQ constructed by Kantor [15] from the generalized hexagon $H(q)$ of order q associated with the group $G_2(q)$. In the first case O is $\mathcal{D}(2)$ (a non-degenerate conic) while in the second case O is $\mathcal{D}(4)$ (see, for example, [25]). We remark that the hypothesis that the GQ of order (q^2,q) is of the type $GQ(\mathcal{C})$ is necessary since $T_2(\mathcal{D}(\sigma))$, where $\sigma \in \mathrm{AutGF}(q)$ satisfies $\sigma^2 \equiv 2 \pmod{q-1}$, occurs as a subquadrangle of order q of $T_3(\Omega)^\wedge$ where Ω is a Tits ovoid.

Penttila and Storme [37] have partially extended this result to cover the case of a subquadrangle of the form $T_2(O_i)$ where O_i is a monomial oval, but so far have been unable to completely classify the resulting GQ.

4   GQ with subGQ of order q and ovoids

In this section we explore a further connection between GQ of order (q^2,q) and ovoids. The connection arises because whenever a GQ of order (q^2,q) has a subquadrangle of order q, there are spreads subtended in the subquadrangle. Thus there are ovoids subtended in the dual subquadrangle.

We also remark that further progress on the classification of ovoids in $\mathrm{PG}(3,q)$ would have consequences on the classification of the GQ of order

(q^2, q). Thus knowledge of the ovoids in the dual subquadrangle can lead to information about the GQ. Let S be a GQ of order (q^2, q), q even, with a classical subquadrangle. It follows that the ovoids subtended in the dual subquadrangle are ovoids of $PG(3, q)$; so if these are classified then it should be possible to classify the possible GQ S. Partial results in this direction are due to Thas and Payne [46] and Brown [5].

Theorem 4.1 [46, 5] Let S be a GQ of order (q, q^2), for q even, which has a subquadrangle isomorphic to $W(q)$. If each ovoid subtended by S in $W(q)$ is either an elliptic quadric or a Tits ovoid, then S is isomorphic to $Q^-(5, q)$ (and in this case each subtended ovoid is an elliptic quadric).

If the conjecture that the only ovoids of $W(q)$ are the elliptic quadrics and the Tits ovoids, then every GQ of order (q, q^2), q even, which contains a subGQ isomorphic to $W(q)$ is isomorphic to $Q^-(5, q)$. In particular, this is true for $q \leq 32$. On the other hand, if there is another such GQ then there are new ovoids in $PG(3, q)$.

In the case q odd, there are interesting non-classical examples of GQ of order (q^2, q), such as the Kantor GQs (see, for example, Bader [1]). Again, Brown has made some progress on classification in this case. The duals of the Roman GQ (here $q = 3^r, r \geq 2$) also have classical subquadrangles of order q ([46]).

Acknowledgement: This work was supported by the Australian Research Council.

References

[1] L. Bader, Flocks of cones and generalized hexagons, *Advances in Finite Geometries and Designs*, J.W.P. Hirschfeld et al. (eds.), Isle of Thorns 1990, Oxford University Press, Oxford, 1991, pp. 7–18.

[2] L. Bader, G. Lunardon and J.A. Thas, Derivation of flocks of quadratic cones, *Forum Math.* **2** (1990), 163–174.

[3] A. Barlotti, Un'estensione del teorema di Segre-Kustaanheimo, *Boll. Un. Mat. Ital.* **10** (1955), 96–98.

[4] A. Barlotti, Some topics in finite geometrical structures, Institute of Statistics Mimeo Series No. 439, University of North Carolina, 1965.

[5] M.R. Brown, Generalized quadrangles of order (q, q^2), q even, containing W(q) as a subquadrangle, *Geom. Dedicata* **56** (1995), 299–306.

[6] F. Buekenhout, ed., *Handbook of incidence geometry: buildings and foundations* North-Holland, Amsterdam, 1995.

[7] W. Cherowitzo, T. Penttila, I. Pinneri and G.F. Royle, Flocks and ovals, *Geom. Dedicata* **60** (1996), 17–37.

[8] F. De Clerck, H. Gevaert and J.A. Thas, Flocks of a quadratic cone in $PG(3, q), q \leq 8$, *Geom. Dedicata* **26** (1988), 215–230.

[9] F. De Clerck and C. Herssens, Flocks of the quadratic cone in $PG(3, q)$, for q small, *Reports of the CaGe project* **8** (1992), 1–75.

[10] P. Dembowski, *Finite Geometries*, Springer-Verlag, Berlin, Heidelberg, 1968.

[11] G. Fellegara, Gli ovaloidi di uno spazio tridimensionale di Galois di ordine 8, *Atti. Accad. Naz. Lincei Rend.* **32** (1962), 170–176.

[12] H. Freudenthal, Une étude de quelques quadrangles généralisés, *Ann. Mat. Pura Appl. (4)* **102** (1975), 109–133.

[13] D.G. Glynn, The Hering classification for inversive planes of even order, *Simon Stevin* **58** (1984), 319–353.

[14] D.G. Glynn, C.M. O'Keefe, T. Penttila and C.E. Praeger, Ovoids and monomial ovals, *Geom. Dedicata* **59** (1996), 223–241.

[15] W.M. Kantor, Generalized quadrangles associated with $G_2(q)^*$, *J. Combin. Th. Ser. A* **29** (1980), 212–219.

[16] W.M. Kantor, Some generalized quadrangles with parameters (q^2, q), *Math. Zeit.* **192** (1986), 45–50.

[17] N. Knarr, A geometric construction of generalized quadrangles from polar spaces of rank three, *Resultate Math.* **21** (1992), 332–334.

[18] H. Luneburg, *Translation planes*, Springer-Verlag, Berlin Heidelberg, 1980.

[19] C.M. O'Keefe, Ovoids in $PG(3, q)$: a survey, *Discrete Math.* **151** (1996), 175–188.

[20] C.M. O'Keefe and T. Penttila, Ovoids of $PG(3, 16)$ are elliptic quadrics, *J. Geom.* **38** (1990), 95–106.

[21] C.M. O'Keefe and T. Penttila, Ovoids of $PG(3, 16)$ are elliptic quadrics, II, *J. Geom.* **44** (1992), 140–159.

[22] C.M. O'Keefe and T. Penttila, Symmetries of arcs, *J. Combin. Theory Ser. A* **66** (1994), 53–67.

[23] C.M. O'Keefe and T. Penttila, Ovoids with a pencil of translation ovals, *Geom. Dedicata* **62** (1996), 19–34.

[24] C.M. O'Keefe and T. Penttila, Ovals in translation hyperovals and ovoids, submitted.

[25] C.M. O'Keefe and T. Penttila, Herds containing a translation oval, in preparation.

[26] C.M. O'Keefe and T. Penttila, Subquadrangles of generalized quadrangles of order (q^2, q), in preparation.

[27] C.M. O'Keefe, T. Penttila and G.F. Royle, Classification of ovoids in $PG(3, 32)$, J. Geom. **50** (1994), 143–150.

[28] G. Panella, Caratterizzazione delle quadriche di uno spazio (tridimensionale) lineare sopra un corpo finito, Boll. Un. Mat. Ital. **10** (1955), 507–513.

[29] S.E. Payne, Generalized quadrangles as group coset geometries, Congress. Numer. **29** (1980), 717–734.

[30] S.E. Payne, A new infinite family of generalized quadrangles, Congress. Numer. **49** (1985), 115–128.

[31] S.E. Payne, An essay on skew translation generalized quadrangles, Geom. Dedicata **32** (1989), 93–118.

[32] S.E. Payne, T. Penttila and G.F. Royle, Building a cyclic q-clan, submitted.

[33] S.E. Payne and C.E. Maneri, A family of skew-translation generalized quadrangles of even order, Congress. Numer. **36** (1982), 127–135.

[34] S.E. Payne and J.A. Thas, Finite Generalized Quadrangles, Pitman, London, 1984.

[35] T. Penttila and C.E. Praeger, Ovoids and translation ovals London Math. Soc., to appear.

[36] T. Penttila and G.F. Royle, BLT-sets for small odd q, in preparation.

[37] T. Penttila and L. Storme, Monomial flocks, in preparation.

[38] O. Prohaska and M. Walker, A note on the Hering type of inversive planes of even order, Arch. Math. **28** (1977), 431–432.

[39] B. Segre, On complete caps and ovaloids in three-dimensional Galois spaces of characteristic two, Acta Arith. **5** (1959), 315–332.

[40] J.J. Seidel, Strongly regular graphs with $(-1, 1, 0)$ adjacency matrix having eigenvalue 3, Lin. Alg. Appl. **1** (1968), 281–298.

[41] E.E. Shult, Characterizations of certain classes of graphs, J. Combin. Theory Ser. B **13** (1972), 142–167.

[42] J.A. Thas, On 4-gonal configurations with parameters $r = q^2 + 1$ and $k = q + 1$, part I, Geom. Dedicata **3** (1974), 365–375.

[43] J.A. Thas, Generalized quadrangles and flocks of cones, *European J. Combin.* **8** (1987), 441–452.

[44] J.A. Thas, Generalized quadrangles of order (s, s^2), I, *J. Combin. Theory Ser. A* **67** (1994), 140–160.

[45] J.A. Thas and H. Van Maldeghem, Generalized quadrangles and the axiom of Veblen, preprint 1996.

[46] J.A. Thas and S.E. Payne, Spreads and ovoids in finite generalized quadrangles, *Geom. Dedicata* **52** (1994), 227–253.

Christine M. O'Keefe, Department of Pure Mathematics, The University of Adelaide, Australia 5005.
e-mail: cokeefe@maths.adelaide.edu.au

Tim Penttila, Department of Mathematics, University of Western Australia, Australia 6907.
e-mail: penttila@maths.uwa.edu.au

Uniform parallelisms of $PG(3,3)$

Alan R. Prince

Abstract

We show that there are no regular parallelisms of $PG(3,3)$ but that
there are many parallelisms consisting entirely of subregular spreads
of index one.

1 Introduction

We denote by $PG(3,q)$ the 3-dimensional projective space over the field of q elements. A *spread* of $PG(3,q)$ is a set of q^2+1 lines which partition the points of the space. A *parallelism* or *packing* of $PG(3,q)$ is a collection of $q^2 + q + 1$ spreads which partition the lines of the space. A *regular* spread is one which contains the regulus determined by any three of its lines. A *subregular* spread of *index* one is a spread obtained from a regular spread by replacing a single regulus in the spread by its opposite. A *regular* parallelism is one consisting only of regular spreads. A *uniform* parallelism is one consisting entirely of spreads which are projectively equivalent.

The situation regarding the known parallelisms is as follows. For $q = 2$, the parallelisms are completely known: there are two up to projective equivalence (see [5, Theorem 17.5.6]). For $q > 2$, $PG(3,q)$ has at least two projectively inequivalent parallelisms consisting of one regular spread and q^2+q subregular spreads of index one (see [1],[2] and [5, Theorem 17.4.2]). There is a regular parallelism of $PG(3,8)$ discovered by Denniston [4]. This example is actually *cyclic*, in other words the spreads involved are permuted cyclically by some collineation of the space. Denniston [3] has shown that there are no cyclic parallelisms of $PG(3,3)$ and $PG(3,4)$. The author [8] has determined all the parallelisms of $PG(3,3)$, which are invariant under a collineation of order 5. There are seven, up to projective equivalence.

The main purpose of this paper is to demonstrate that there are many uniform parallelisms of $PG(3,3)$. There are only two orbits of spreads of $PG(3,3)$ under the action of the collineation group $PGL(4,3)$. One orbit consists of the regular spreads and the other consists of the subregular spreads of index one. Thus, there are correspondingly two possible types of uniform parallelism. We prove the following result.

Theorem 1.1 (a) *There are no regular parallelisms of* $PG(3,3)$.

(b) *There are at least 54 projectively inequivalent uniform parallelisms of* $PG(3,3)$ *consisting of subregular spreads of index one.*

We remark that Lunardon has confirmed to the author that there is an error in [6]. Thus, the problem of the existence of regular parallelisms in $PG(3,q)$ for q odd, $q > 3$, is still open.

2 Preliminary Results

The concrete model we take for $PG(3,3)$ is as follows. We consider the field $\mathbf{F}_{81}$ generated over $\mathbf{F}_3$ by a primitive element θ satisfying $\theta^4 = \theta + 1$. Then, $\mathbf{F}_{81}$ is a 4-dimensional vector space over $\mathbf{F}_3$ and we take its 1-spaces, 2-spaces and 3-spaces to be the points, lines and planes of $PG(3,3)$. The 40 points of $PG(3,3)$ may be represented by θ^i for $i = 0, 1, 2, \cdots, 39$. Denoting the point represented by θ^i simply by i, the 130 lines of $PG(3,3)$ are:

L					L					L					L					L				
L1	0	1	4	13	L14	1	2	5	14	L27	1	7	12	33	L40	1	8	24	26	L53	1	10	37	38
L2	0	2	24	17	L15	2	3	6	15	L28	2	19	26	4	L41	2	22	12	32	L54	2	33	29	13
L3	0	3	12	39	L16	3	5	27	20	L29	3	38	32	24	L42	3	10	26	28	L55	3	4	7	16
L4	0	5	26	34	L17	5	6	9	18	L30	5	30	28	12	L43	5	33	32	36	L56	5	24	19	13
L5	0	6	32	11	L18	6	27	35	1	L31	6	16	36	26	L44	6	4	28	21	L57	6	12	38	17
L6	0	27	28	31	L19	27	9	25	2	L32	27	13	21	32	L45	27	24	36	23	L58	27	26	30	39
L7	0	9	36	37	L20	9	35	14	3	L33	9	17	23	28	L46	9	12	21	8	L59	9	32	16	34
L8	0	35	21	29	L21	35	25	15	5	L34	35	39	8	36	L47	35	26	23	22	L60	35	28	13	11
L9	0	25	23	7	L22	25	14	20	6	L35	25	34	22	21	L48	25	32	8	10	L61	25	36	17	31
L10	0	14	8	19	L23	14	15	20	6	L36	14	11	10	23	L49	14	28	22	33	L62	14	21	39	37
L11	0	15	22	38	L24	15	20	1	9	L37	15	31	33	8	L50	15	36	10	4	L63	15	23	34	29
L12	0	20	10	30	L25	20	18	2	35	L38	20	37	4	22	L51	20	21	33	24	L64	20	8	11	7
L13	0	18	33	16	L26	18	1	3	25	L39	18	29	24	10	L52	18	23	4	12	L65	18	22	31	19

L					L					L					L					L				
L66	1	11	21	31	L79	1	16	23	39	L92	1	17	19	34	L105	1	22	30	36	L118	1	28	29	32
L67	2	31	23	37	L80	2	13	8	34	L93	2	39	38	11	L106	2	10	16	21	L119	2	36	7	28
L68	3	37	8	29	L81	3	17	22	11	L94	3	34	30	31	L107	3	33	13	23	L120	3	21	19	36
L69	5	29	22	7	L82	5	39	10	31	L95	5	11	16	37	L108	5	4	17	8	L121	5	23	38	21
L70	6	7	10	19	L83	6	34	33	37	L96	6	31	13	29	L109	6	24	39	22	L122	6	8	30	23
L71	27	19	33	38	L84	27	11	4	29	L97	27	37	17	7	L110	27	12	34	10	L123	27	22	16	8
L72	9	38	4	30	L85	9	31	24	7	L98	9	29	39	19	L111	9	26	11	33	L124	9	10	13	22
L73	35	30	24	16	L86	35	37	12	19	L99	35	7	34	38	L112	35	32	31	4	L125	35	33	17	10
L74	25	16	12	13	L87	25	29	26	38	L100	25	19	11	30	L113	25	28	37	24	L126	25	4	39	33
L75	14	13	26	17	L88	14	7	32	30	L101	14	38	31	16	L114	14	36	29	12	L127	14	24	34	4
L76	15	17	32	39	L89	15	19	28	16	L102	15	30	37	13	L115	15	21	7	26	L128	15	12	11	24
L77	20	39	28	34	L90	20	38	36	13	L103	20	16	29	17	L116	20	23	19	32	L129	20	26	31	12
L78	18	34	36	11	L91	18	30	21	17	L104	18	13	7	39	L117	18	8	38	28	L130	18	32	37	26

We use the same labelling of the lines as in [8] (the groups show the cycles in the action of a collineation of order 13).

There are a total of 8424 spreads of $PG(3,3)$. Under the action of the collineation group $PGL(4,3)$, these split into two orbits of lengths 2106 and 6318, consisting of the regular spreads and the subregular spreads of index one, respectively (see [8]). The translation plane corresponding to a subregular spread of index one is the Hall plane of order 9 (see [7]). The spreads in each orbit were generated by computer and stored.

3 Regular parallelisms

In this section, we consider regular parallelisms. Since the regular spreads form a single orbit under the action of the collineation group $PGL(4,3)$, we may assume, in determining regular parallelisms up to projective equivalence, that some fixed regular spread S is in the parallelism. The natural choice to take for S, in our context, is the regular spread corresponding to the subspaces $\mathbf{F}_9 \cdot \theta^i$ of $\mathbf{F}_{81}$ for $i = 0, 1, 2, \cdots, 9$. This is the spread consisting of the lines $i, 10 + i, 20 + i, 30 + i$ for $i = 0, 1, 2, \cdots, 9$. These are the lines L12, L21, L31, L41, L66, L97, L98, L107, L117, L127 listed in the previous section. A computer search shows that there are 1080 regular spreads disjoint to S (as sets of lines). In order to avoid duplication in the search, the next step is to determine the orbits of the stabiliser of S in $PGL(4,3)$ in its action on these 1080 spreads.

Denote by τ the collineation of $PG(3,3)$ induced by the map $x \mapsto x\theta$ of $\mathbf{F}_{81}$. Note that τ permutes the 40 points cyclically and, therefore, is a Singer cycle of $PG(3,3)$. Let σ denote the collineation of $PG(3,3)$ induced by the Frobenius automorphism $x \mapsto x^3$ of $\mathbf{F}_{81}$. Both τ and σ leave S invariant. We need another collineation fixing S, in order to generate the full stabiliser in $PGL(4,3)$.

The field $\mathbf{F}_{81}$ is a 2-dimensional vector space over its subfield $\mathbf{F}_9$ with basis $1, \theta$. Moreover, θ^{10} is a primitive element for $\mathbf{F}_9$. Thus, any element of $\mathbf{F}_{81}$ has coordinates (x, y), where $x, y \in \mathbf{F}_9$, with respect to the basis $1, \theta$. The linear map of $\mathbf{F}_{81}$ (viewed as a vector space over its subfield $\mathbf{F}_9$) defined by $(x, y) \mapsto (x, x\theta^{10} + y)$ leaves invariant the set of ten subspaces $\mathbf{F}_9 \cdot \theta^i$ of $\mathbf{F}_{81}$ for $i = 0, 1, 2, \cdots, 9$. This map is still linear if we view $\mathbf{F}_{81}$ as a vector space over $\mathbf{F}_3$ and its action on the basis $1, \theta, \theta^2, \theta^3$ is

$$1 \mapsto \theta^3 + \theta^2, \quad \theta \mapsto \theta, \quad \theta^2 \mapsto \theta^3 - \theta^2 - \theta - 1, \quad \theta^3 \mapsto \theta^3 + \theta.$$

The induced collineation ρ of $PG(3,3)$, which has order 3, leaves S invariant. The collineations τ, σ, ρ generate the full stabiliser of S in $PGL(4,3)$.

It is easy to compute the action of the generators τ, σ, ρ on the 130 lines of $PG(3,3)$ and, following this, the action on the 1080 regular spreads disjoint to S. Then, an iterative procedure establishes that $< \tau, \sigma, \rho >$ has two orbits on the set of 1080 spreads of lengths 360 and 720.

If a regular parallelism, involving S, contains a spread from the orbit of length 720 then we may assume that it contains some fixed representative of that orbit. A computer search then yields that there are 527 regular spreads disjoint to both S and this fixed representative. Another computer search shows that there is no subset of size 11 of this set of 527 spreads, which consists of pairwise disjoint spreads. Thus, there is no regular parallelism which contains S and a representative from the orbit of size 720.

In any regular parallelism involving S, all the remaining spreads must, therefore, be in the orbit of size 360. We may suppose that some fixed representative of this orbit is involved in the parallelism. There are 180 spreads in this orbit disjoint to both S and this fixed representative. Applying the same computer programme searching for a subset of size 11 consisting of pairwise disjoint spreads, with these 180 spreads as input, yields again that no such subset exists. We conclude:

- **there are no regular parallelisms of PG(3,3).**

4 Uniform parallelisms

In this section, we consider uniform parallelisms which are not regular. Such a parallelism must consist entirely of subregular spreads of index one. The subregular spreads of index one form a single orbit under the action of $PGL(4,3)$ and hence we may assume that some fixed representative T of this orbit is involved in the uniform parallelism. In the computations, the choice we take for T is one of the spreads invariant under a collineation of order 5 listed in [8], namely: L1, L15, L36, L43, L46, L58, L65, L99, L103, L113. This choice is really quite arbitrary.

The stabiliser of T in the collineation group $PGL(4,3)$ is of order $5! \cdot 2^4$ (see [8]). There are 3160 subregular spreads disjoint to T and the stabiliser of T acts on this set of spreads. We would like to determine the orbit lengths for this action. However, this is more difficult than the analogous case in the previous section as it is not so easy to find generators for the stabiliser. By using orbit invariants, we can partition the 3160 spreads into sets which are unions of orbits in the following way.

For each of the 3160 spreads, we calculate the number of spreads, amongst the 3160, disjoint to the given one. This is the number of subregular spreads disjoint to both T and the given subregular spread. It turns out that there are only six distinct values for this parameter, and these values together with the number of spreads for which the parameter has this value are given in the following table.

parameter value	1569	1506	1499	1500	1502	1503
number of spreads	16	240	320	760	864	960

Note that spreads in the same orbit under the action of the stabiliser of T must have the same value for this parameter, so that the set of spreads with a given value of the parameter is a union of orbits. We now consider each of these sets in turn. For each spread of the set, we calculate the number of spreads in the set disjoint to the given spread. This parameter is also an orbit invariant and produces a finer subdivision into sets, which are unions of orbits, given in the following table.

size of set	16	240		320		760			864		960
parameter value	10	90	136	161	171	348	343	356	401	413	463
size of subset	16	120	120	160	160	120	160	480	384	480	960

A further iteration of this procedure does not produce any finer subdivision. Label these subsets of spreads $O_1, O_2, \cdots, O_{11}$ as in the following table:

set	O_1	O_2	O_3	O_4	O_5	O_6	O_7	O_8	O_9	O_{10}	O_{11}
size of set	16	120	120	120	160	160	160	384	480	480	960
parameter value	10	348	90	136	343	161	171	401	356	413	463

We know that each of the sets O_i is a union of orbits but we cannot be certain that each actually is an orbit. In what follows, we proceed as we would if these sets were orbits. This is sufficient to achieve our aim of finding many examples of projectively inequivalent uniform parallelisms but we cannot be completely sure that our enumeration is complete.

Let $A_i = O_1 \cup O_2 \cup \cdots O_i$ for $i = 1, 2, \cdots, 11$. For each A_i, we choose a representative element from the set O_i and let D_i be the set of spreads in A_i which are disjoint to it. The size of D_i is independent of the choice of the representative, which is further evidence that the O_i may be orbits. We then search for 11-sets in D_i consisting of pairwise disjoint spreads. Any such 11-set together with the spread T and the representative from O_i forms a parallelism. For each parallelism, we compute a parameter λ, defined below. The results are summarised in the following table.

set	$|D_i|$	parallelisms	λ
A_1	10	0	-
A_2	64	0	-
A_3	70	0	-
A_4	214	84	64,72,80
A_5	221	42	38,46,60
A_6	377	216	36,40,48,52,56,60,76
A_7	383	24	50,54
A_8	542	170	22,23,24,25,26,28,30,32,34,38,39,42,48
A_9	828	3872	8,10-24,26-32,34,35,36,38,39,40,42, 44,46,48,50,52,53,54,56,58,60,64,68,72
A_{10}	1062	14772	6,8-40,42,44,46,48,50,52, 53,54,56,58,60,64,68,72,76,80
A_{11}	1503	235472	5-40,42,44-48,50,52,53,54,56,58,60,68,76

For any parallelism, the parameter λ is defined to be the number of spreads meeting each constituent spread of the parallelism in at most one line. Clearly, projectively equivalent parallelisms have the same value of λ.

The six largest values of λ which occur are 60,64,68,72,76,80. For each of these values, the following table gives one of the parallelisms found with this particular value of λ.

In total, there are 54 different values of λ which occur in the parallelisms we have found. Since projectively equivalent parallelisms have the same value of λ we have established the following, which completes the proof of the main theorem:

- there are at least 54 projectively inequivalent uniform parallelisms of $PG(3,3)$ consisting of subregular spreads of index one.

References

[1] A. Beutelspacher, On parallelisms in finite projective spaces, *Geom. Dedicata* **3** (1974), 35-40.

[2] R.H.F. Denniston, Some packings of projective spaces, *Rend. Accad. Naz. Lincei* **52** (1972), 36-40.

[3] R.H.F. Denniston, Packings of $PG(3,q)$, *Finite Geometric Structures and their Applications* (Bressanone, 1972), Cremonese, 1973, 195-199.

[4] R.H.F. Denniston, Cyclic packings of the projective space of order 8, *Rend. Accad. Naz. Lincei* **54** (1973), 373-377.

PARALLELISM 1, $\lambda = 80$									
1	15	36	43	46	58	65	99	103	113
2	17	37	47	55	62	90	100	110	118
3	21	32	38	39	92	101	111	119	122
4	18	51	52	61	68	88	89	93	124
5	20	27	50	56	67	77	87	91	123
6	22	40	41	63	72	95	104	120	125
7	16	35	54	70	75	79	112	117	128
8	23	28	29	33	64	74	82	83	105
9	24	30	34	71	81	96	106	127	130
10	25	42	45	57	59	66	69	102	126
11	14	44	48	73	78	97	98	107	129
12	26	31	49	76	80	84	85	86	121
13	19	53	60	94	108	109	114	115	116

PARALLELISM 2, $\lambda = 76$									
1	15	36	43	46	58	65	99	103	113
2	18	38	48	89	94	104	111	114	121
3	21	28	32	33	39	64	83	101	105
4	23	34	51	67	70	72	74	81	118
5	16	54	61	79	86	115	117	124	127
6	20	69	90	92	106	122	126	128	130
7	26	44	56	63	88	93	123	125	129
8	19	40	49	55	57	78	82	102	116
9	17	41	50	66	68	71	73	75	77
10	25	30	31	35	53	76	84	85	107
11	22	27	42	45	80	91	95	98	112
12	14	37	52	59	60	87	97	109	120
13	24	29	47	62	96	100	108	110	119

PARALLELISM 3, $\lambda = 72$									
1	15	36	43	46	58	65	99	103	113
2	17	37	47	55	62	90	100	110	118
3	21	28	32	33	39	64	83	101	105
4	18	51	52	61	68	88	89	93	124
5	20	27	50	56	67	77	87	91	123
6	22	40	41	63	72	95	104	120	125
7	16	35	54	70	75	79	112	117	128
8	23	29	38	74	82	92	111	119	122
9	24	60	71	94	106	108	109	114	130
10	25	42	45	57	59	66	69	102	126
11	14	44	48	73	78	97	98	107	129
12	26	31	49	76	80	84	85	86	121
13	19	30	34	53	81	96	115	116	127

PARALLELISM 4, $\lambda = 68$									
1	15	36	43	46	58	65	99	103	113
2	16	27	34	44	63	100	101	124	130
3	14	35	50	73	96	97	111	116	117
4	18	33	37	38	39	74	88	93	120
5	23	77	85	86	87	105	106	107	108
6	20	48	52	54	90	92	95	109	115
7	22	40	41	84	89	94	104	121	125
8	26	31	49	56	64	67	72	76	110
9	24	28	29	60	82	83	91	114	123
10	25	42	45	57	59	66	69	102	126
11	19	30	51	68	70	75	78	79	112
12	17	47	55	61	62	71	80	118	129
13	21	32	53	81	98	119	122	127	129

PARALLELISM 5, $\lambda = 64$									
1	15	36	43	46	58	65	99	103	113
2	17	37	47	55	62	90	100	110	118
3	18	39	49	61	72	80	95	115	116
4	22	33	68	71	104	105	106	112	128
5	14	34	51	52	87	89	94	97	124
6	20	38	40	54	70	74	76	78	121
7	26	44	56	63	88	93	123	125	129
8	19	30	50	64	92	101	107	109	130
9	23	41	53	73	77	96	108	111	120
10	25	42	45	57	59	66	69	102	126
11	16	27	31	48	60	67	91	98	127
12	21	28	32	79	81	83	85	114	117
13	24	29	35	75	82	84	86	119	122

PARALLELISM 6, $\lambda = 60$									
1	15	36	43	46	58	65	99	103	113
2	16	37	44	78	79	86	87	88	124
3	14	35	50	73	96	97	111	116	117
4	22	33	68	71	104	105	106	112	128
5	19	30	34	38	39	92	101	107	115
6	17	40	41	55	62	63	90	100	125
7	23	47	51	70	74	93	94	108	118
8	26	31	49	56	64	67	72	76	110
9	24	28	29	60	82	83	91	114	123
10	25	42	45	57	59	66	69	102	126
11	18	48	52	54	75	77	85	95	120
12	20	27	61	80	84	89	109	121	130
13	21	32	53	81	98	119	122	127	129

[5] J.W.P. Hirschfeld, *Finite Projective Spaces of Three Dimensions*, Oxford University Press, Oxford, 1985.

[6] G. Lunardon, On regular parallelisms in $PG(3,q)$, *Discrete Math.* **51** (1984), 229-235.

[7] H.Lüneburg, *Translation Planes*, Springer-Verlag, Berlin, 1980.

[8] A.R. Prince, Parallelisms of $PG(3,3)$ invariant under a collineation of order 5, to appear.

Alan R. Prince, Department of Mathematics, Heriot-Watt University, Edinburgh EH14 4AS, United Kingdom.
e-mail: a.r.prince@ma.hw.ac.uk

Double-fives and partial spreads in $PG(5,2)$

Ron Shaw

Abstract

A double-five of planes is a set ψ of 35 points in $PG(5,2)$ which admits two distinct decompositions $\psi = \alpha_1 \cup \alpha_2 \cup \alpha_3 \cup \alpha_4 \cup \alpha_5 = \beta_1 \cup \beta_2 \cup \beta_3 \cup \beta_4 \cup \beta_5$ into a set of five mutually skew planes such that $\alpha_r \cap \beta_r$ is a line, for each r, while $\alpha_r \cap \beta_s$ is a point, for $r \neq s$. In a recent paper, [Sh96], a construction of a double-five was given, starting out from a (suitably coloured) icosahedron, and some of its main properties were described. The present paper deals first of all with some further properties of double-fives. In particular the existence of an invariant symplectic form is demonstrated and some related duality properties are described.

Secondly the relationship of double-fives to partial spreads of planes in $PG(5,2)$ is considered. The α-planes, or equally the β-planes, of double-fives provide the only examples of maximal partial spreads. It is shown that one of the planes of a non-maximal partial spread of five planes is always privileged, and this fact is seen to give rise to a nice geometric construction of an overlarge set of nine 3-(8,4,1) designs having automorphism group $\Gamma L_2(8)$.

1 Double-fives of planes in $PG(5,2)$: a recap

Our chief concern will be with a vector space $V = V(6,2)$ of dimension 6 over $GF(2)$, and the associated 5-dimensional projective space $\mathbf{P}V = PG(5,2)$, which will be identified with $V \backslash \{0\}$.

A *double-five of planes* is a 35-point set ψ in $PG(5,2)$ which admits two distinct decompositions

$$\psi = \alpha_1 \cup \alpha_2 \cup \alpha_3 \cup \alpha_4 \cup \alpha_5 = \beta_1 \cup \beta_2 \cup \beta_3 \cup \beta_4 \cup \beta_5$$

into a set of five planes such that, for $r \neq s$,

$$\alpha_r \cap \alpha_s = \emptyset \qquad \beta_r \cap \beta_s = \emptyset$$
$$\alpha_r \cap \beta_r = \text{a line } \lambda_r \qquad \alpha_r \cap \beta_s = \text{a point } n_{rs}$$

In [7] a double-five was constructed starting out from a regular icosahedron $\mathfrak{I}$. Under the action of the rotational symmetry group $G_0(\mathfrak{I}) \cong A_5$ of $\mathfrak{I}$ the 15 pairs of opposite edges of $\mathfrak{J}$ form 5 blocks of 3 mutually perpendicular pairs, and these blocks are coloured using the "colours" $1, 2, 3, 4, 5$. The 20 faces of $\mathfrak{J}$ form 5 blocks of 4 in two enantiomorphic ways, and the faces are correspondingly coloured, using the same colours $1, 2, 3, 4, 5$, in two ways, one "left-handed" and one "right-handed". If a face has colour r in the left-handed scheme, the opposite face is assigned colour r in the right-handed scheme. If the 20 faces of $\mathfrak{J}$ are $\{f_{rs}, r \neq s\}$, where the face f_{rs} has colour r in the left-handed scheme and colour s in the right-handed scheme, then, see [7, Figs. 1,2], the edge-colouring can be chosen such that the two faces adjoining an edge of colour t are f_{pq} and f_{rs} for some permutation $pqrst$ of 12345. The 60 rotational symmetries of $\mathfrak{J}$ correspond to the 60 even permutations of the colours $1, 2, 3, 4, 5$.

In the icosahedral construction of a double-five, the 6 vertices $\{B_1, \ldots, B_6\}$ of a particular *polar cap* of $\mathfrak{J}$ are mapped on to a basis $\mathcal{B} = \{b_1, \ldots, b_6\}$ for $V(6, 2)$, and the 6 vertices of the antipodal cap on to a basis $\overline{\mathcal{B}} = \{\overline{b}_1, \ldots, \overline{b}_6\}$, with $b_i, \overline{b}_i$ arising from an antipodal pair of vertices, and with the two bases being related by $\overline{b}_i = b_i + u$, where $u = \sum_{i=1}^{6} b_i$.

Lemma 1.1 *([7]) Let $X_1, \ldots, X_6$ be the vertices of any one of the twelve polar caps of $\mathfrak{J}$. Then $\{x_1, \ldots, x_6\}$ is a basis for $V(6, 2)$ and $\sum_{i=1}^{6} x_i = u$.* $\square$

An edge XY of $\mathfrak{J}$ is mapped on to the sum $x + y$ of the corresponding vectors, and similarly a face XYZ on to $x + y + z$. Since $2u = 0$, a pair of opposite edges map on to the same vector. In contrast, a pair of opposite faces map on to a pair of vectors $f, \overline{f}$, where $\overline{f} = f + u$. Consider the 35 points ψ of $PG(5, 2) = V(6, 2) \backslash \{0\}$ which arise from the 15 pairs of opposite edges and the 20 faces of $\mathfrak{J}$. The colouring guides us to arrange these 35 points as an array

$$\psi = \begin{pmatrix} \lambda_1 & n_{12} & n_{13} & n_{14} & n_{15} \\ n_{21} & \lambda_2 & n_{23} & n_{24} & n_{25} \\ n_{31} & n_{32} & \lambda_3 & n_{34} & n_{35} \\ n_{41} & n_{42} & n_{43} & \lambda_4 & n_{45} \\ n_{51} & n_{52} & n_{53} & n_{54} & \lambda_5 \end{pmatrix} . \tag{1.1}$$

Here n_{rs} denotes the point arising from the face f_{rs} of $\mathfrak{J}$ and λ_r denotes the 3 points arising from the 3 pairs of opposite edges of $\mathfrak{J}$ having colour r. Since the face f_{rs} is opposite to the face f_{sr}, note that $n_{rs} + n_{sr} = u$, $r \neq s$.

Theorem 1.2 *([7]) The 35-set $\psi \subset PG(5, 2)$ in Eq. (1.1) is a double-five, with the 7-set of the rth row forming a plane α_r and the 7-set of the sth column forming a plane β_s. Indeed, if for $ijklmn$ any permutation of 123456 we define $b_{ij} = b_i + b_j = \overline{b}_i + \overline{b}_j$, $\overline{b}_{ij} = b_{ij} + u = \overline{b}_i + b_j$, $b_{ijk} = b_i + b_j + b_k$, then, with the same labelling conventions as in [7], the 35 points of ψ are explicitly*

$$\psi = \begin{pmatrix} 34\,\overline{52}\,16 & 124 & 123 & 236 & 246 \\ 356 & 45\,\overline{13}\,26 & 235 & 234 & 346 \\ 456 & 416 & 51\,\overline{24}\,36 & 341 & 345 \\ 451 & 516 & 526 & 12\,\overline{35}\,46 & 452 \\ 513 & 512 & 126 & 136 & 23\,\overline{41}\,56 \end{pmatrix} . \tag{1.2}$$

(Here ij, $\overline{ij}$, ijk are abbreviations for b_{ij}, $\overline{b}_{ij}$, b_{ijk}.)   □

Theorem 1.3 *([7]) A double-five ψ of planes is a projectively unique configuration in $PG(5,2)$, and it determines uniquely:*

1) the five planes α_r and the five planes $\beta_r, r = 1, \ldots, 5$, there being no other planes lying on ψ;

2) the five solids (that is $PG(3,2)$'s) $\sigma_r = join(\alpha_r, \beta_r)$;

3) the set $\{\lambda_1, \ldots, \lambda_5\}$ of five skew lines, where $\lambda_r = \alpha_r \cap \beta_r$;

4) a privileged hyperplane $\varpi = \varpi(\psi)$, the even, or invariant, hyperplane of ψ, spanned by the five lines λ_r;

5) a parabolic quadric $\mathcal{P}_4 \subset \varpi = PG(4,2)$, consisting of the 15 "diagonal points" $\lambda_1 \cup \ldots \cup \lambda_5$ of ψ;

6) a privileged point $u = u(\psi) \in \varpi$, the nucleus of ψ, which satisfies $n_{rs} + n_{sr} = u$, $r \neq s$, and which is the nucleus of $\mathcal{P}_4$;

7) the set $\{n_{rs} \equiv \alpha_r \cap \beta_s; \quad r \neq s\}$ of twenty "off-diagonal" points;

8) the symmetry group $G(\psi)$, isomorphic to that $G(\mathfrak{I})$ of $\mathfrak{I}$, given by $G(\psi) = G_0(\psi) \times Z_2 \subset GL_6(2)$, with $G_0(\psi) \cong A_5$, and $Z_2 = \langle J \rangle$, where $Jb_i = \overline{b}_i$.   □

The set ψ is a *set of hyperbolic type*, that is it has the intersection property $|\psi \cap \pi| \in \{15, 19\}$ for any hyperplane $\pi \subset PG(5,2)$. It was shown in [7] that use instead of the 15 pairs of opposite edges and the 12 vertices of the icosahedron yields a set $\phi \subset PG(5,2)$ of 27 points of *elliptic type*, satisfying, that is, $|\phi \cap \pi| \in \{11, 15\}$ for any hyperplane $\pi \subset PG(5,2)$. The set ϕ shares with ψ the same icosahedral symmetry group: $G(\phi) = G(\psi) \cong G(\mathfrak{I})$. The sets ψ and ϕ are of Tonchev type 3b, see [9, Table I].

2   Further Properties

2.1 $G(\psi)$-invariant symplectic form

Let $d(\ ,\ )$ be the distance function for the icosahedral graph $\Gamma(\mathfrak{I})$; in particular $d(X, Y)$ is 1 if the vertices X, Y of $\mathfrak{I}$ are neighbours, and 3 if they are antipodal.

Lemma 2.1 *There exists a non-degenerate $G(\psi)$-invariant symplectic form $x.y$ on $V(6,2)$ such that, for $x, y \in \mathcal{B} \cup \overline{\mathcal{B}}$,*

$$x.y = \begin{cases} 0, & \text{if } d(X,Y) \in \{0,2\}, \\ 1, & \text{if } d(X,Y) \in \{1,3\}, \end{cases} \quad (2.3)$$

where X, Y are the vertices of $\mathfrak{I}$ corresponding to the vectors x, y.

Each of the five planes α_r, and each of the five planes β_r, of ψ is self-polar with respect to the symplectic polarity on $PG(5,2)$ determined by (2.3).

Proof Consider the bilinear symplectic form defined by Eq. (2.3) for x, y restricted to belong to the basis $\mathcal{B} = \{b_1, \ldots, b_6\}$. It is easily checked to be non-degenerate.

Using bilinearity it easily follows that $u.x = 1$, $x \in \mathcal{B}$, and hence that Eq. (2.3) indeed holds good for $x, y \in \mathcal{B} \cup \bar{\mathcal{B}}$. Knowing now that Eq. (2.3) holds for X, Y any of the 12 vertices of $\mathfrak{I}$, the invariance of the symplectic form $x.y$ under $G(\psi) \cong G(\mathfrak{I})$ is manifest.

That α_r and β_r are self-polar follows easily from the nature of the colouring of $\mathfrak{I}$, see [7, Figs. 1,2]. For example, suppose that the face f_{rs} of $\mathfrak{I}$ has vertices ABC, and so corresponds to the vector $a + b + c \in V(6,2)$, and let ABD be the other face sharing the edge AB. Then one of the edges AD, BD has colour r and the other colour s. Consider $(a + d).(a + b + c)$. From Eq. (2.3), $a.(a + b + c) = 0 + 1 + 1 = 0$, and $d.(a + b + c) = 1 + 1 + 0 = 0$. Hence $(a + d).(a + b + c) = 0$, and similarly $(b + d).(a + b + c) = 0$. Thus $x.n_{rs} = 0$ for $x \in \lambda_r$ and for $x \in \lambda_s$.   □

Corollary 2.2 *Under the action of the group $G(\psi)$ the 63 hyperplanes of $PG(5,2)$ fall into five orbits, of lengths* $20, 12, 15, 15, 1$.

Proof   The bilinear form $x.y$ pairs a point $a \in PG(5,2)$ with its polar hyperplane $a.x = 0$. So hyperplane orbits follow from point orbits. But, under the action of the group $G(\psi)$, the 63 points of $PG(5,2)$ fall into five orbits, namely $\{n_{rs}\}$, $\mathcal{B} \cup \bar{\mathcal{B}}$, $\mathcal{P}_4$, $\overline{\mathcal{P}_4} (= u + \mathcal{P}_4)$, $\{u\}$, of respective lengths $20, 12, 15, 15, 1$.   □

Remarks   (i) The more detailed results given in [7, Theorem 3.9] are also more easily derived by use of the symplectic form $x.y$. For example, the point orbit $\mathcal{B} \cup \bar{\mathcal{B}}$ yields the $G(\psi)$-orbit of 12 hyperplanes with equations

$$b_r.x = 0, \quad b_6.x = 0, \quad \bar{b_r}.x = 0, \quad \bar{b_6}.x = 0,$$

where $r \in \{1, 2, 3, 4, 5\}$. With the labelling as in [7], it follows from Eq. (2.3) that these 12 hyperplanes have the equations

$$x_{r-1} + x_{r+1} = x_6, \quad \sum_{r=1}^{5} x_r = 0, \quad x_{r-2} + x_r = x_{r+2}, \quad x_6 = 0.$$

(ii) The singleton orbits $\{u\}$, $\{\varpi\}$ form a polar pair, the invariant hyperplane ϖ having equation $u.x = 0$.

(iii) It was noted in [7] that the 10 planes α_r, β_r of the double-five $\psi \subset PG(5,2)$ automatically give rise to 10 planes α_r^0, β_r^0 of a dual double-five $\psi^* \subset PG(5,2)^*$, this following from the incidence properties without appeal to any metrical considerations. (Here $\alpha_r^0 \subset PG(5,2)^*$ denotes the subspace consisting of those nonzero linear forms which are zero on each point of α_r.) Now the scalar product (2.3) gives rise to a linear isomorphism $f : V \to V^*$ defined by $a \mapsto f_a$, where $f_a(x) = a.x$. Consequently the plane $\alpha_r \subset PG(5,2)$ gives rise to a plane $\alpha_r^* = \{f_a : a \in \alpha_r\} \subset PG(5,2)^*$. In fact, because α_r is self-polar, α_r^* coincides with α_r^0; similarly β_r^* coincides with β_r^0. However the new fact arising from the existence of the invariant symplectic form is that, by restriction, f provides us with ten linear isomorphisms $\alpha_r \to \alpha_r^0, \beta_r \to \beta_r^0$. An explicit expression for ψ^* then follows from Eq. (1.2), in agreement with that in [7].

(iv) Except for the square of the linear form $u.x$, no quadratic form on V is invariant under $G_0(\psi)$. To see this, note that the $G_0(\psi)$-orbits of points coincide with the $G(\psi)$-orbits and recall the intersection properties of point orbits with hyperplanes as given in [7, Eq. 3.21]. For example, the 35-point set $\mathcal{B} \cup \overline{\mathcal{B}} \cup \overline{\mathcal{P}}_4$ can not have a quadratic equation, since it has even intersection with some hyperplanes.

2.2   Cubic equation $\Psi = 0$ of double-five ψ

The above bilinear form $x.y$ is *over*-determined by laying down, as in Eq. (2.3), its values on the *double basis* $\mathcal{B} \cup \overline{\mathcal{B}}$. However the virtue of such an "over-definition", once it has been justified, is the resulting manifest invariance under the icosahedral symmetry group $G(\psi)$. We give now another example of over-determination. (For still another example, see [7, Theorem 3.4, Proof].)

The double-five ψ has, see [7], equation $\Psi = 0$, with Ψ, satisfying $\Psi(0) = 0$, of degree 3. Let B, T be the first, second polarizations of Ψ:

$$
\begin{aligned}
B(x,y) &= \Psi(x+y) + \Psi(x) + \Psi(y) \\
T(x,y,z) &= B(x, y+z) + B(x,y) + B(x,z) \\
&= \sum_{(\lambda,\mu,\nu) \in GF(2)^3} \Psi(\lambda x + \mu y + \nu z).
\end{aligned}
$$

For the moment restrict x, y, z to be elements of $\mathcal{B} \cup \overline{\mathcal{B}}$; it then follows that

$$
\begin{aligned}
\Psi(x) &= 1, \\
B(x,y) &= \begin{cases} 0, & \text{if } d(X,Y) \in \{0,1\}, \\ 1, & \text{if } d(X,Y) \in \{2,3\}, \end{cases} \\
T(x,y,z) &= \begin{cases} 1, & \text{if } X, Y, Z \text{ define a triple of acute type}, \\ 0, & \text{otherwise}. \end{cases}
\end{aligned}
$$

The cubic Ψ is determined by the values of Ψ, B, T on a *basis*, and so the foregoing values on the double basis $\mathcal{B} \cup \overline{\mathcal{B}}$ over-determine Ψ, but have the virtue of bringing out the icosahedral symmetry $G(\psi)$ of Ψ, B, T. For *any* $x \in V$ we thus have, in agreement with [7, Theorem 5.4],

$$
\Psi(x) = \sum \beta_i x_i + \sum_{i<j} \beta_{ij} x_i x_j + \sum_{i<j<k} \beta_{ijk} x_i x_j x_k,
$$

where, in our labelling of the icosahedron, the nonzero coefficients are

$$
\begin{aligned}
\beta_i &= \Psi(b_i) = 1, \quad i = 1, 2, \ldots, 6, \\
\beta_{ij} &= B(b_i, b_j) = 1 \quad ij = 13, 24, 35, 41, 52, \\
\beta_{ijk} &= T(b_i, b_j, b_k) = 1 \quad ijk \in \Omega,
\end{aligned}
$$

where, as in [7], $\Omega = \{126, 236, 346, 456, 516, 124, 235, 341, 452, 513\}$.

Remarks   (i) Given Ψ, and therefore its polarizations B, T, then the nucleus of ψ is that point u which satisfies $T(x, y, u) = 0$, for all $x, y \in V$ and the invariant hyperplane has equation $B(u, x) = 0$. (Query: *Can one determine the invariant bilinear form x.y in a simple way from* Ψ?)

(ii) The cubic Ψ^* for the dual double-five ψ^* is seen to be

$$\Psi^* = \xi_6 + \sum_{1 \leq r < s \leq 5} \xi_r \xi_s + \sum_{ijk \in \Omega^*} \xi_i \xi_j \xi_k,$$

where $\Omega^* = \{136, 246, 356, 416, 526, 123, 234, 345, 451, 512\}$ and where the coordinates ξ_i are dual to the x_i.

3   The projective line $PG(1, 8)$

3.1   Introduction

Of relevance to our discussion, in sections 4.2, 4.3, of symmetry aspects of spreads, and of partial spreads, in $PG(5, 2)$, will be certain properties of the 9-point projective line $\mathcal{S} = PG(1, 8)$, acted upon sharply 3-transitively by the simple group $PGL_2(8) \cong L_2(8)$ of order $9.8.7 = 504$. Actually it is the action of the larger group $P\Gamma L_2(8)$ which will be especially relevant. In this connection recall that the field $\mathbb{F} = GF(8)$ has automorphism group $\mathfrak{A} = \text{Aut}(\mathbb{F})$ isomorphic to Z_3, generated by the Frobenius automorphism $\phi : \lambda \mapsto \lambda^\phi = \lambda^2$. Relative to any fixed basis $\{e_1, e_2\}$ for $V(2, 8)$, the ϕ-linear map with coordinate expression $(x_1, x_2) \mapsto (x_1^\phi, x_2^\phi)$ yields an element T of the extended group $P\Gamma L_2(8) \cong L_2(8) \rtimes Z_3$ which is of order 3. Moreover the subgroup $\langle T \rangle \cong Z_3$ generated by T consists precisely of those elements of $P\Gamma L_2(8)$ which keep fixed each of the three points $\langle e_1 \rangle, \langle e_2 \rangle, \langle e_1 + e_2 \rangle \in \mathcal{S}$ — the other six points of $\mathcal{S}$ being permuted by T in two 3-cycles. (Thus $P\Gamma L_2(8)$ acts $\frac{7}{2}$-transitively on $\mathcal{S}$.)

Elements of $PGL_2(q)$ of order 2 are termed involutions. Recall that associated with any tetrad $\{P_0, P_1, P_2, P_3\}$ of four distinct points of a projective line $PG(1, q)$ are the three mutually commuting involutions $J_i, i = 1, 2, 3$, which effect the following permutations of the four points:

$$J_1 : P_0 \rightleftharpoons P_1, P_2 \rightleftharpoons P_3; \quad J_2 : P_0 \rightleftharpoons P_2, P_3 \rightleftharpoons P_1; \quad J_3 : P_0 \rightleftharpoons P_3, P_1 \rightleftharpoons P_2. \quad (3.4)$$

In the cases when $q = 2^h$ is even, the following facts are known, see [3]:

Lemma 3.1 *(i) there is a single class of $q^2 - 1$ involutions;*

(ii) each involution has just one fixed point; moreover the $q - 1$ involutions fixing a given point mutually commute, and so they comprise the non-identity elements of a subgroup $\cong (Z_2)^h$ of $PGL_2(q)$;

(iii) the three involutions J_i in (3.4) share the same fixed point P, called the associated *point of the tetrad $\{P_0, P_1, P_2, P_3\}$.*   □

In the case $q = 8$ there are thus 63 involutions, each of the 9 points of S being the common fixed point of 7 commuting involutions. In fact, since 2 is prime to the order of $\mathfrak{A}$, the only elements of $P\Gamma L_2(8)$ of order 2 are these 63 involutions $\in PGL_2(8)$. It is easy to check, see [3], that each tetrad T of distinct points of $PG(1,8)$ admits as cross-ratios each of the six values of $GF(8)\backslash\{0,1\}$, and hence that any two (unordered) tetrads are projectively equivalent. Of especial interest will be another property peculiar to $PG(1,8)$:

Lemma 3.2 *The associated point P of a tetrad T is also the associated point of the residual tetrad $T' = S \backslash (T \cup \{P\})$.*

Proof Choose any involution J with fixed point P, and let its 2-cycles be $(P_r P_{r'})$, $r = 0, 1, 2, 3$. Then J is one of the three involutions associated with the tetrad $B = \{P_0, P_{0'}, P_1, P_{1'}\}$, and is also one of the three involutions associated with the tetrad $B' = \{P_2, P_{2'}, P_3, P_{3'}\}$. So the property holds for the tetrad B, and hence for any tetrad. (See [3] for a different proof). □

3.2 Stabilizer subgroups

By lemma 3.2, the three involutions J_i in (3.4), which are associated with the tetrad $T = \{P_0, P_1, P_2, P_3\}$ and whose common fixed point is P, must coincide with the three involutions associated with the residual tetrad $T' = \{P_0', P_1', P_2', P_3'\} = S \backslash (T \cup \{P\})$. Consequently, after a suitable relabelling of the points of the residual tetrad T', the three involutions are, in an obvious abbreviated notation,

$$J_1 : (01)(23)(0'1')(2'3'), \quad J_2 : (02)(31)(0'2')(3'1'), \quad J_3 : (03)(12)(0'3')(1'2'). \quad (3.5)$$

Let us denote by $\Gamma\mathcal{G}_P$ the subgroup of $P\Gamma L_2(8)$ which fixes the point P, and by $\Gamma\mathcal{G}(T)$ and $\Gamma\mathcal{G}(T')$ the subgroups of $P\Gamma L_2(8)$ which stabilize (setwise) the tetrads T and T'. Subgroups $\mathcal{G}_P, \mathcal{G}(T)$ and $\mathcal{G}(T')$ of $PGL_2(8)$ are analogously defined. Let $\Gamma\mathcal{G}(T, T')$ be the subgroup of $P\Gamma L_2(8)$, indeed of $\Gamma\mathcal{G}_P$, which fixes the $4 + 1 + 4$ partition $T \cup \{P\} \cup T'$ of S, and let the subgroup $\mathcal{G}(T, T')$ of $\mathcal{G}_P$ be analogously defined. Observe that $\Gamma\mathcal{G}(T)$ is of index 2 in $\Gamma\mathcal{G}(T, T')$, and $\mathcal{G}(T)$ is of index 2 in $\mathcal{G}(T, T')$.

Theorem 3.3 *(i) The 7 involutions fixing the point P are $J_1, J_2, J_3, J, J_{1'}, J_{2'}, J_{3'}$, where, for a suitable labelling of the points of T', J_i, $i = 1, 2, 3$, are as in (3.5), where $J_{i'} = JJ_i$, $i = 1, 2, 3$, and where J effects the permutation $P_r \rightleftharpoons P_{r'}$, $r \in \{0, 1, 2, 3\}$. Thus*

$$J : (00')(11')(22')(33'), \quad J_{1'} : (01')(23')(0'1)(2'3),$$
$$J_{2'} : (02')(31')(0'2)(3'1), \quad J_{3'} : (03')(12')(0'3)(1'2). \quad (3.6)$$

(ii) The following hold for the subgroups as defined above:

$$\Gamma\mathcal{G}(T) = \Gamma\mathcal{G}(T') \subset \Gamma\mathcal{G}_P, \quad \mathcal{G}(T) = \mathcal{G}(T') \subset \mathcal{G}_P, \quad (3.7)$$
$$\Gamma\mathcal{G}_P \cong ((Z_2)^3 \rtimes Z_7) \rtimes Z_3, \quad \mathcal{G}_P \cong (Z_2)^3 \rtimes Z_7, \quad (3.8)$$
$$\Gamma\mathcal{G}(T) \cong A_4 \cong (Z_2)^2 \rtimes Z_3, \quad \mathcal{G}(T) \cong (Z_2)^2. \quad (3.9)$$

(iii) Of the four involutions (3.6), just one centralizes $\Gamma\mathcal{G}(\mathcal{T})$, thus accounting for the Z_2 subgroup in the first of the following isomorphisms:

$$\Gamma\mathcal{G}(\mathcal{T}, \mathcal{T}') \cong A_4 \times Z_2, \qquad \mathcal{G}(\mathcal{T}, \mathcal{T}') \cong (Z_2)^3. \qquad (3.10)$$

Proof   (i) Consider, for $i = 1, 2, 3$, the involution K_i which effects the interchanges $(00')(ii')$. Then, by lemma 3.1 (iii), K_i must have the same fixed point P as the involution J_i, since the latter effects $(0i)(0'i')$. Since K_1, K_2, K_3 all fix P and effect $(00')$, they must coincide, and hence have the effect $(00')(11')(22')(33')$. Thus J in (3.6) is indeed an involution, and hence so are the $J_{i'}$, since $J_{i'} = JJ_i = J_iJ$.

(ii) Concerning (3.7), any $T \in \Gamma\mathcal{G}(\mathcal{T})$ necessarily normalizes the set of three involutions (3.5), and so fixes the point P. Concerning (3.9), note that an element of $\Gamma\mathcal{G}(\mathcal{T})$ can not induce a single transposition on $\mathcal{T}$, such as $(0)(1)(23)$, since an involution can not have two fixed points. On the other hand a 3-cycle, such as $(0)(321)$, can be induced semilinearly, but not linearly. For if the cross-ratio $\{0, 1; 2, 3\} = \alpha$ satisfies $\alpha^3 = 1 + \alpha$, note that $\{0, 3; 1, 2\} = (1 + \alpha)/\alpha = \alpha^2 = \{0, 1; 2, 3\}^\phi$, and $\{0, 2; 3, 1\} = \alpha^4 = \{0, 3; 1, 2\}^\phi$. Hence there exists $T \in \Gamma L_2(8)$, necessarily ϕ-linear, which induces $(0)(321)$, and hence a ϕ^2-linear element T^2 which induces $(0)(123)$. (If instead $\alpha^3 = 1 + \alpha^2$, then it is $(0)(123)$ which is induced ϕ-linearly.) Consequently $\Gamma\mathcal{G}(\mathcal{T}) \cong A_4$, with normal subgroup $\mathcal{G}(\mathcal{T}) = \{I, J_1, J_2, J_3\} \cong (Z_2)^2$. Concerning (3.8), the group $PGL_2(8)$ contains 36 Z_7 subgroups, a particular Z_7 keeping fixed a particular duad of points out of the $\binom{9}{2} = 36$ duads; the 8 duads which contain P account for the 8 Z_7's in the structure $(Z_2)^3 \rtimes Z_7$ of $\mathcal{G}_P$. The 7 involutions of the normal subgroup $(Z_2)^3$ of $\mathcal{G}_P$ are precisely those exhibited in (3.5), (3.6).

(iii) Each of the four involutions (3.6) defines a bijection $\mathcal{T} \leftrightarrow \mathcal{T}'$. However the fact that $\Gamma\mathcal{G}(\mathcal{T})$ acts as A_4 simultaneously upon the points of both $\mathcal{T}$ and $\mathcal{T}'$ picks out one of these four involutions as preferred, in that it alone centralizes $\Gamma\mathcal{G}(\mathcal{T})$. For let the labelling of the points of the two tetrads be such that P_r and $P_{r'}, r = 0, 1, 2, 3$, are both fixed by the same Z_3 subgroup of $\Gamma\mathcal{G}(\mathcal{T})$. With this choice of labelling the first involution J in (3.6) will centralize each Z_3 subgroup, and hence centralize $\Gamma\mathcal{G}(\mathcal{T})$. (The involutions $J_{i'} = JJ_i$ do not then centralize, since J_i lies in $\Gamma\mathcal{G}(\mathcal{T}) \cong A_4$.)   □

Remark   The 56 elements of order 3 lying in $PGL_2(8)$ form a single class $\mathcal{C}$. Since only the identity element of $PGL_2(8)$ fixes three points, each $T_0 \in \mathcal{C}$ necessarily permutes the nine points of $\mathcal{S}$ in three 3-cycles. So each of the 28 Z_3 subgroups of $PGL_2(8)$ gives rise to a 3+3+3 partition of $\mathcal{S}$, say $\mathcal{S} = \mathcal{Q}_1 \cup \mathcal{Q}_2 \cup \mathcal{Q}_3$. (In fact such a partition is stabilized not only by a Z_3 subgroup $\langle T_0 \rangle \subset PGL_2(8)$ but also by three Z_3 subgroup $\langle T_i \rangle \subset P\Gamma L_2(8)$, where T_i, $i = 1, 2, 3$, fixes each point of the triad $\mathcal{Q}_i$.) The $\binom{9}{3} = 84$ triads of points of $\mathcal{S}$ thus sort themselves out into 28 triples of the above kind $\{\mathcal{Q}_1, \mathcal{Q}_2, \mathcal{Q}_3\}$, cf. [3]. They are the 28 1-factors of a 1-factorization of the complete 3-uniform 9-hypergraph K_9^3, and, since this 1-factorization has automorphism group $P\Gamma L_2(8)$ of order 1512, it must be isomorphic to the first one in the census of Mathon & Rosa, see [5, Table 1]. The subgroup of $P\Gamma L_2(8)$ which stabilizes one of these $3+3+3$ partitions of S is a maximal subgroup $\cong Z_9 \rtimes Z_6$, cf. [2].

4 Spreads and double-fives in $PG(5,2)$

4.1 Spreads, partial spreads and double-fives

For the vector space $V = V(6,2)$ a *partial spread* is a collection $\Sigma = \{W\}$ of 3-dimensional subspaces $W \subset V$ such that $V = W \oplus W'$ for each distinct pair $W, W' \in \Sigma$. Equivalently we may deal with the corresponding set $\{PW\}$ of pairwise skew planes of the projective space $PV = PG(5,2) = V\backslash\{0\}$. If $|\Sigma| = 9$, then Σ is a (complete) *spread* — the 9 planes yielding a partition of the 63 points of $PG(5,2)$: $9 \times 7 = 63$. Let Σ_9 denote a spread for V, and denote by Σ_k the partial spread formed by choosing a k-subset of Σ_9. Of course Σ_k extends to a bigger partial spread Σ_{k+1} (indeed to the complete spread Σ_9). A partial spread which does not extend is termed *maximal*.

Lemma 4.1 *Each double-five ψ provides us with two examples, say Ψ_5, Ψ_5', of maximal partial spreads of size 5 — arising from the two sets of 5 skew planes lying on ψ.*

Proof We need to show that there the complementary 28-set ψ^c contains no planes. Suppose to the contrary that $\gamma \subset \psi^c$ for some plane γ. Now the intersection of γ with each of the 5 solids $\sigma_r = \text{join}(\alpha_r, \beta_r)$ is non-empty. However $u \notin \gamma \cap \sigma_r$, for otherwise $\gamma \cap \varpi$ would contain a line through the nucleus u of $\mathcal{P}_4$ — and such a line is tangent to $\mathcal{P}_4$, contradicting $\gamma \subset \psi^c$. Hence the intersection $\gamma \cap \sigma_r$ is a point of $\overline{\lambda}_r(= u + \lambda_r)$. So γ has 5 points $\gamma \cap \sigma_r$ lying on ϖ, and so $\gamma \subset \varpi$. But then γ has non-empty intersection with $\mathcal{P}_4$, again contradicting $\gamma \subset \psi^c$. $\square$

¿From now it will help to view our space $V(6,2)$ as a direct sum $V = X \oplus X$ of two copies of a space $X = V(3,2)$, and to write the general element $v \in V$ in the form $v = (x,y)$, $x,y \in X$. We write $W_0 = \{(x,0) : x \in X\}$ and $W_\infty = \{(0,y) : y \in X\}$, so that V is the internal direct sum $W_\infty \oplus W_0$. The planes in $PG(5,2)$ which are skew to both PW_∞ and PW_0 are those with equation $y = Cx$ for some $C \in GL(X) \cong GL(3,2)$, that is planes of the form PW_C, where

$$W_C = \{(x, Cx) : x \in X\}, \qquad C \in GL(X). \tag{4.11}$$

In using this model for V, the following elementary facts concerning $GL(3,2)$ will be of use.

Lemma 4.2 *(i) An element $A \in GL(X)$ is fixed-point-free on $PX = PG(2,2) = X\backslash\{0\}$ if and only if it has order 7. Such elements fall into two conjugacy classes, with A, A^2, A^4 lying in one class, and their inverses A^6, A^5, A^3 lying in the other. Elements of one class have characteristic polynomial $t^3 + t + 1$, and elements of the other have characteristic polynomial $t^3 + t^2 + 1$.*

(ii) If A is of order 7 then its commutant $[A] \subset L(X, X)$ is a field $\mathbb{K}$ of order 8:

$$\mathbb{K} = \{0\} \cup \mathbb{K}^\times \cong GF(8), \qquad where \ \mathbb{K}^\times = \langle A \rangle \cong Z_7. \qquad (4.12)$$

The normalizer of $\langle A \rangle$ is a maximal subgroup $F_{21} \cong Z_7 \rtimes Z_3$ of $GL(X)$.

(iii) For a given A of order 7 there are precisely 7 solutions $B \in GL(X)$ of the equations

$$A^7 = B^7 = (A^{-1}B)^7 = I, \qquad AB \neq BA, \qquad (4.13)$$

namely $\{A^r B_0 A^{-r} : r = 0, 1, ..., 6\}$, where B_0 is any particular solution of (4.13). Moreover A is not conjugate to B.

(iv) An element $A \in GL(X)$ has a line of fixed points in $PG(2, 2)$ if and only if it has order 2, and it has a single fixed point if and only if it has order 3 or 4.

(v) For any solution B of (4.13) the order of $A^{-r}B$ is 7, 3, 4, 2, 4, 3 according as $r = 1, 2, 3, 4, 5, 6$, respectively. In particular, only for $r = 4$ does $A^{-r}B$ have a line of fixed points in $PG(2, 2)$.   □

Using part (i) of the lemma, along with elementary linear algebra, one sees, cf. [8], that any partial spread of 4 subspaces is projectively equivalent to the partial spread

$$\Sigma_4 = \{W_\infty, W_0, W_I, W_A\}, \quad where \ A^7 = I, \ and \ A^3 = A + I. \qquad (4.14)$$

Note that Σ_4 extends to the spread

$$\Sigma_9 = \{W_\infty, W_0, W_{A^r}; \ r = 0, 1, ..., 6\}. \qquad (4.15)$$

Now from lemma 4.1 we know that Σ_4 must also possess an extension to a maximal partial spread, say

$$\Psi_5 = \{W_\infty, W_0, W_I, W_A, W_B\}. \qquad (4.16)$$

By using lemma 4.2 we see that there are *precisely* 7 ways to extend Σ_4 to a partial spread Ψ_5 of 5 planes, other than to a $\Sigma_5 \subset \Sigma_9$, namely by choosing B in (4.16) to be one of the 7 solutions of equations (4.13), *all 7 such extensions thus being projectively equivalent.* It follows that the supporting 35-set $\psi \subset PG(5, 2)$ of Ψ_5 in (4.16) must be a double-five; moreover all maximal partial spreads are of this kind, namely of size 5 (cf. [4]) and arising from a double-five. Also one sees that the extension of a Σ_4 to a Σ_9 is unique, and there is just one kind of spread, representable in the form (4.15) for some choice of A. Hence there is also only one kind of non-maximal partial spread Σ_5, namely of the form $\Sigma_9 \setminus \Sigma_4$.

Remark The supporting 35-set $\psi \subset PG(5, 2)$ of Ψ_5 in (4.16) must admit a partition into another maximal partial spread Ψ_5'. Nevertheless it is not easy to prove this directly from Ψ_5 as given in (4.16).

Let Σ_4' be any partial spread of size 4, and let Σ_9 be its unique extension to a spread. It follows from the foregoing that, adopting projective terminology, the only planes skew to the 4 planes of Σ_4' are the 5 planes of $\Sigma_5 = \Sigma_9 \setminus \Sigma_4'$ together with 7 planes Y_i, such that $\Sigma_4' \cup \{Y_i\}$ is a Ψ_5 for each $i = 0, 1, \dots, 6$. Thus for any non-maximal partial spread Σ_5 of 5 planes it follows that its supporting 35-set $S_5 \subset PG(5,2)$ possesses precisely 7 further internal planes Y_i, $i = 0, 1, \dots 6$. We enquire, how do the 7 planes Y_i intersect the 5 planes of Σ_5?

Theorem 4.3 *Any non-maximal partial spread Σ_5 of 5 planes possesses a privileged member, say $W_* \in \Sigma_5$, such that each of the 7 further internal planes Y_i of S_5 meets W_* in a line and meets each of the four other planes $W \in \Sigma_5$ in a point. The 7 lines $Y_i \cap W_*$ are distinct, as are the 28 points $Y_i \cap W$, $i = 0, 1, \dots 6$, $W \in \Sigma_5 \setminus \{W_*\}$.*

Proof   Consider the partial spread $\Sigma_5 = \{W_{A^r} : r = 2, 3, 4, 5, 6\} = \Sigma_9 \setminus \Sigma_4'$, where $\Sigma_4' = \{W_\infty, W_0, W_I, W_A\}$. By lemma 4.2, parts (iv), (v), the plane W_B meets $W_* = W_{A^4}$ in a line and meets W_{A^r} in a point for $r = 2, 3, 5, 6$. The remaining assertions are easily seen to hold.   □

Let $\mathcal{G}(\Sigma_r) \subset GL(V) \cong GL_6(2)$ denote the symmetry group of the partial spread Σ_r and let $\mathcal{G}(S_r) \subset GL(V)$ be that of its supporting point set S_r. On account of the foregoing facts concerning the $5 + 7 = 12$ internal planes of S_5, it follows that $\mathcal{G}(\Sigma_5) = \mathcal{G}(S_5)$. (Also $\mathcal{G}(\Sigma_r) = \mathcal{G}(S_r)$ for $r < 5$, since the only internal planes of S_r are those of Σ_r.) Clearly any element $T \in \mathcal{G}(\Sigma_5)$ must fix the privileged plane $W_* \in \Sigma_5$, and permute the remaining 4 planes of Σ_5 amongst themselves.

Theorem 4.4 *Given a partial spread $\Sigma_5 \subset \Sigma_9$, with privileged plane W_*, set $\Sigma_4' = \Sigma_9 \setminus \Sigma_5$; thus $\Sigma_9 = \Sigma_4 \cup \{W_*\} \cup \Sigma_4'$, where $\Sigma_5 = \Sigma_4 \cup \{W_*\}$. Then W_* is also the privileged plane of the partial spread $\Sigma_5' = \Sigma_4' \cup \{W_*\}$.*

Proof   Consider $\Sigma_9 = \Sigma_4 \cup \{W_{A^4}\} \cup \Sigma_4'$, where the partial spreads Σ_4, Σ_4' are taken to be $\Sigma_4 = \{W_\infty, W_0, W_I, W_A\}$, $\Sigma_4' = \{W_{A^2}, W_{A^3}, W_{A^5}, W_{A^6}\}$. Consider the element $L \in GL(V)$ defined by $L(x,y) = (A^3 y, A^4 x)$, $x, y \in X$. Observe that $L^2 = I_V$ and that $L \in \mathcal{G}(\Sigma_9)$, since L effects the following permutation ρ_L of the 9 members of Σ_9:

$$\rho_L = (\infty 0)(IA)(A^2 A^6)(A^3 A^5). \tag{4.17}$$

(Here, and elsewhere, we use the labels ∞, 0 and A^r as stand-ins for the elements W_∞, W_0 and W_{A^r}, $r = 0, 1, \dots, 6$, of the spread Σ_9.) Consider the partial spreads $\Sigma_5 = \Sigma_4 \cup \{W_{A^4}\}$, $\Sigma_5' = \Sigma_4' \cup \{W_{A^4}\}$. From (4.17) we see that L lies in both $\mathcal{G}(\Sigma_5)$ and $\mathcal{G}(\Sigma_5')$, but leaves fixed (in fact pointwise) only the plane W_{A^4}. Consequently W_{A^4} is the privileged plane of both Σ_5 and Σ_5'.   □

Corollary 4.5 $\mathcal{G}(\Sigma_4) = \mathcal{G}(\Sigma_5) = \mathcal{G}(\Sigma_4') = \mathcal{G}(\Sigma_5')$.   □

Remarks   (i) The plane W_* in the theorem will also be referred to as the *associated* plane of Σ_4; it is also the associated plane of Σ_4'.

(ii) As noted previously, for $r \leq 5$, $\mathcal{G}(\Sigma_r) = \mathcal{G}(S_r)$. In contrast, $\mathcal{G}(\Psi_5) \cong A_5$ is a proper subgroup of $\mathcal{G}(\psi) \cong A_5 \times Z_2$.

(iii) Given any $\Sigma_4^{(1)} \subset \Sigma_9$ and $\Sigma_4^{(2)} \subset \Sigma_9$, then there exists $T \in \mathcal{G}(\Sigma_9)$, not just $T \in GL(V)$, sending $\Sigma_4^{(1)}$ to $\Sigma_4^{(2)}$. This follows because the extension of a Σ_4 to a Σ_9 is unique.

4.2   Symmetry groups

For a partial spread $\Sigma_r \subset \Sigma_9$, we denote by $\mathcal{G}(\Sigma_r, \Sigma_9)$ the subgroup of $\mathcal{G}(\Sigma_9)$ which stabilizes Σ_r, and by $\mathcal{G}(\dot{\Sigma}_r, \Sigma_9)$ the subgroup which fixes each element $W \in \Sigma_r$. Then, for $\Sigma_9 = \Sigma_r \cup \Sigma_{9-r}$, we have $\mathcal{G}(\Sigma_r, \Sigma_9) = \mathcal{G}(\Sigma_{9-r}, \Sigma_9)$. Also, since a Σ_4 has a unique extension to a Σ_9, we have $\mathcal{G}(\Sigma_r, \Sigma_9) = \mathcal{G}(\Sigma_r)$ for $r \geq 4$.

For $C \in GL(X)$ we define $D_C \in GL(V)$ by $D_C = C \oplus C$, that is by $D_C(x, y) = (Cx, Cy)$. Observe that any $T \in GL(V)$ which fixes each of W_∞, W_0, W_I is of the form $T = D_C$ for some $C \in GL(X)$, and that T will also fix each of the other subspaces W_{A^r}, $r = 1, 2, \ldots, 6$, of the spread Σ_9 if and only if C centralizes A, that is if and only if $C \in \langle A \rangle \cong Z_7$. The subgroup

$$N = \langle D_A \rangle = \{D_{A^r}, r = 0, 1, \ldots, 6\} \cong Z_7 \qquad (4.18)$$

of $\mathcal{G}(\Sigma_9)$ is thus the kernel of the permutation representation ρ of $\mathcal{G}(\Sigma_9)$ on the 9-set Σ_9; it is a normal subgroup of $\mathcal{G}(\Sigma_9)$, and also of each of the groups $\mathcal{G}(\Sigma_r, \Sigma_9)$. The associated quotient groups are denoted $\mathcal{G}_1(\Sigma_9) = \mathcal{G}(\Sigma_9)/N$ and $\mathcal{G}_1(\Sigma_r, \Sigma_9) = \mathcal{G}(\Sigma_r, \Sigma_9)/N$.

In the same vein, any $T \in \mathcal{G}(\dot{\Sigma}_3, \Sigma_9)$, where $\Sigma_3 = \{W_\infty, W_0, W_I\}$, is of the form $T = D_C$ for some C belonging to the normalizer F_{21}, see lemma 4.2, of $\langle A \rangle$ in $GL(X)$. Thus

$$\mathcal{G}(\dot{\Sigma}_3, \Sigma_9) \cong F_{21} \cong Z_7 \rtimes Z_3, \qquad \mathcal{G}_1(\dot{\Sigma}_3, \Sigma_9) \cong Z_3. \qquad (4.19)$$

If S is any element of F_{21} of order 3 which induces the squaring automorphism of $\langle A \rangle : SAS^{-1} = A^2$, then the permutation ρ_{D_S} induced on Σ_9 by D_S is $\rho_{D_S} = (AA^2A^4)(A^6A^5A^3)$. Incidentally, since a 3-point stabilizer $\mathcal{G}(\dot{\Sigma}_3, \Sigma_9)$ thus has no element of order 2, it follows that every involution $J \in \mathcal{G}(\Sigma_9)$ necessarily fixes precisely one element of Σ_9, and permutes the other eight in four 2-cycles. If $R(x, y) = (x + y, x)$, then R is seen to be another element of $\mathcal{G}(\Sigma_3, \Sigma_9)$ of order 3, yielding the permutation $\rho_R = (\infty 0 I)(AA^4A^2)(A^6A^5A^3)$. It follows that $T = RD_S = D_SR \in \mathcal{G}(\Sigma_3, \Sigma_9)$, given by $T(x, y) = (Sx + Sy, Sx)$, is also of order 3, and induces $\rho_T = (\infty 0 I)(A^3A^5A^6)$.

For $\Sigma_4 = \{W_\infty, W_0, W_I, W_A\}$ we already know two elements of the group $\mathcal{G}(\Sigma_4)$, namely the preceding element T, of order 3, and the involution L of equation (4.17). These two generate any even permutation of the 4-set Σ_4. On the other hand it is easy to see that no element $G \in \mathcal{G}(\Sigma_4)$ can induce an odd permutation. For suppose G were to fix W_∞, W_0, but interchange W_I, W_A. Then $G(x, y) = (Cx, ACy)$ for some $C \in GL(X)$ satisfying $ACA = C$, contradicting the fact that A is not conjugate to A^{-1}. So

$$\mathcal{G}_1(\Sigma_4) \cong A_4 \cong (Z_2)^2 \rtimes Z_3, \quad \text{and hence } \mathcal{G}(\Sigma_4) \cong (Z_7 \times (Z_2)^2) \rtimes Z_3. \qquad (4.20)$$

But, as remarked after corollary 4.5, $\mathcal{G}_1(\Sigma_9)$ acts transitively on the $\binom{9}{4} = 126$ partial spreads of kind Σ_4 in Σ_9. Hence $\mathcal{G}_1(\Sigma_9)$ has order $126 \times |A_4| = 1512$, and $\mathcal{G}(\Sigma_9)$ has order $1512 \times 7 = 10584$. As a permutation group on the 9-set Σ_9, $\mathcal{G}_1(\Sigma_9)$ is generated by the two permutations $(\infty 0 I)(A^3 A^5 A^6)$ and $(\infty A^5)(A A^3)(A^2 0)(A^4 A^6)$. In fact, as we will see below in section 4.3, we have the group isomorphisms

$$\mathcal{G}(\Sigma_9) \cong \Gamma L_2(8) \cong (Z_7 \times L_2(8)) \rtimes Z_3, \qquad \mathcal{G}_1(\Sigma_9) \cong P\Gamma L_2(8) \cong L_2(8) \rtimes Z_3 \,. \quad (4.21)$$

Let $\mathcal{G}(\Sigma_4, \Sigma_4')$ denote the subgroup of $\mathcal{G}(\Sigma_9)$ consisting of all those elements which preserve the $4+1+4$ partition $\Sigma_9 = \Sigma_4 \cup \{W_*\} \cup \Sigma_4'$. The group $\mathcal{G}(\Sigma_4, \Sigma_4')$ contains $\mathcal{G}(\Sigma_4) = \mathcal{G}(\Sigma_4')$ as a subgroup of index 2, and in fact it has a direct product structure:

$$\mathcal{G}(\Sigma_4, \Sigma_4') \cong \mathcal{G}(\Sigma_4) \times Z_2, \qquad \mathcal{G}_1(\Sigma_4, \Sigma_4') \cong \mathcal{G}_1(\Sigma_4) \times Z_2 \,. \quad (4.22)$$

To see this, consider the $4+1+4$ partition $\Sigma_9 = \Sigma_4 \cup \{W_I\} \cup \Sigma_4'$ with

$$\Sigma_4 = \{W_\infty, W_A, W_{A^2}, W_{A^4}\}, \qquad \Sigma_4' = \{W_0, W_{A^6}, W_{A^5}, W_{A^3}\} \,. \quad (4.23)$$

Defining $J' \in GL(V)$ by $J'(x, y) = (A^3 x + Ay, Ax + A^3 y)$, we see that J' is an involution which fixes W_I and effects the permutation $(\infty A^2)(A A^4)(0 A^5)(A^6 A^3)$. So $J' \in \mathcal{G}(\Sigma_4)$, and the associated plane of Σ_4 in (4.23) is indeed W_I. Another element of $\mathcal{G}(\Sigma_4)$ is D_S: $D_S(x, y) = (Sx, Sy)$, where $SAS^{-1} = A^2$, since, as noted previously, D_S induces the permutation $(A A^2 A^4)(A^6 A^5 A^3)$. Now an element of $\mathcal{G}(\Sigma_4, \Sigma_4')$ which interchanges Σ_4, Σ_4' is the involution J defined by $J(x, y) = (y, x)$ with effect $(\infty 0)(A A^6)(A^2 A^5)(A^4 A^3)$. Now note that J commutes with both J' and D_S, and hence with the A_4 subgroup of $\mathcal{G}(\Sigma_4)$ which is generated by J' and D_S. Hence $\mathcal{G}(\Sigma_4, \Sigma_4') = \mathcal{G}(\Sigma_4) \times \langle J \rangle$, as claimed in (4.22). Observe that a $4+1+4$ partition $\Sigma_9 = \Sigma_4 \cup \{W_*\} \cup \Sigma_4'$ thus comes along with a built-in linear bijection J, which map the 28 points of $\mathcal{S}_4$ on to the 28 points of $\mathcal{S}_{4'}$, and which commutes with every element of $\mathcal{G}(\Sigma_4) = \mathcal{G}(\Sigma_4')$.

Note that $\mathcal{G}(\Sigma_4, \Sigma_4'), \mathcal{G}_1(\Sigma_4, \Sigma_4')$ are subgroups of $\mathcal{G}(\{W_*\}, \Sigma_9), \mathcal{G}_1(\{W_*\}, \Sigma_9)$, respectively. These latter in fact have the structures

$$\mathcal{G}(\{W_*\}, \Sigma_9) \cong \{Z_7 \times ((Z_2)^3 \rtimes Z_7)\} \rtimes Z_3, \qquad \mathcal{G}_1(\{W_*\}, \Sigma_9) \cong ((Z_2)^3 \rtimes Z_7) \rtimes Z_3 \quad (4.24)$$

These isomorphisms can perhaps be best appreciated from the perspective of section 4.3 below. Nevertheless from our present viewpoint it is easy to exhibit the elements of the normal subgroup $H \cong (Z_2)^3$ of $\mathcal{G}(\{W_I\}, \Sigma_9)$ explicitly. Define the element $U \in L(V, V)$ by $U(x, y) = (x+y, x+y)$; thus $U^2 = 0$, and $\mathrm{im} U = \ker U = W_I$. Define

$$J(K) = I_V + D_K U, \qquad K \in \mathbb{K}. \quad (4.25)$$

Then $J(K)^2 = I_V$ and $J(K)J(K') = J(K+K') = J(K')J(K)$, so that $K \mapsto J(K)$ is an isomorphism of the additive abelian group of $\mathbb{K}$ onto a subgroup $H \cong (Z_2)^3$ of $\mathcal{G}(\Sigma_9)$. The 7 involutions $J(K)$, $K \neq 0$, are precisely those involutions in $\mathcal{G}(\Sigma_9)$ which fix W_I, and so $H = \{J(K) : K \in \mathbb{K}\}$ is a normal subgroup of $\mathcal{G}(\{W_I\}, \Sigma_9)$. Note that the involutions all fix W_I pointwise, not merely setwise. Explicitly the

7 involutions $J(A^r)$, $r = 0, 1, \ldots, 6$, induce the following permutations on the 9-set Σ_9:

$$
\begin{aligned}
J(I) &: (\infty 0)(AA^6)(A^2 A^5)(A^4 A^3), & J(A) &: (\infty A^2)(AA^4)(0A^5)(A^6 A^3), \\
J(A^2) &: (\infty A^4)(AA^2)(0A^3)(A^6 A^5), & J(A^3) &: (\infty A^5)(AA^3)(A^2 0)(A^4 A^6), \\
J(A^4) &: (\infty A)(A^2 A^4)(0A^6)(A^5 A^3), & J(A^5) &: (\infty A^6)(A0)(A^2 A^3)(A^4 A^5), \\
J(A^6) &: (\infty A^3)(AA^5)(A^2 A^6)(A^4 0) .
\end{aligned}
\tag{4.26}
$$

Note that $\{I, J(A), J(A^2), J(A^4)\}$ is the $(Z_2)^2$ subgroup of $\mathcal{G}(\Sigma_4)$, and that $J(I)$ and $J(A)$ are our previous J and J'.

It is not hard to see that $\mathcal{G}_1(\Sigma_2, \Sigma_9)$ is isomorphic to $Z_7 \rtimes Z_6$. In the case that $\Sigma_2 = \{W_\infty, W_0\}$, the element $(x, y) \mapsto (x, Ay)$ of $\mathcal{G}(\Sigma_9)$ effects the 7-cycle $(IAA^2A^3A^4A^5A^6)$. It should also be noted that $\mathcal{G}(\Sigma_9)$ contains elements of order 9, such as $Q : (x, y) \mapsto (y, x + A^2 y)$, which has effect $(\infty A^2 A^3 A I A^6 A^4 A^5 0)$. Consequently $D_A Q (= Q D_A)$ is a Singer cycle, that is an element of $GL_6(2)$ of order 63.

4.3   The $V(2, 8)$ view of Σ_9

On account of lemma 4.2(ii) we may view $V(6, 2) = X \oplus X$ as a 2-dimensional vector space $V(2, 8)$ over the field $\mathbb{K} = \{0\} \cup \langle A \rangle \cong GF(8)$. Choosing any nonzero $e \in X$, set $e_1 = (e, 0) \in W_\infty$ and $e_2 = (0, e) \in W_0$; then $\{e_1, e_2\}$ is a basis for $V(2, 8)$. The nine 1-dimensional subspaces of $V(2, 8)$ are then $W_\infty = \langle e_1 \rangle$, $W_0 = \langle e_2 \rangle$ and $W_{A^r} = \langle e_1 + A^r e_2 \rangle$, $r = 0, 1, \ldots, 6$; viewing these as 3-dimensional subspaces over the subfield $\{0, I\} \cong GF(2)$, they constitute the spread $\Sigma_9 = \{W_\infty, W_0, W_{A^r}$; $r = 0, 1, \ldots, 6\}$ in $V(6, 2)$ previously dealt with. Now each $T \in \Gamma L_2(8)$ induces a permutation of the elements of Σ_9, and since T is $GF(2)$-linear, $\Gamma L_2(8)$ is thereby a subgroup of $\mathcal{G}(\Sigma_9)$. The isomorphisms (4.21) follow: $\mathcal{G}(\Sigma_9) \cong \Gamma L_2(8)$, since both groups have order 10584, and hence, on quotienting by $N \cong \langle A \rangle \cong \mathbb{K}^\times$, we obtain $\mathcal{G}_1(\Sigma_9) \cong P\Gamma L_2(8)$.

Clearly, various other aspects of (non-maximal) partial spreads Σ_r in $V(6, 2)$ dealt with earlier can be viewed advantageously in terms of $V(2, 8)$ and $PG(1, 8)$. The associated point P of a tetrad $\mathcal{T}$ of points of $PG(1, 8)$ becomes the associated plane W_* of a $\Sigma_4 \subset \Sigma_9$. The 7 involutions of theorem 3.3(i) are, if $P = \langle e_1 + e_2 \rangle = W_I$, those of equation (4.26). The structure of the groups $\Gamma \mathcal{G}_P, \Gamma \mathcal{G}(\mathcal{T})$ and $\Gamma \mathcal{G}(\mathcal{T}, \mathcal{T}')$ in equations (3.8), (3.9) and (3.10) yields that of the groups $\mathcal{G}_1(\{W_*\}, \Sigma_9), \mathcal{G}_1(\Sigma_4)$ and $\mathcal{G}_1(\Sigma_4, \Sigma_4')$ in equations (4.24), (4.20) and (4.22).

Nevertheless, it should be noted that the $V(2, 8)$, and concomitant $PG(1, 8)$, perspectives are attractive and simple *only if they confine their view to the 9 subspaces of a particular spread* Σ_9. At the projective level of $PG(1, 8)$ only 9 of the 1395 planes of $PG(5, 2)$ are visible; at the vector space level of $V(2, 8)$, the remaining 1386 3-dimensional subspaces of $V(6, 2)$, that is those which are not 1-dimensional subspaces of $V(2, 8)$, are there, but not very visible. Certainly it seems hard to believe that a $V(2, 8)$ view-point could be helpful in discussing a double-five. In particular it should be noted that the symmetry group A_5 of a maximal partial spread Ψ_5 has order $2^2.3.5$, yet 5 does not divide the order of $\Gamma L_2(8)$. In a related vein, the geometrical property, see theorem 4.3, which picks out the privileged member of a Σ_5, or equally the associated plane of a Σ_4, is not visible in the $PG(1,8)$ treatment of the associated point of a tetrad.

4.4   An overlarge set of 3-(8,4,1) designs

We end on a combinatorial note, by showing how the properties of the projective line $PG(1,8)$, as described in section 3, give rise to a simple construction of an overlarge set of 3-(8,4,1) designs (as defined below). We also provide details of the construction in the richer geometrical setting of a spread Σ_9 of planes in $PG(5,2)$.

In terms of the complete graph $K(8)$ on the 8 points $S\backslash\{P\}$, with a transposition (rs) viewed as an edge $\{r,s\}$, note that each of the 7 involutions in (3.5), (3.6) defines a 1-factor, and the set of 7 involutions exhibits a 1-factorization of $K(8)$. These 7 involutions, with fixed point P, define a $3-(8,4,1)$ design $\mathcal{D}(P)$ on the 8 points $S\backslash\{P\}$, $S=PG(1,8)$, as follows: a tetrad of points is a block of the design $\mathcal{D}(P)$ if and only if its associated involutions have P as fixed point. In terms of the labels $0,1,2,3,0',1',2',3'$ for the 8 points, as used in theorem 3.3, the 14 blocks of the design $\mathcal{D}(P)$ are

$$0123, \ 010'1', \ 020'2', \ 030'3', \ 012'3', \ 023'1', \ 031'2',$$
$$0'1'2'3', \ 232'3', \ 313'1', \ 121'2', \ 0'1'23, \ 0'2'31, 0'3'12 . \qquad (4.27)$$

Here the 7 blocks which contain 0 are given in the first line of the display, the remaining 7 blocks being their complements. (The 3-(8,4,1) design is unusual in that the complement of a block is a block.) Equivalently described, the 7 complementary pairs $\mathcal{B}, \mathcal{B}'$ of blocks of $\mathcal{D}(P)$ are those for which $\mathcal{G}(\mathcal{B}, \mathcal{B}')$ is the normal subgroup $\cong (Z_2)^3$ of $\mathcal{G}_P$, see theorem 3.3. Any three points Q_1, Q_2, Q_3 of $S\backslash\{P\}$ lie in a unique block of $\mathcal{D}(P)$, namely $\{Q_1, Q_2, Q_3, Q_4\}$ where $Q_4 = L(Q_3)$, with L that involution which fixes P and interchanges Q_1, Q_2.

Varying P over the nine points of S, we have a set $\{\mathcal{D}(P) : P \in PG(1,8)\}$ of nine 3-(8,4,1) designs. These designs are mutually disjoint, since a tetrad whose involutions have P as fixed point is a block only of the design $\mathcal{D}(P)$. (The numbers tally: each design contributes 14 blocks, thus accounting for all the $\binom{9}{4} = 9 \times 14 = 126$ tetrads on the line $PG(1,8)$.) Such a set of 9 mutually disjoint 3-(8,4,1) designs is termed an *overlarge set*, see [1], [6]. In [1], Breach & Street showed that, up to isomorphism, there are just two overlarge sets, L_1 and L_2, of 3-(8,4,1) designs, with automorphism groups of orders 216 and 1512 respectively. The above overlarge set $\{\mathcal{D}(P) : P \in PG(1,8)\}$ clearly has automorphism group $P\Gamma L_2(8)$, and so is isomorphic to the second overlarge set L_2 of Breach & Street. (Incidentally, the 28 resolution classes displayed in [1] are the 28 partitions of S discussed at the end of section 3.2.) The above construction of an L_2, using the properties of the 9-point projective line, would appear to be the simplest possible. Observe, from theorem 3.3 (iii), that the overlarge set has the following feature: each of the 63 pairs $\{B, B'\}$ of blocks comes along with a privileged bijection $B \leftrightarrow B'$, arising from that unique involutory automorphism J which commutes with every automorphism which preserves the pair $\{B, B'\}$.

In a $PG(5, 2)$ setting the construction of an overlarge set, based on a 9-set Σ_9 of planes of a spread, proceeds as follows. For each plane $W_* \in \Sigma_9$ the blocks of a 3-(8,4,1) design $\mathcal{D}(W_*)$, based on the 8-set $\Sigma_9 \backslash \{W_*\}$, are defined as follows. Choose a line $\lambda \subset W_*$, and to each plane $Y \neq W_*$ of $PG(5, 2)$ which contains λ associate the block

$$\Sigma_4^Y = \{W \in \Sigma_9 : |W \cap Y| = 1\}. \tag{4.28}$$

There are 14 such planes Y, and the associated 14 blocks come along in complementary pairs $\Sigma_4^Y, \Sigma_4^{Y'}$ — since, for each Y, we have $\Sigma_9 = \Sigma_4^Y \cup \{W_*\} \cup \Sigma_4^{Y'}$ where Y' is the image of Y under that involution $J \in \mathcal{G}(\Sigma_9)$ which, see equation (4.22), centralizes $\mathcal{G}(\Sigma_4^Y)$. By theorem 4.3 the blocks (4.28) are the same, independent of the choice of line λ.

Acknowledgements. The author was aided in this research by the award of an Emeritus Fellowship by the Leverhulme Trust. He thanks J. Conway, T. S. Griggs, J.W.P. Hirschfeld and A. Rosa for useful discussions, and especially A. Rosa for the provision of references on overlarge sets. He is also grateful to N.A. Gordon for certain accuracy checks carried out using MAGMA.

References

[1] D.R. Breach & A.P. Street, Partitioning sets of quadruples into designs II, *J. Combin. Math. Combin. Comput.* **3** (1988), 41-48.

[2] J.H. Conway, R.T. Curtis, S.P. Norton, R.A. Parker & R.A. Wilson *Atlas of Finite Groups*, Clarendon Press, Oxford, 1985.

[3] J.W.P. Hirschfeld, *Projective Geometries over Finite Fields*, Clarendon, Oxford, 1979.

[4] H. Lüneburg, *Translation Planes*, Springer-Verlag, New York, 1980.

[5] R. Mathon & A. Rosa, A census of 1-factorizations of K_9^3: solutions with group of order > 4, *Ars Combin.* **16**(1983), 129-147.

[6] M.J. Sharry & A.P. Street, Cyclic and rotational overlarge sets of some small Steiner systems, *Utilitas Math.* **44**(1993), 115-130.

[7] R. Shaw, Icosahedral sets in $PG(5, 2)$, *European J. Combin.* **18**(1997), 315-339.

[8] R. Shaw, Configurations of planes in $PG(5, 2)$, contribution to the conference *Combinatorics '96*, Assisi, 1996, submitted for publication in the *Proceedings*.

[9] V.D. Tonchev, Quasi-symmetric designs, codes, quadrics, and hyperplane sections, *Geom. Dedicata* **48** (1993), 295-308.

Ron Shaw, School of Mathematics, University of Hull, Hull HU6 7RX, United Kingdom.
email: r.shaw@maths.hull.ac.uk

Rank three geometries with simplicial residues

E. E. Shult

Abstract

This is a survey of all known rank three geometries belonging to a string diagram of type (c^*, c)-geometry. There are three types of objects: points, lines, and blocks subject to axioms imposed by the diagram. There are several other formulations described here which are more convenient for presenting certain of the examples. All examples fall into these six classes:

1. Simplicial type, which can easily be characterized.

2. Fischer spaces with no affine planes.

3. Orthogonal types, whose points and lines are exterior points and tangent lines of certain low-dimensional quadrics.

4. Hall type, determined by alternating multilinear forms over the field of two elements.

5. Affine type, whose points are vectors in some d-dimensional space over the integers mod 2. Here, blocks are not subspaces.

6. A few special examples determined by coherent pairs: the construction of odd type of Cameron and Fisher, and two examples of Blokhuis and Brouwer.

1 Introduction

All geometries in this report are connected and every flag (that is, a clique of the multipartite incidence graph) lies in a **chamber** (that is, a clique of the incidence graph containing one object from each part).

1.1   The definition

This survey concerns geometries with three sorts of objects, which we will call **points** $(\mathcal{P})$, **lines** $(\mathcal{L})$, and **blocks**, $(\mathcal{B})$, which, as an incidence geometry belongs to a specific **diagram**, (c^*, c):

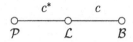

which is a way of encoding these axioms:

(CC1) *Any two distinct lines incident with a common point p are both incident with a unique common block.*

(CC2) *Any two lines incident with a common block are incident with a unique common point.*

(CC3) *Given a point p incident with a block B, there are exactly two lines incident with both.*

(CC4) (The String Property) *Any point incident with a line of a block B is also incident with B.*

There is an apparent duality in the diagram, and a real duality in the axioms. If the words "point" and "block" are transposed throughout these axioms, the axioms (CC1) and (CC2) are interchanged, while (CC3) and (CC4) are self-dual.

It follows that two distinct points are together incident with at most one line, otherwise these two points are incident with two lines, and by (CC1) and (CC4) together, the entire quartette of objects is incident with a unique common block B. But lines of a block are incident with a unique common point by (CC2). Thus $(\mathcal{P}, \mathcal{L})$ is a partial linear space.

Consequently, the lines and blocks can be identified with their sets of incident points, and (by (CC4)), incidence is inclusion among these sets.

By duality, two distinct blocks are incident with at most one line in common.

At times we will consider two further properties of these geometries:

(S) (The Subspace Axiom) *A block is a subspace of the point-line geometry. Specifically, this means that if a line L is incident with two points of a block, then the line itself (and hence all points incident with it) are incident with that block.*

The dual of this axiom asserts that if two blocks are incidence with a common line L and a common point, then that point is incident with L. But this is implied by (S) for if $\{p\} \cup L \subseteq B_1 \cap B_2$, there is a line N of B_1 on p, and if p were not on L, then N would meet L at a second point q, by (CC2). By (S), N would be incident with B_2, so $B_1 = B_2$ by (CC1). Thus (S) implies its dual. By symmetry of the diagram, (S) is a self-dual axiom.

REMARK: The condition (S) is equivalent to the assertion that the three points of any triangle in the collinearity graph together lie in a unique block.

A second property is

(I) (The intersection property.) *The intersection of two distinct blocks (regarded as sets of points) is either a common incident line, a point or the empty set.*

Property (I) is even stronger than property (S). Suppose (I) holds, and two distinct points x and y belong to some common line L, and a block B. By the diagram geometry assumption that all flags lie in a chamber, L must lie in some block B'. If L were not incident with B then $B' \neq B$. But then by (I), $B' \cap B = L$ and L is incident with B, a contradiction. Thus $|L \cap B| \geq 2$ implies $L \subseteq B$, so (S) holds.

But in general (S) does not imply (I). Even assuming (S), it is conceivable that two blocks may intersect at two or more pair-wise non-collinear points, without impairing the property that blocks are subspaces. Indeed, most of our examples possess property (S) but not (I).

Assuming that we have now acquired some familiarity with the way these axioms work, we leave the reader to observe that (I) is a self-dual property.

1.2   Parameters.

Suppose now that $\Gamma = (\mathcal{P}, \mathcal{L}, \mathcal{B})$ is a finite string geometry of points, lines and blocks. (The word "string" here simply means that if a point and block are incident with a common line, then they are incident with each other.) We say that Γ **has parameters** if the following is true:

There exist integers k and ℓ such that

1. *Each line is incident with k points.*
2. *Each line is incident with ℓ blocks.*

Recall, that by assumption, all diagram geometries are connected.

Lemma 1.1 *(Hughes) Every finite (c^*, c)-geometry has parameters.*

Suppose now, Γ has parameters k and ℓ. Then

- Each block contains $k + 1$ lines.

- Each block contains $k(k + 1)/2$ points.

- Each point lies in $\ell + 1$ lines.

- Each point lies on $\ell(\ell + 1)/2$ blocks.

1.3   Alternative Formulations

1.3.1   Points with two line classes.

Let $(\mathcal{P}, \mathcal{L})$ be a connected incidence system of points and lines, the lines regarded as sets of points. Suppose there is a partition of the lines into two classes $\mathcal{L}_1$ and $\mathcal{L}_2$ such that

(LL1) If $(A, B) \in \mathcal{L}_1 \times \mathcal{L}_1 \cup \mathcal{L}_2 \times \mathcal{L}_2$, then $|A \cap B| = 0$ or 1.

(LL2) If $(A, B) \in \mathcal{L}_1 \times \mathcal{L}_2$, then $|A \cap B| = 0$ or 2.

(LL3) Any pair of distinct points is $\mathcal{L}_1$-collinear if and only if it is $\mathcal{L}_2$-collinear.

In addition to the incidence of points and lines inherited from $(\mathcal{P}, \mathcal{L})$, let us say that a line L from $\mathcal{L}_1$ is incident with a line N of $\mathcal{L}_2$ if and only if $|L \cap N| = 2$. The result is a rank three incidence geometry $\Gamma = (\mathcal{L}_1, \mathcal{P}, \mathcal{L}_2)$. Then given $N \in \mathcal{L}_2$, by (LL2) and (LL3), the elements of $\mathcal{L}_1$ and $\mathcal{P}$ incident with N are bijective with the 2-subsets and 1-subsets, respectively, of N, with the same incidence; thus this residue is a c^*-geometry. By symmetry of the assumptions affecting the partition $\mathcal{L} = \mathcal{L}_1 + \mathcal{L}_2$, the residue of a line $L \in \mathcal{L}_1$ is a geometry of type c. If $(L, N) \in \mathcal{L}_1 \times \mathcal{L}_2$, and L and N are incident with a common point, then they intersect at two points and so are incident with each other. Thus, as remarked in Pasini ([**P**] p. 220, exercise 7.24),

Theorem 1.2 *The geometry* $\Gamma = (\mathcal{L}_1, \mathcal{P}, \mathcal{L}_2)$ *is a* (c^*, c)*-geometry.*

1.3.2   Locally grid graphs

A **simple graph** $G = (V, E)$ is an undirected graph without loops or multiple edges. The subgraph induced on the set of neighbors of vertex v is denoted $G(v)$. A **grid graph** is a connected graph whose vertices can be coordinatized by a cartesian product $A \times B$, such that two vertices are adjacent if and only if they share a common left coordinate or a common right coordinate, but not both. A simple graph is a **locally grid graph** if and only if, for every vertex v, $G(v)$ is the graph of a grid. In a locally grid graph, any two maximal cliques intersect in at most two points. Moreover, if C is the collection of all maximal cliques of such a graph, then the graph $(C, \sim)$ of the relation of two maximal cliques intersecting in exactly one vertex, has at most two connected components. Of course if one of the grids $G(v)$ is not square, two components must occur.

Lemma 1.3 *Suppose G is a locally grid graph whose graph of maximal cliques $(C, \sim)$ has two connected components, C_1 and C_2. Then setting two cliques to be incident if they intersect at two vertices, and vertex-clique incidence the obvious one, (C_1, V, C_2) is a (c^*, c)-geometry with the subspace property (S).*

Proof The clique-classes C_i play the role relative to V of that of the line classes L_i relative to P in the Axioms (LL1)–(LL3). So by Theorem 1.2, a (c^*, c)-geometry results.

Suppose p is a vertex of the locally grid graph G and that $\{p\} = C_1 \cap C_1'$, for cliques C_1 and C_1' in C_1. Then p must be incident with any clique $C_2 \in C_2$ which is incident with both C_1 and C_1', because otherwise $G(p)$ would not be a grid. In the language of points, lines and blocks, this says that if a line is incident with two points which are incident with a block, then that line is incident with that block. But this is the subspace property.   □

1.3.3   Coherent pairs in $(0, 2)$-block configurations

Let P be a set of *points* and let B be a collection of subsets of P whose elements are called *blocks*. Here, (P, B) is defined to be a $(0, k)$-**configuration** if and only if

(Bk) *Any two blocks of B intersect at either 0 or k points.*

For any two distinct points a and b, let B_{ab} be the set of blocks which contain a but do not contain b. Clearly, if $k \geq 1$, there is an induced $(0, k-1)$-configuration $(P - \{a, b\}, B_{ab})$. If x and y are in $P - \{a, b\}$, we say that x **and** y **are** B_{ab}-**collinear** if and only if there is a block in B containing $\{a, x, y\}$ but not b.

The key axioms appear in the next definition. Two distinct points $\{a, b\}$ of a $(0, 2)$-configuration are said to be a **coherent pair** if and only if

(CO1) *Any two points $x, y \in P - \{a, b\}$ are B_{ab}- collinear if and only if they are also B_{ba}-collinear.*

(CO2) *The $(0,1)$-system $(\mathcal{P} - \{a,b\}, \mathcal{B}_{ab})$ is connected.*

REMARK: Note that in view of (CO1), axiom (CO2) could be replaced just as well by the assertion that $(\mathcal{P} - \{a,b\}, \mathcal{B}_{ba})$ is connected. In either case axiom (CO2) is not really a restriction. If we had begun with a $(0,2)$-configuration $(\mathcal{P}, \mathcal{B})$ and a pair of distinct points a and b satisfying axiom (CO1) (that is, a "precoherent pair", if you will), then one may replace $\mathcal{B}$ by $\mathcal{B}' := \mathcal{B}_{ab} \cup \mathcal{B}_{ba}$, and $\mathcal{P}$ by $\mathcal{P}' := \{a,b\} \cup X$, where X is a connected component of the point-collinearity graph of $(\mathcal{P} - \{a,b\}, \mathcal{B}_{ab})$ to obtain a coherent pair $\{a,b\}$ of the $(0,2)$-system $(\mathcal{P}', \mathcal{B}')$.

The following is nearly immediate:

Theorem 1.4   (i) *If $\{a,b\}$ is a coherent pair of the $(0,2)$-configuration $(\mathcal{P}, \mathcal{B})$, then $(\mathcal{B}_{ab}, \mathcal{P} - \{a,b\}, \mathcal{B}_{ba})$ is a (c^*,c)-geometry.*

(ii) *Conversely, if $(\mathcal{P}, \mathcal{L}_1, \mathcal{L}_2)$ satisfies the axioms (LL1) – (LL3) , and if " 1 " and " 2 " are two new objects not in $\mathcal{P}$, then, setting*

$$
\begin{aligned}
\mathcal{B}_{12} &= \{\{1\} \cup L | L \in \mathcal{L}_1\} \\
\mathcal{B}_{21} &= \{\{2\} \cup N | N \in \mathcal{L}_2\} \\
\mathcal{P}' &= \mathcal{P} \cup \{1,2\} \\
\mathcal{B}' &= \mathcal{B}_{12} \cup \mathcal{B}_{21},
\end{aligned}
$$

it is true that $\{1,2\}$ is a coherent pair of the $(0,2)$-configuration $(\mathcal{P}', \mathcal{B}')$.

The only advantage in approaching this subject through coherent pairs of $(0,2)$-configurations is perhaps a psychological one. A person seeking (c^*,c)-geometries should be on the alert for systems of points and blocks whose blocks meet each other in zero or two points.

A few miscellaneous examples from Section 4 seem best described from this point of view.

1.4   The structure of a block, and their models

A finite block consists of $k+1$ lines $\{L_o, \ldots, L_k\}$, each having k points. Each point of the block lies on exactly two of the lines and so may be indexed by the indices of those two lines. Thus we can let p_{ij} be the name of the unique point in $L_i \cap L_j$. Since, also any two lines of a block meet at a point, all 2-subsets $\{i,j\}$ drawn from $\{0,1,\ldots,k\}$ arise to index the points. Thus we have a "standard model" of a block. For a given parameter k, the standard block B_k has elements of a $k+1$-set $\Omega_k := \{0,1,\ldots,k\}$ as **lines**, and all 2-sets of Ω_k as **points**.

Any other model of a block with $k + 1$ lines is equivalent to the one above. The usefulness of having other models for blocks is that they suggest how the standard model might be embedded into other geometries. Each of the known examples displays such a model which suggests the particular example. This being so, it is illuminating if we began each of the examples described below with the model for a block which it displays.

2 The simplicial (or Johnson) examples

One can dualize the standard model of a block by taking complements of the sets involved. In this model, the points are the $(k - 1)$-subsets of a $(k + 1)$-subset $\Omega_k = \{0, \ldots, k\}$, and the lines are the k-subsets of Ω_k. Incidence is containment.

Now let X be an n-set, $n > k + 1$. The points, $\mathcal{P}$, are the $(k - 1)$-subsets of X; the lines, $\mathcal{L}$, are the k-subsets of X, and the blocks, $\mathcal{B}$, are the $(k + 1)$-subsets of X. Proper containment among these subsets is incidence. Then $(\mathcal{P}, \mathcal{L}, \mathcal{B}) := S(n, k)$ is clearly a (c^*, c)-geometry with parameters $(n, n - k)$. Obviously the dual geometry is obtained by taking complements of all the relevant subsets in X, and so is of the same type, having parameters $(n-k, k)$.

It is easy to see that two points represented by $(k - 1)$-sets are collinear in the (c^*, c)-geometry , if and only if they meet at a $(k - 2)$-set, and so the point-collinearity graph is the distance regular Johnson graph, $J(n, k - 1)$. In particular, two points are at distance d from one another if and only their corresponding $(k - 1)$-sets intersect at a $(k - d - 1)$-set.

Now it should be clear that blocks are subspaces, and the intersection of (the point-shadows of) two distinct blocks cannot contain two non-collinear points. Thus the intersection property (I) holds.

In fact one has

Theorem 2.1 *Suppose* $\Gamma = (\mathcal{P}, \mathcal{L}, \mathcal{B})$ *is a finite* (c^*, c)-*geometry. Then* Γ *is simplicial type if and only if it satisfies the intersection condition* (I).

This is remarked in Exercise 7.26 of [P] and proved in some detail in [S2]. The latter proof employs Theorem 9.1.3 and Corollary 9.1.4 of [BCN, pp. 156-7]. The proof of the theorem comes from Blokhuis and Brouwer [BB] (see also Dowling [D] and Moon [M]). The Corollary follows from Sprague [Sp] and Tůma [Tu].

Corollary 2.2 *Let* J *be a finite* (c^*, c)-*geometry of simplicial type.*

(i) J *is simply connected. This means that if* $\kappa : \Gamma \to J$ *is a 2-cover as rank-3 geometries, then* κ *is an isomorphism.*

(ii) *If $\kappa : J \to \Gamma$ is a 2-cover, then κ is an isomorphism–that is any simplicial (c^*, c)-geometry is not a 2-cover of any further geometry.*

The corollary is proved in [**S2**], using the Theorem 9.1.3 (*loc. cit.*).

This is probably a good place to note that the difference between the (c^*, c)-geometries and the multiply extended grid graphs studied in [**MP**] produces different meanings to the notions of "simply connected" and "homomorphism". In the case of the multiply-extended grid graphs corresponding to (c^*, c)-geometries of simplicial type the situation is different in these respects:

- The graphs are *not* distinguished by the intersection property.

- There is a relatively easy proof that they are simply connected, using only the cycle structure of the Johnson graph (see page 255 of [**MP**]).

- The graphs do possess nontrivial homomorphic images (p. 256 of [**MP**]) in contrast to the Corollary 2.2.

3   The examples of Fischer type

The (c^*, c)-geometries of this section are characterized by the fact that one of the parameters k or ℓ is 3. To keep from describing both a geometry and its dual, we assume here that it is the parameter k which is equal to three – that is, there are just three points on each line. The model for a block is very simple: it is the dual affine plane of order 2.

3.1   Blocks are subspaces.

Our first observation is

Lemma 3.1 *Any (c^*, c)-geometry with $k = 3$ satisfies the subspace condition (S).*

Proof Suppose B is a block and suppose L is a line of the (c^*, c)-geometry with $B \cap L = \{a, a'\}$ so the two points are not collinear by a line of B. Thus without loss of generality we may take L to be $\{a, a', d\}$. Let A be a line of B on point a. By axiom (CC1), the lines $A = \{a, b, c\}$ and L are incident with a unique block B'. In B', a' on L is either collinear to b or c – say c. But in B, as the figure shows, a' is collinear to *both* b and c. Thus by partial linearity, B and B' share a common line on c and a'. But they also share line A on c. Then (CC1) forces $B = B'$, against the assumption that L was not incident with B. Thus no such L is so positioned relative to a block, so (S) holds. □

3.2   The connection with Fischer spaces

We thus see that the $(\mathcal{P}, \mathcal{L})$-truncation of a (c^*, c)-geometry with $k = 3$ is a partial linear space with three points per line in which the *subspace* generated by two distinct intersecting lines is a dual affine plane. This is a special instance of a Fischer space.

A **Fischer space**, is a point-line incidence system satisfying

(F1) Each line has exactly three points, any two of them intersecting in at most one point.

(F2) The subspace generated by any two intersecting lines is either the affine plane of order 3 or the dual affine plane of order 2.

For each point p of a Fischer space, there is a permutation t_p of the points which fixes point p and all points which are not collinear with p, but transposes the two remaining points in each line on p. It is clear that t_p stabilizes each subspace generated by a pair of lines on p, and is in fact an automorphism of the Fischer space. Setting G to be the group of automorphisms of $(\mathcal{P}, \mathcal{L})$ generated by the set $T = \{t_p | p \in \mathcal{P}\}$, one sees that T is a class of **3-transpositions** of G – that is, a conjugacy class of involutions of G any two of which generate $\mathrm{Sym}(3)$ or the elementary group of order four. Conversely, if T is a class of 3-transpositions one recovers a Fischer space, $(T, \mathcal{L})$, by declaring each triplet of involutions of T within a $\mathrm{Sym}(3)$ to be a line. A great deal is known about 3-transposition groups, and we can utilize this in our discussion of (c^*, c)-geometries of Fischer type.

As we have seen, the point-line truncation, of a (c^*, c)-geometry Γ of Fischer type is just a Fischer space without subspaces which are affine planes of order 3. The class T of 3-transpositions takes blocks to blocks and so are automorphisms of the full rank three (c^*, c)-geometry. Thus by definition, G being a subgroup of the automorphism group of Γ, acts faithfully on Γ. Since G in general is primitive on points, G has a trivial center. This, together with the absence of the affine planes forces $O_3(G) = 1$ – that is, G has no non-trivial normal 3-groups. It can, however possess normal 2-subgroups. If we factor out the largest normal 2-subgroup, $O_2(G)$, the absence of affine planes forces

$$G/O_2(G) \simeq \left\{ \begin{array}{l} \mathrm{Sym}(n), \\ \mathrm{Sp}(2n, 2), \text{ or} \\ O^\epsilon(2n, 2), \epsilon = \pm. \end{array} \right.$$

The group homomorphism $G \to G/O_2(G)$ is reflected in the geometry. One does not actually obtain a geometry homomorphism, since blocks may collapse to a line: for example, for a block B, $\mathrm{Aut}(B) \simeq \mathrm{Sym}(4)$, but if the Klein four-subgroup (that is $O_2(\text{Aut}(B)) \leq O_2(G)$, one obtains such a collapse for B. But it does induce a homomorphism of the Fischer spaces

$$(T, \mathcal{L}) \to (\bar{T}, \bar{\mathcal{L}}),$$

where $\bar{T}$ and $\bar{\mathcal{L}}$ are the images of the 3-transposition class T and the images of the class of Sym(3)'s induced by $G \to O_2(G)$. This is because the fibre above each point is a coclique in the collinearity graph of the domain Fischer space.

When $G/O_2(G)$ is larger then Sym(3) or Sym(4), the three groups presented are indeed (c^*, c)-geometries, but they are homomorphic images of the (c^*, c)-geometries defined by G which are not 2-coverings of the geometry, because the fate of some blocks is that they must become lines. So these homomorphisms involve an asymmetry between points and blocks not revealed by the $c*c$ diagram. Nonetheless, we may say that we almost understand these geometries: roughly speaking they are obtained by blowing up the one of the four 3-transposition geometries defined by Sym(n) (actually a (c^*, c)-geometry of simplicial type using 2-subsets as points), Sp($2n, 2$) and the two orthogonal groups $O^+(2n, 2)$ and $O^-(2n, 2)$ (in the last three cases, 3-transpositions are the transvections), by uniformly replacing each line by a (c^*, c)-geometry which is a larger transverse design.

The exact nature of the normal 2-groups involved (where many of the elements of order 3 in the Fischer space lines (the Sym(3)'s) must act without fixed points) is a difficult subject. The reader is referred to the papers of J. I. Hall for the most definitive description of these normal 2-subgroups.

4   The Examples of Orthogonal Type

There are five infinite families here. The cases $q \equiv 3(\text{mod } 4)$) of the first example below are in Pasini's recent book, *Diagram Geometries* (Ex. 7.28, pg. 221 of [**P**]), and the first two examples are in Section 2.4 of [**MP**]. All five of the geometries presented below are chamber-transitive and satisfy the subspace condition (S). The last example yields a new locally-grid graph whose automorphism group is transitive on "flags" – that is ordered triples (v, e, c) where c is a maximal clique, e is one of its edges, and v is a vertex incident with that edge.

4.1 The block models

Following the custom set at the beginning, we begin with a description of the pertinent models for a block. There are two models, both involving a non-degenerate quadratic form Q on a vector space V over a finite field $GF(q)$.

Let $B : V \times V \to GF(q)$ be the associated bilinear form. (In characteristic 2, where B is a symplectic form, the radical of B is not allowed to contain any nonzero singular vectors).

MODEL ONE: Suppose dim $V = 3$. The singular 1-spaces comprise a conic C of the projective plane $\pi := \mathbf{P}(V) \simeq PG(2,q)$. If q is odd then the points of the plane exterior to C are of two types: (1) the $q(q+1)/2$ points of "hyperbolic type" which live on two tangent lines of the conic, and the $q(q-1)/2$ points of "elliptic type" which lie on no tangent lines. The Lines of our block are the $q+1$ lines tangent to the conic. The Points of our block are the points of hyperbolic type. Each point is on exactly two such lines, and any two tangent lines meet at a unique point (though each may contain many points). Thus this system of points and lines is a rank 2 geometry of type c^*, and so is a model for a block.

MODEL TWO: Now suppose dim $V = 4$ with no restriction on the parity of q. We suppose that Q is an "elliptic quadric". Each projective point (1-space) p of O lies in a unique tangent plane

$$\pi_p := p^\perp = \{r = \text{ 1-space of } V | B(p,r) = 0\}.$$

Two such tangent planes meet at an exterior line – that is a 2-subspace of V containing no 1-space representing a point of the ovoid. Conversely, the "perp" of an exterior line must be a hyperbolic 2-space containing exactly two singular 1-spaces representing points of the ovoid. One way to summarize this is by saying that the $q^2(q^2+1)/2$ lines exterior to O and the $1+q^2$ tangent planes comprise a geometry of type c^*, and so form a block.

4.2   Three examples with blocks from the first model.

4.2.1   The examples in $Q^\epsilon(3,q)$.

Let q be an odd prime power, F the field of q elements, and let V be a 4-dimensional vector space over F. Let $Q : V \to F$ be a non-degenerate quadratic form. With some abuse of notation we also let Q denote the projective variety of the zero-set of Q, called the **quadric**, — that is, the collection of all singular 1-spaces of V. There is a sign $\epsilon = \pm$ which is "+" if V possesses totally singular 2-dimensional subspaces (Q is then a *hyperbolic quadratic*) and is "−" otherwise (when Q is called an *elliptic quadric*). There are two classes of non-singular 1-spaces, distinguished according to whether they are generated by vectors v, for which $Q(v)$ is or is not a square in F. Both classes contain the same number of 1-spaces, and we pick one of these, call it $\mathcal{N}$, to be the **points** of our geometry. The **lines** are the 2-dimensional subspaces of the form $\langle n, s \rangle$, where $n \in \mathcal{N}$ and s is a totally singular 1-space in $n^\perp$. There are also two classes of non-degenerate 3-subspaces of V. In one of these classes, the projective points from $\mathcal{N}$ form the set of hyperbolic points of the projective plane of the three space – these are the 3-spaces which form the **blocks** of the geometry. Containment among subspaces is incidence. Thus the blocks form a c^*-geometry via Model One, above. The lines and blocks containing a point n, is isomorphic to the geometry of singular 1-spaces, and hyperbolic 2-subspaces of the 3-space $n^\perp$ — that is, the 1- and 2-subsets of the set $n^\perp \cap Q$ of singular 1-spaces orthogonal to n. Thus we obtain a (c^*, c)-geometry.

4.2.2   An example from the quadric $Q(4, q)$.

Here, the **points** are the class $\mathcal{N}^-$ of non-singular 1-spaces n such that $(n^\perp \cap Q)$ is an elliptic quadric of $PG(3, q)$. The lines and blocks are exactly as define in the previous example: **lines** are degenerate 2-spaces containing q 1-spaces from $\mathcal{N}^-$, and **blocks** are non-degenerate 3-spaces whose $\mathcal{N}^-$-points lie on two tangents of the conic. This too is a (c^*, c)-geometry.

4.2.3   Parameters and other invariants of the these geometries.

It is clear that in these three geometries, each line is incident with exactly q points. Given a line, say $\langle n, s \rangle$, $s \in Q$ and $n \in \mathcal{N}$ or $\mathcal{N}^-$ as is appropriate, a block on this line is obtained by adjoining a second singular subspace $s' \in n^\perp \cap Q - \{s\}$ to form the 2-space (plane) $\langle n, s, s' \rangle$. Thus each line lies on q blocks in the first two cases, and q^2 blocks in the third.

In all three cases, there are two sorts of points at distance two (in the point- collinearity graph) from a point n. Say $x \in \Delta^+(n)$ if and only if $\langle n, x \rangle$ is a hyperbolic 2-space, and let $x \in \Delta^-(n)$ if $\langle n, x \rangle$ is an anisotropic 2-subspace of V. (Note that if $q > 5$, these cannot both be a single orbit under the action of the stabilizer of n in the orthogonal group, since in one of these sets (depending upon the parity of $(q - 1)/2$) there are elements of $\mathcal{N}$ (or $\mathcal{N}^-$) which are orthogonal to n as well as those that are not.)

Letting δ denote a sign, $\pm$, let β^δ be the number of blocks containing the distance-two pair of points (n, x) when $x \in \Delta^\delta(n)$. This is a constant, because each hyperbolic line lies in a constant number of planes in the class $\mathcal{B}$ of all projective planes π for which $\mathcal{N} \cap \pi$ (or $\mathcal{N}^- \cap \pi$) are hyperbolic points, and similarly, each anisotropic line lies in a constant number of planes from $\mathcal{B}$. It is easy to calculate that in the geometries obtained from $Q^\epsilon(3, q)$,

$$\beta^\delta = (q - \delta\epsilon)/2.$$

For the geometry obtained from $Q(4, q)$, we have

$$\beta^+ = \beta^- = q(q - 1)/2.$$

Moreover, if $x \in \Delta^\delta(n)$, any two blocks containing the two points x and n, intersect at *the same coclique of c^δ points* where

$$c^\delta = (q - \delta)/2,$$

namely, the 1-subspaces of $\langle x, n \rangle$ which are points of the geometry, that is, in $\mathcal{N}$ (or $\mathcal{N}^-$). Thus we have:

(BI) (Block Intersection Axiom) *the intersection of the point-shadows of two distinct blocks containing two non-collinear points x and y, is the same set of c pairwise non-collinear points, for any two distinct blocks containing x and y.*

Note that the local "mu" parameter, the number of points collinear with both x and n, is $4\beta^\delta$ (4β) and the number of lines on x carrying these points is $2\beta^\delta$ (2β). The cardinalities of the two distance-two classes can be determined by calculating the number of lines which are respectively secant and exterior to the quadric, and which lie on point n. All of these results are compiled in the following tables:

Geometry	order (k, ℓ)	β^+	β^-	c^+	c^-
$Q^-(3, q)$	(q, q)	$(q+1)/2$	$(q-1)/2$	$(q-1)/2$	$(q+1)/2$
$Q^+(3, q)$	(q, q)	$(q-1)/2$	$(q+1)/2$	$(q-1)/2$	$(q+1)/2$
$(Q(4, q), \mathcal{N})$	(q, q^2)	$q(q-1)/2$	$q(q-1)/2$	$(q-1)/2$	$(q+1)/2$

Geometry	$|\mathcal{P}|$	$|\Delta^+(n)|$	$|\Delta^-(n)|$
$Q^-(3, q)$	$q(q^2+1)/2$	$(q^2-q)(q-3)/4$	$(q^2+q)(q-1)/4$
$Q^+(3, q)$	$q(q^2-1)/2$	$(q^2+q)(q-3)/4$	$(q^2-q)(q-1)/4$
$(Q(4, q), \mathcal{N})$	$q^2(q^2-1)/2$	$q(q^2+1)(q-3)/4$	$q(q^2+1)(q-1)/4$

4.2.4 A two-parameter family of strongly regular graphs.

It is evident that the point collinearity graphs of these geometries always have a "lambda" value determined by the geometry parameters k and ℓ. But in the last case, $(Q(4, q), \mathcal{N})$, the "mu" parameter is uniform as well.

The same phenomenon occurs when we replace the class $\mathcal{N}^-$ by $\mathcal{N}^+$ in $Q(4, q)$ or if we replace the "4" by any even projective dimension. In fact in **[S2]**, the following is proved:

Theorem 4.1   (i) *The point-line geometry $\Gamma_n^\epsilon(q) := (\mathcal{N}^\epsilon, \mathcal{L}^\epsilon)$ (obtained from a class of exterior points, $\mathcal{N}^\epsilon$), and their incident tangent lines, $\mathcal{L}^\epsilon$, of the quadric $Q(2n, q)$, q odd) is a partial linear space (called a* **projective tangent space** *in* **[CP]**).

(ii) *If two distinct lines L_1 and L_2 of $\Gamma_n^\epsilon(q)$ intersect at a point p, then the subspace that they generate is either an affine plane of order q, or a block in the sense of Section 2 (defining blocks for (c^*, c)-geometries).*

(iii) *The point-collinearity graph of $\Gamma_n^\epsilon(q)$ is a strongly regular graph with these parameters:*

$\epsilon = -.$

$$
\begin{aligned}
v &= q^n(q^n - 1)/2 \\
k &= (q^{n-2} + \ldots q + 1)(q^n + 1)(q - 1) \\
\lambda &= 2q^{2n-2} - q^n + q^{n-1} - 2 \\
\mu &= 2q^{n-1}(q^{n-1} - 1)
\end{aligned}
$$

$\epsilon = +.$

$$
\begin{aligned}
v &= q^n(q^n + 1)/2 \\
k &= (q^{n-1} + \ldots + q + 1)(q^{n-1} + 1)(q - 1) \\
\lambda &= 2q^{2n-2} + q^n - q^{n-1} - 2 \\
\mu &= 2q^{n-1}(q^{n-1} + 1)
\end{aligned}
$$

REMARK: These strongly regular graphs yield symmetric designs with $k + 1$ points per block when $q = 3$. This seems to be an observation of Arunas Rudvalis ([**Ru**]) around 1968. The fact that strongly regular graphs occur in even projective dimension for all odd $q > 3$, may also have been known to him. In any event it seems worthwhile to record or revive attention to this phenomenon, whichever the case may be.

4.3 Geometries whose blocks are based on the second model.

Again, we assume $Q : V \to F$ is a non-degenerate quadratic form, where $F = GF(q)$ is finite. The *points* of our geometry Γ comprise the set $\mathcal{A}$ of all anisotropic 2-dimensional subspaces of V (in characteristic 2 this means 2-spaces without singular 1-subspaces). The *lines* are 3-dimensional subspaces of the form $A \perp s$ where A is an anisotropic 2-space, s is a singular 1-space and "$\perp$" denotes the direct sum of subspaces which are orthogonal to one another. *Blocks*, then, are $PG(3, q)$'s which meet the quadric Q in an elliptic quadric. In order to get any two lines on a point of the geometry on a point A to generate a block, it is necessary and sufficient that $A^\perp \cap Q$ be a rank 1 polar space — either a non-degenerate plane or an elliptic $PG(3, q)$. This requirement leaves us with two cases: $Q = Q(4, q)$ and $Q = Q^+(5, q)$.

Note that there is no requirement here that q be odd. Basically, this is because our model of a block makes no such requirement in this case. Suppose q is even, and dim $V = 5$. Then the radical of the associated bilinear form B is a 1-dimensional non-singular space r. Our points are 2-subspaces A containing no singular 1-subspaces. and so cannot be degenerate with respect to B. In particular, they do not contain r. For example, when $q = 2$, there are exactly 20 such spaces, and the (c^*, c)-geometry is also that of simplicial type $J(4, 6)$ or the dual of one of those of Fischer type based on the transpositions of Sym(6).

4.3.1 Common features of the geometries of the second type.

The number k of points on a line is always q^2, and the number ℓ of blocks on a line is q when $Q = Q(4, q)$ (when $A^\perp$ is a plane) and q^2 when $Q = Q^+(5, q)$ (when $A^\perp \cap Q$ is an elliptic quadric).

These two geometries have point-diameter at least three. The elements of $\mathcal{A}$ form a 6- or 7-class association scheme depending on whether $Q = Q^+(5, q)$ or $Q(4, q)$, respectively. Let A_1 and A_2 be anisotropic 2-subspaces of V representing two points of Γ. Only these relationships are possible:

(1) $A_1 = A_2$.

(2) $W = \langle A_1, A_2 \rangle$ is a degenerate 3-space. In this case A_1 and A_2 are collinear points of Γ.

(3,4) $W = \langle A_1, A_2 \rangle$ is a non-degenerate 3-space of V. If $Q = Q^+(5, q)$, there is only one isometry class of such 3-spaces in the general orthogonal group, and if $Q = Q(4, q)$, there are two such classes. (This fact must be argued separately for each parity of q.)

(5) $\langle A_1, A_2 \rangle$ is an elliptic 4-space (i.e. a block). Here, clearly $\beta(A_1, A_2) = 1$.

(6) $\langle A_1, A_2 \rangle$ is a hyperbolic 4-space.

(7) $W = \langle A_1, A_2 \rangle$ is a 4-space with a non-trivial radical $R(W)$ (with respect to the bilinear form B). Then dim $R(W) = 1$ or 2 according as q is odd or even. In the former case $W = U \perp R(W)$, where U is a non-degenerate 3-space. The discriminant of U is determined if $Q = Q(4, q)$ and if $Q = Q^+(5, q)$ the two possibilities for the discriminant give 4-spaces conjugate in $GO^+(5, q)$. In the latter case, when q is even, $W = U \perp s$, where s is a singular 1-space, and U has a (non-singular) radical $R(U)$ with $R(W) = R(U) \perp s$, the direct sum of two perpendicular B-isotropic 1-spaces. Again, there is only one isometry class of such 4-spaces W.

In cases (3), (4), and (5), A_1 and A_2 are at distance two in the collinearity graph for Γ, for they lie in at least one common block. In case (5-7), they are at distance three in the collinearity graph. Moreover, in all cases that A_1 and A_2 are at distance two and do not lie in a unique block in Γ, $\dim W = 3$, so that as blocks correspond to vector spaces of dimension 4, the block intersection property (BI) holds.

4.4   Dualities among geometries of orthogonal type

The results proved in [S2] can be summarized as follows:

- The (c^*, c)-geometries whose points are one of the two classes of exterior points of the quadric $Q^\epsilon(3, q)$, where ϵ is fixed and q is odd, is closed under taking dualities. (The dualities do not always result from the "perp" map.)

- The (c^*, c)-geometry whose points are the class $\mathcal{N}^-$ of "elliptic" exterior points of the quadric $Q(4, q)$ is dual to the (c^*, c)-geometry $Q(4, q)\mathcal{A}$ whose points are the lines exterior to the quadric, provided q is odd. When q is even, the dual of Γ_2 seems not to have an interpretation as subspaces of an orthogonal geometry.

- The (c^*, c)-geometry whose points are the exterior lines of the quadric $Q^+(5, q)$, is self-dual under the "perp" mapping.

5   Examples of Hall type

A class of locally grid graphs constructed from attenuated spaces over $GF(2)$ appears in Section 2.7 of the Meixner-Pasini census ([MP]), with an attribution to the author of this survey article. However, the latter was only the messenger, for these were perhaps the only flag-transitive examples of a far more general class of (c^*, c)-geometries of whose existence the author first learned from a conversation with J. I. Hall in Fargo, North Dakota ([H]). I have therefore taken the liberty of distinguishing this class with his name.

5.1   The description of a block

Here, our model of a block is very simple: The points and lines of a block are the affine lines and affine points of an affine geometry $AG(d + 1, 2)$, of dimension $d + 1$ over the field of two elements. Since this is a linear space with line size two, a block is a c^* geometry with $2^d - 1$ points on each line.

5.2   The definition of the geometry

In order to approach the (c^*, c)-geometries , we must first describe a **Grassmann space**. Fix a vector space V of finite dimension n over a field F. Let d be an integer with $2 \leq d \leq n-2$. The Grassmann space $A_{n-1,d}(F)$ is a geometry of points and lines: its points are the d-dimensional vector subspaces of V; its lines are incident pairs (A, B) of respective vector space dimensions $(d-1, d+1)$. A point is incident with a line if the d-space represented by the point lies between the pair of spaces represented by the line. Thus if F is the finite field $GF(q)$, there are $q+1$ points on each line.

A Grassmann space is always a partial linear space, that is two distinct points have *at most* one common incident line. The distance between two points is always $d-i$, where i is the vector space dimension of the intersection of the two d-subspaces represented by these points. Thus, if $d \leq n/2$, (as may be supposed by passing to the dual space of V if necessary) the diameter of the point-collinearity graph is always d.

Recall that a **subspace** of any geometry of points and lines is just a subset X of points with the property that any line which has at least two of its points in X in fact has all of its incident points in X. A subspace is called a **singular subspace** if and only if any two of its points are collinear.

In the case of Grassmann space, there are two distinguished classes $\mathcal{A}_1$ and $\mathcal{A}_2$ of maximal singular subspaces with these properties:

(G1) Any singular subspace lies in some member of $\mathcal{A}_1$ or $\mathcal{A}_2$, and any line lies in just one of each.

(G2) If $A, B \in \mathcal{A}_i$, then $A \cap B = \emptyset$ or a point, $i = 1, 2$.

(G3) If $(A, B) \in \mathcal{A}_1 \times \mathcal{A}_2$, then the set of points of $A \cap B$ is either the empty set or the set of points of a line.

The maximal singular subspaces here are all projective spaces: the elements of $\mathcal{A}_1$ are $PG(n-d)$'s, those of $\mathcal{A}_2$ are $PG(d)$'s. The two classes $\mathcal{A}_1$ and $\mathcal{A}_2$ are respectively bijective with the $(d-1)$-subspaces and $(d+1)$-subspaces of V. Be aware that the axioms (G1)-(G3) alone do not characterize Grassmann spaces.

A **geometric hyperplane** of a point-line geometry is a proper subspace which intersects each line non-trivially. Let H be any geometric hyperplane of the Grassmann space $A_{n-1,d}(F)$ considered above. There is then an induced geometry $A_H := (\mathcal{P}', \mathcal{L}')$ whose points are the points of the Grassmann space which do *not* lie in the hyperplane H, and whose lines are those lines of the Grassmann space which do not lie in H. This geometry has two systems of maximal singular subspaces consisting of the sets

$$\mathcal{L}_i := \{A - H \mid A \in \mathcal{A}_i, A - H \neq \emptyset\}, i = 1, 2.$$

The elements of $\mathcal{L}_1$ are affine spaces $AG(n - d, F)$, while those in $\mathcal{L}_2$ are affine spaces $AG(d, F)$. Each line of A_H lies in exactly one space of each type. Now, when $F = GF(2)$, all lines of A_H have exactly two points and $(\mathcal{L}_1, \mathcal{P}', \mathcal{L}_2)$ satisfies the axioms (LL1)-(LL3) of Section 1.4. Thus

$$(\mathcal{P}, \mathcal{L}, \mathcal{B}) = (\mathcal{L}_1, \mathcal{P}', \mathcal{L}_2)$$

is a (c^*, c)-geometry whose blocks are as described in the first subsection of this section.

Thus for each choice of integers (n, d) with $1 < d \leq n/2$ and $n > 3$ and choice of geometric hyperplane H, there is a (c^*, c)-geometry with parameters $(k, \ell) = (2^{n-d} - 1, 2^d - 1)$.

5.3 The attenuated case

There is at least one choice of H for which A_H is flag-transitive, the case that A_H is a so-called **attenuated space**. Here the (c^*, c)-geometry has a particularly simple description:

Again let V be an n-dimensional vector space over the field $GF(2)$. Fix an $n - d$-dimensional subspace D, $1 < d < n - 1$. Our points $(\mathcal{P})$ are all $(d-1)$-dimensional vector subspaces of V which meet D at the zero subspace. Lines $(\mathcal{L})$ are all d-dimensional subspaces of V which complement D. Finally, blocks $(\mathcal{B})$ are all $(d + 1)$-dimensional subspaces which intersect D in a 1-space. Letting incidence be containment, one obtains the (c^*, c)-geometry, $\Gamma = (\mathcal{P}, \mathcal{L}, \mathcal{B})$.

It admits the full stabilizer of the subspace D in $\Gamma L(V)$ as a flag transitive group of automorphisms. When $n = 2d$, this became the flag-transitive locally grid graph of the Meixner-Pasini census ([**MP**],Section 2.7, p. 264).

5.4 Properties of the Hall geometries

Clearly Γ has the subspace property (S).

The following is shown in [**S2**]:

- Except for the attenuated case, the block intersection property (BI) fails.

- There are no standard quotients in the sense of Cuypers and Pasini ([**CP**]).

- All the hyperplanes are defined by alternating trilinear forms ([**R**] or [**S1**]).

As far as is known, only the attenuated case admits a flag-transitive group.

6   Geometries of Affine Type

The geometries described in this section do not satisfy the property (S), that
blocks are a subspace. In fact the truncation to points and lines is an affine
space.

6.1   The planar case

Here, the model for a block is obtained by removing a line and all its points
from a block defined by a line-hyperoval.

Let Π be a Desarguesian projective plane of order q, a power of 2. Then
there is a dual hyperoval $\mathcal{H}' = \{N_0, N_1, \ldots N_{q+1}\}$, of $q+2$ lines, no three on a
common point. Then the point union U of these lines consists of $(q + 2)(q +
1)/2$ points, each lying on exactly two of the lines of $\mathcal{H}'$. Thus $(U, \mathcal{H}')$ is a
geometry of type c^* and so is already a model for a block. Now remove line
N_0 and all its points from Π to form an affine plane $A_0 = (V_0, \mathcal{L}_0)$. Setting
$L_i = N_i - N_0$, $i = 1, \ldots, q + 1$, $\mathcal{L}'_0 = \{L_1, \ldots, L_{q+1}\}$, and $U_0 = U - L_0$, we
see that $B_0 = (U_0, \mathcal{L}'_0)$ is also a c^* geometry with $(q + 1)q/2 = |U_0|$ points,
and $q + 1$ lines, each being a representative of a unique parallel class of A_0.

We form the system $\mathcal{B}_0$ of all vector translates of B_0 in A_0: that is

$$\mathcal{B}_0 = B_0^T = \{B_0 + v | v \in V_0\}.$$

Note that if v is a non-zero vector of V_0 and $B \in \mathcal{B}_0$, then $B \neq B + v$. For
otherwise, if L were a line incident with the block B, then $L + v$ would also
be a line of B, so $L = L + v$, as L and $L + v$ would intersect otherwise. But
this cannot be true for all lines L of B, since translation by v would fix the
point of intersection of two such lines, against the fact that the translation
group T is regular on the vectors of V_0.

Thus we see that $|V_0| = |\mathcal{B}| = q^2$. Then if $\mathcal{L}_0$ is the full set of $q(q + 1)$
lines of $A_0 \simeq AG(2, q)$, we see that every pair of 1-spaces lies on at least
one translated block $B_0 + v$ containing 0, since the lines of B_0 have one
representative of each parallel class of A_0. On the other hand, $|V| = |\mathcal{B}|$
shows that vector 0 lies on exactly $|B_0| = q(q + 1)/2$ blocks.

It follows that with containment among subsets of V_0 as incidence, $\Gamma_0 =
(V_0, \mathcal{L}_0, \mathcal{B}_0)$ is a (c^*, c)-geometry with parameters $(k, \ell) = (q, q)$.

Suppose now that the hyperoval of lines $\{L_0, L_1, \ldots, L_{q+1}\}$ which was
used to define the geometry Γ_0, consisted of a dual conic and its nucleus L_0
Then this hyperoval admits the group $SL(2, q) \simeq O_3(q)$, transitive on the
lines of the dual conic, while stabilizing L_0. It follows that $SL(2, q)$ acts on
$A_0 \simeq AG(2, q)$, fixing block B_0, while acting doubly transitive on its lines and
transitive on its points. Thus the (c^*, c)-geometry $\Gamma_0 = (A_0, \mathcal{L}_0, \mathcal{B}_0)$ admits
the collineation group $T \cdot SL(2, q)$ as a flag-transitive group.

6.2   The general case

We can now use Γ_0 to construct even larger (c^*, c)-geometries of affine type. Suppose V is an n-dimensional vector space over $GF(q)$ where $q = 2^d$ and $n \geq 3$. Let $V_0, V_1, \ldots, V_m$ be the collection of all 2-dimensional subspaces of V, so

$$m + 1 = \left[\begin{array}{c} n \\ 2 \end{array} \right]_q ,$$

the Gaussian coefficient. For each V_i, the translates of its 1-spaces in V_i forms the set $\mathcal{L}_i$, of lines of an affine plane $A_i = (V_i, \mathcal{L}_i)$. On each A_i we construct a block B_i of $1 + q$ lines and $q(q + 1)/2$ points using a hyperoval of lines of the associated projective plane Π_i, obtained by adding the "line at infinity". Of course this can be done in various ways, depending on the orbit of hyperovals and the choice of the removed line L_0, but we choose at least one such way for each i to obtain a block B_i of A_i. Then as before, set $\mathcal{B}_i = \{B_i + a | a \in A_i\}$ so that

$$\Gamma_i = (V_i, \mathcal{L}_i, \mathcal{B}_i)$$

is a (c^*, c)-geometry for each A_i.

Now set

$$\mathcal{B} = \{B_i + v | v \in V, i = 0, \ldots, m\},$$

the set of all V-translates of the B_i. Finally, let $\mathcal{L}$ be the V-translates of all the lines in all the $\mathcal{L}_i$. Then of course $A = (V, \mathcal{L})$ is the affine space $AG(n, q)$.

We form the incidence geometry

$$\Gamma = (V, \mathcal{L}, \mathcal{B}).$$

with containment of vector shadows as incidence. Then Γ admits the full group T of all translations by vectors of V as a point-regular group of automorphisms. Also, any block $B \in \mathcal{B}$ spans an affine plane subspace $A(B)$ of A which is a translate of a particular V_i. Thus setting $\mathcal{F}_i = B_i^T$, we see that the block set $\mathcal{B}$ partitions into $m + 1$ sets $\mathcal{F}_0, \ldots, \mathcal{F}_m$. Thus blocks in distinct $\mathcal{F}_i$ can only meet in the empty set, a point, two points or a line, since the affine plane subspaces which they generate can intersect at the empty set, a point or an affine line which is either a line of both blocks, or meets the points of one block in at most two points. Two distinct blocks from the same $\mathcal{F}_i$ share no common points. Put another way

(*) *Two intersecting lines of $\mathcal{L}$ are incident with at most one block of $\mathcal{B}$.*

Now if N and M are two lines on the zero vector (that is, they are distinct 1-subspaces of V), then $\langle N, M \rangle_V = V_i$, for some i, and N and M are together incident with a unique block of the (c^*, c)-geometry Γ_i. Since T is transitive on V, the same holds for every pair of intersecting lines. Thus

Theorem 6.1 $\Gamma = (V, \mathcal{L}, \mathcal{B})$ *is a* (c^*, c)-*geometry with parameters.*

$$k = q \ \text{and} \ \ell = \begin{bmatrix} n \\ 2 \end{bmatrix}_q - 1 = q(1 + q + \cdots + q^{n-2}).$$

6.3   A flag-transitive case, and an isomorphism theorem

Let us call a (c^*, c)-geometry **of classical affine type** if $\dim V = 2$, and $\Gamma = (V, \mathcal{L}, \mathcal{B})$ has its translation class of blocks defined by a dual conic $\{L_1, \ldots, L_{q+1}\}$. Then as observed earlier, Γ has q^2 points and blocks, $q(q+1)$ lines, and the automorphism group of Γ is a group G, which is the semidirect product of the additive group of V by $SL(2, q)$ in its natural action on V. We denote this group $Qd(q)$ so that it is easily distinguished from the geometry $AG(2, q)$ on which it acts. The classical construction introducing this paragraph shows that there are at least two conjugacy classes of complements (isomorphic to $SL(2, q)$) to V in $Qd(q)$ – in fact there are q of them. This group acts naturally on the affine plane $\mathbf{A} = AG(2, q)$ whose points are the vectors of V, and whose lines $\mathcal{L}$ are the translates of all 1-spaces of V. A p-Sylow group P stabilizes just one parallel class Π, acts regularly on the lines of Π. Let M be the unique normal subgroup of index q in P fixing every line of Π.

Theorem 6.2 *Let* $G = Qd(q)$ *with* $O_p(G) = V$, $\mathbf{A}$ *and* P *defined as above with* q *even. Let* H *and* K *be any two non-conjugate complements of* V *in* G.

(i) *The coset geometry* $\Gamma(G; H, P, K)$ *is isomorphic to the unique* (c^*, c)-*geometry of classical affine type with parameters* (q, q).

(ii) *Any* (c^*, c)-*geometry with parameters* $(k, \ell) = (q, q)$, q *even, which admits* $Qd(q)$ *as a flag-transitive group, is of classical affine type.*

(iii) *The Cameron-Fisher construction of even type (see* [**MP**], *sec. 2.6, or* [**CF**]*) is a* (c^*, c)-*geometry of classical affine type.*

This is proved in [**S2**].

7   Examples drawn from coherent pairs and elsewhere

7.1   The cases derived from designs.

Assume the $(0, 2)$-configuration $(V, \mathcal{D})$ is a $2 - (v, k, \lambda)$-design. We have

Lemma 7.1 *If $k > 3$, and the $2 - (v, k, \lambda)$-design $(V, \mathcal{D})$ contains a coherent pair, then $(k + 1)/2 < \lambda \leq k - 1$. In particular, no reasonable biplane can possess a coherent pair.*

This is proved in [S2]. An interesting example occurs when the $(0, 2)$-block configuration is the Steiner system for the Mathieu group on 22 letters. The resulting (c^*, c)-geometry has parameters $(4, 4)$ and admits $Qd(4)$ as a flag-transitive group. It is thus of classical affine type.

A slight variation on the coherent-pair theme allows a presentation of the Fisher-Cameron construction of odd type, described in [CF] and in [MP, sec. 2.5].

Finally in [BB], Brouwer and Blokhuis constructed two (c^*, c)-geometries having 32 points, 40 lines and 32 blocks, one of which might be a 2-fold cover of the unique geometry of classical affine type with $q = 4$.

8   Final remarks

What are the prospects for classifying the (c^*, c)-geometries ? The examples of the Sections 5 and 6 demonstrate that such a classification problem is at least as difficult as (i) the classification of all alternating d-linear forms in n variables over $GF(2)$, or (ii) the classification of all hyperovals of $PG(2, q)$, q even.

It therefore seems more realistic to look at the classification problem with either (1) the assumption of flag-transitivity, or (2) the assumption of special geometric properties. Flag transitivity implies two doubly transitive actions, one on the set of lines on a point, the other on the set of blocks on a line. In both of these actions a special 2-graph is preserved.

Acknowledgement Research for this article was supported by a grant from the U. S. National Science Foundation. The author warmly thanks the organizers for the opportunity to participate in this historic first Pythagorean Conference. I especially thank Prof. S. Magliveras and his family for arranging such an unparalleled environment for work.

REFERENCES

[BB]   A.Blokhuis and A. E. Brouwer, Locally 4 by 4 grid graphs,
       J. Graph Theory **13** (1989), 229-244.

[BCN] A. E. Brouwer, A. M. Cohen, and A. Neumaier, *Distance-regular graphs*, Springer Verlag, Berlin, Heidelberg, 1989.

[CF]   P. Cameron and P. Fisher, Small extended generalized quadrangles, *European J. Combin.* **11** (1990), 403-413.

[CP]   H. Cuypers and A. Pasini, Locally polar geometries with affine planes, *European J. Combin.* **13** (1992), 39-57.

[D]   T. A. Dowling, A characterization of the T_m graph, *J. Combin. Theory Ser. A* **6** (1969), 251-263.

[H]   J. I. Hall, Personal communication, Fargo, North Dakota, 1993.

[M]   A. Moon, A characterization of the graphs of the Johnson schemes $G(3k, k)$ and $G(3k + 1)$, *J. Combin. Theory Ser. B* **33** (1982), 213-221.

[MP]   T. Meixner and A. Pasini, A census of multiply-extended grid graphs, *Finite Geometries and Combinatorics.* London Math. Soc. Lecture Notes No. 191, DeClerck F. et al, (eds.) Cambridge University Press, Cambridge 1993, pp. 249-268.

[P]   A. Pasini, *Diagram Geometries*, Oxford University Press, Oxford, 1994.

[R]   M. Ronan, Embeddings and hyperplanes of discrete geometries, *European J. Combin.* **8** (1987), 179-185.

[Ru]   A. Rudvalis, Personal communication, 1968.

[S1]   E. E. Shult, Geometric hyperplanes of embeddable Grassmannians, *J. Algebra* **145** (1992), 55-82.

[S2]   E. E. Shult, A survey of c^*c geometries. Preprint, 1996, 53 pp.

[Sp]   A. Sprague, Pasch's axiom and projective spaces, *Discrete Math.* **33** (1981), 79-87.

[Tu]   J. Tůma, A structure theorem for lattices of generalized partitions, *Contributions to Lattice Theory*, Colloq. Math. Soc. Janos Bolyai **33**, A. P. Huhn and E. T. Schmidt (eds.), North Holland, Amsterdam, 1983.

E. E. **Shult**, Department of Mathematics, Kansas State University, Manhattan, KS 66502, U.S.A.
e-mail: shult@math.ksu.edu

Generalized quadrangles and the Axiom of Veblen

J. A. Thas   *H. Van Maldeghem*

Abstract

If x is a regular point of a generalized quadrangle $\mathcal{S} = (P, B, I)$ of order $(s, t), s \neq 1$, then x defines a dual net with $t + 1$ points on any line and s lines through every point. If $s \neq t, s > 1, t > 1$, then $\mathcal{S}$ is isomorphic to a $T_3(O)$ of Tits if and only if $\mathcal{S}$ has a coregular point x such that for each line L incident with x the corresponding dual net satisfies the Axiom of Veblen. As a corollary we obtain some elegant characterizations of the classical generalized quadrangles $Q(5, s)$. Further we consider the translation generalized quadrangles $\mathcal{S}^{(p)}$ of order $(s, s^2), s \neq 1$, with base point p for which the dual net defined by L, with p I L, satisfies the Axiom of Veblen. Next there is a section on Property (G) and the Axiom of Veblen, and a section on flock generalized quadrangles and the Axiom of Veblen. This last section contains a characterization of the TGQ of Kantor in terms of the Axiom of Veblen. Finally, we prove that the dual net defined by a regular point of $\mathcal{S}$, where the order of $\mathcal{S}$ is (s, t) with $s \neq t$ and $s \neq 1 \neq t$, satisfies the Axiom of Veblen if and only if $\mathcal{S}$ admits a certain set of proper subquadrangles.

1   Introduction

For terminology, notation, and results concerning finite generalized quadrangles and not explicitly given here, see the monograph of Payne and Thas [11], which is henceforth denoted FGQ.

Let $\mathcal{S} = (P, B, I)$ be a (finite) generalized quadrangle (GQ) of order $(s, t), s \geq 1, t \geq 1$. So $\mathcal{S}$ has $v = |P| = (1 + s)(1 + st)$ points and $b = |B| = (1 + t)(1 + st)$ lines. If $s \neq 1 \neq t$, then $t \leq s^2$ and, dually, $s \leq t^2$; also $s + t$ divides $st(1 + s)(1 + t)$.

There is a point-line duality for GQ (of order (s, t)) for which in any definition or theorem the words "point" and "line" are interchanged and the parameters s and t are interchanged. Normally, we assume without further notice that the dual of a given theorem or definition has also been given.

Given two (not necessarily distinct) points x, x' of $\mathcal{S}$, we write $x \sim x'$ and say that x and x' are *collinear*, provided that there is some line L for which $x \text{ I } L \text{ I } x'$; hence $x \not\sim x'$ means that x and x' are not collinear. Dually, for $L, L' \in B$, we write $L \sim L'$ or $L \not\sim L'$ according as L and L' are concurrent or nonconcurrent. When $x \sim x'$ we also say that x is *orthogonal* or *perpendicular* to x', similarly for $L \sim L'$. The line incident with distinct collinear points x and x' is denoted xx', and the point incident with distinct concurrent lines L and L' is denoted $L \cap L'$.

For $x \in P$ put $x^\perp = \{x' \in P \,\|\, x \sim x'\}$, and note that $x \in x^\perp$. The trace of a pair $\{x, x'\}$ of distinct points is defined to be the set $x^\perp \cap x'^\perp$ and is denoted $\mathrm{tr}(x, x')$ or $\{x, x'\}^\perp$; then $|\{x, x'\}^\perp| = s + 1$ or $t + 1$ according as $x \sim x'$ or $x \not\sim x'$. More generally, if $A \subset P$, A "perp" is defined by $A^\perp = \cap\{x^\perp \,\|\, x \in A\}$. For $x \neq x'$, the *span* of the pair $\{x, x'\}$ is $\mathrm{sp}(x, x') = \{x, x'\}^{\perp\perp} = \{u \in P \,\|\, u \in z^\perp$ for all $z \in x^\perp \cap x'^\perp\}$. When $x \not\sim x'$, then $\{x, x'\}^{\perp\perp}$ is also called the *hyperbolic line* defined by x and x', and $|\{x, x'\}^{\perp\perp}| = s + 1$ or $|\{x, x'\}^{\perp\perp}| \leq t + 1$ according as $x \sim x'$ or $x \not\sim x'$.

2   Regularity

Let $\mathcal{S} = (P, B, \text{I})$ be a finite GQ of order (s, t). If $x \sim x', x \neq x'$, or if $x \not\sim x'$ and $|\{x, x'\}^{\perp\perp}| = t + 1$, where $x, x' \in P$, we say the pair $\{x, x'\}$ is *regular*. The point x is *regular* provided $\{x, x'\}$ is regular for all $x' \in P, x' \neq x$. Regularity for lines is defined dually.

A (finite) *net* of *order* $k \, (\geq 2)$ and *degree* $r \, (\geq 2)$ is an incidence structure $\mathcal{N} = (P, B, \text{I})$ satisfying the following:

(i) each point is incident with r lines and two distinct points are incident with at most one line;

(ii) each line is incident with k points and two distinct lines are incident with at most one point;

(iii) if x is a point and L is a line not incident with x, then there is a unique line M incident with x and not concurrent with L.

For a net of order k and degree r we have $|P| = k^2$ and $|B| = kr$.

Theorem 2.1 (1.3.1 of Payne and Thas [11]) . *Let x be a regular point of the GQ $\mathcal{S} = (P, B, I)$ of order $(s, t), s > 1$. Then the incidence structure with pointset $x^\perp - \{x\}$, with lineset the set of spans $\{y, z\}^{\perp\perp}$, where $y, z \in x^\perp - \{x\}, y \not\sim z$, and with the natural incidence, is the dual of a net of order s and degree $t + 1$. If in particular $s = t > 1$, there arises a dual affine plane of order s. Also, in the case $s = t > 1$ the incidence structure π_x with pointset $x^\perp$, with lineset the set of spans $\{y, z\}^{\perp\perp}$, where $y, z \in x^\perp, y \neq z$, and with the natural incidence, is a projective plane of order s.*

3   Dual nets and the Axiom of Veblen

Now we introduce the *Axiom of Veblen* for dual nets $\mathcal{N}^* = (P, B, I)$.

Axiom of Veblen. *If L_1 I x I $L_2, L_1 \neq L_2, M_1 \not{I} x \not{I} M_2$, and if L_i is concurrent with M_j for all $i, j \in \{1, 2\}$, then M_1 is concurrent with M_2.*

The only known dual net $\mathcal{N}^*$ which is not a dual affine plane and which satisfies the Axiom of Veblen is the dual net $H_q^n, n > 2$, which is constructed as follows : the points of H_q^n are the points of $\mathrm{PG}(n, q)$ not in a given subspace $\mathrm{PG}(n - 2, q) \subset \mathrm{PG}(n, q)$, the lines of H_q^n are the lines of $\mathrm{PG}(n, q)$ which have no point in common with $\mathrm{PG}(n - 2, q)$, the incidence in H_q^n is the natural one. By the following theorem these dual nets H_q^n are characterized by the Axiom of Veblen.

Theorem 3.1 (Thas and De Clerck [14]) *Let $\mathcal{N}^*$ be a dual net with $s+1$ points on any line and $t + 1$ lines through any point, where $t + 1 > s$. If $\mathcal{N}^*$ satisfies the Axiom of Veblen, then $\mathcal{N}^* \cong H_q^n$ with $n > 2$ (hence $s = q$ and $t + 1 = q^{n-1}$).*

4   Generalized quadrangles and the Axiom of Veblen

Consider a GQ $T_3(O)$ of Tits, with O an ovoid of $\mathrm{PG}(3, q)$; see 3.1.2 of FGQ. Here $s = q$ and $t = q^2$. Then the point (∞) is coregular, that is, each line incident with (∞) is regular. It is an easy exercise to check that for each line incident with (∞) the corresponding dual net is isomorphic to H_q^3. Hence for each line incident with the point (∞) the corresponding dual net satisfies the Axiom of Veblen. We now prove the converse.

Theorem 4.1 *Let $\mathcal{S} = (P, B, I)$ be a GQ of order (s, t) with $s \neq t, s > 1$ and $t > 1$. If $\mathcal{S}$ has a coregular point x and if for each line L incident with x the correponding dual net $\mathcal{N}_L^*$ satisfies the Axiom of Veblen, then $\mathcal{S}$ is isomorphic to a $T_3(O)$ of Tits.*

Proof Let L_1, L_2, L_3 be three lines no two of which are concurrent, let M_1, M_2, M_3 be three lines no two of which are concurrent, let $L_i \not\sim M_j$ if and only if $\{i, j\} = \{1, 2\}$ and assume that x I L_1. By 5.3.8 of FGQ it is sufficient to prove that for any line $L_4 \in \{M_1, M_2\}^\perp$ with $L_4 \not\sim L_i, i = 1, 2, 3$, there exists a line M_4 concurrent with L_1, L_2, L_4.

So let $L_4 \in \{M_1, M_2\}^{\perp}$ with $L_4 \not\sim L_i, i = 1, 2, 3$. Consider the line R containing $L_2 \cap M_2$ and concurrent with L_1. Further, consider the line R' containing $M_2 \cap L_4$ and concurrent with L_1. By the regularity of L_1 there is a line $S \in \{M_1, M_3\}^{\perp\perp}$ through the point $L_3 \cap M_2$. Clearly the lines L_1 and S are concurrent. So the line L_1 is concurrent with the lines S, R, R'; also the line M_2 is concurrent with the lines S, R, R'. By the regularity of L_1 the line S belongs to the line $\{R, R'\}^{\perp\perp}$ of the dual net $\mathcal{N}_{L_1}^*$ defined by L_1. Hence the lines $\{R, R'\}^{\perp\perp}$ and $\{M_1, M_3\}^{\perp\perp}$ of $\mathcal{N}_{L_1}^*$ have the element S in common. By the Axiom of Veblen, also the lines $\{M_1, R'\}^{\perp\perp}$ and $\{M_3, R\}^{\perp\perp}$ of $\mathcal{N}_{L_1}^*$ have an element M_4 in common. Consequently M_4 is concurrent with L_1, L_2, L_4. Now from 5.3.8 of FGQ it follows that S is isomorphic to a $T_3(O)$ of Tits. $\square$

Corollary 4.2 *Let S be a GQ of order (s, t) with $s \neq t$, $s > 1$ and $t > 1$.*

(i) *If s is odd, then S is isomorphic to the classical GQ $Q(5, s)$ if and only if it has a coregular point x and if for each line L incident with x the corresponding dual net $\mathcal{N}_L^*$ satisfies the Axiom of Veblen.*

(ii) *If s is even, then S is isomorphic to the classical GQ $Q(5, s)$ if and only if all its lines are regular and if for at least one point x and all lines L incident with x the dual nets $\mathcal{N}_L^*$ satisfy the Axiom of Veblen.*

Proof Let (x, L) be an incident point-line pair of the GQ $Q(5, s)$. By 3.2.4 of FGQ there is an isomorphism of $Q(5, s)$ onto $T_3(O)$, with O an elliptic quadric of $PG(3, s)$, which maps x onto the point (∞). It follows that $\mathcal{N}_L^*$ satisfies the Axiom of Veblen.

Conversely, assume that the GQ S of order (s, t), with s odd, $s \neq t, s > 1$ and $t > 1$, has a coregular point x such that for each line L incident with x the dual net $\mathcal{N}_L^*$ satisfies the Axiom of Veblen. Then by Theorem 4.1 the GQ S is isomorphic to $T_3(O)$. By Barlotti [2] and Panella [9] each ovoid O of $PG(3, s)$, with s odd, is an elliptic quadric. Now by 3.2.4 of FGQ we have $S \cong T_3(O) \cong Q(5, s)$.

Finally, assume that for the GQ S of order (s, t), with s even, $s \neq t, t > 1$, all lines are regular and that for at least one point x and all lines L incident with x the dual nets $\mathcal{N}_L^*$ satisfy the Axiom of Veblen. Then by Theorem 4.1 the GQ S is isomorphic to $T_3(O)$. Since all lines of $S \cong T_3(O)$ are regular, by 3.3.3(iii) of FGQ we finally have $S \cong T_3(O) \cong Q(5, s)$. $\square$

5   Translation generalized quadrangles and the Axiom of Veblen

Let $S = (P, B, I)$ be a GQ of order $(s, t), s \neq 1, t \neq 1$. A collineation θ of S is an *elation* about the point p if $\theta = \mathrm{id}$ or if θ fixes all lines incident with p

and fixes no point of $P - p^\perp$. If there is a group H of elations about p acting regularly on $P - p^\perp$, we say S is an *elation generalized quadrangle* (EGQ) with *elation group* H and *base point* p. Briefly, we say that $(S^{(p)}, H)$ or $S^{(p)}$ is an EGQ. If the group H is abelian, then we say that the EGQ $(S^{(p)}, H)$ is a *translation generalized quadrangle*. For any TGQ $S^{(p)}$ the point p is coregular so that the parameters s and t satisfy $s \leq t$; see 8.2 of FGQ. Also, by 8.5.2 of FGQ, for any TGQ with $s \neq t$ we have $s = q^a$ and $t = q^{a+1}$, with q a prime power and a an odd integer; if s (or t) is even then by 8.6.1(iv) of FGQ either $s = t$ or $s^2 = t$.

In $\mathrm{PG}(2n+m-1, q)$ consider a set $O(n, m, q)$ of $q^m + 1$ $(n-1)$-dimensional subspaces $\mathrm{PG}^{(0)}(n-1, q), \mathrm{PG}^{(1)}(n-1, q), \ldots, \mathrm{PG}^{(q^m)}(n-1, q)$, every three of which generate a $\mathrm{PG}(3n-1, q)$ and such that each element $\mathrm{PG}^{(i)}(n-1, q)$ of $O(n, m, q)$ is contained in a $\mathrm{PG}^{(i)}(n+m-1, q)$ having no point in common with any $\mathrm{PG}^{(j)}(n-1, q)$ for $j \neq i$. It is easy to check that $\mathrm{PG}^{(i)}(n+m-1, q)$ is uniquely determined, $i = 0, 1, \ldots, q^m$. The space $\mathrm{PG}^{(i)}(n+m-1, q)$ is called the *tangent space* of $O(n, m, q)$ at $\mathrm{PG}^{(i)}(n-1, q)$. For $n = m$ such a set $O(n, n, q)$ is called a *generalized oval* or an $[n-1]$-*oval* of $\mathrm{PG}(3n-1, q)$; a generalized oval of $\mathrm{PG}(2, q)$ is just an oval of $\mathrm{PG}(2, q)$. For $n \neq m$ such a set $O(n, m, q)$ is called a *generalized ovoid* or an $[n-1]$-*ovoid* or an *egg* of $\mathrm{PG}(2n+m-1, q)$; a $[0]$-ovoid of $\mathrm{PG}(3, q)$ is just an ovoid of $\mathrm{PG}(3, q)$.

Now embed $\mathrm{PG}(2n+m-1, q)$ in a $\mathrm{PG}(2n+m, q)$, and construct a point-line geometry $T(n, m, q)$ as follows.

Points are of three types :

(i) the points of $\mathrm{PG}(2n+m, q) - \mathrm{PG}(2n+m-1, q)$;

(ii) the $(n+m)$-dimensional subspaces of $\mathrm{PG}(2n+m, q)$ which intersect $\mathrm{PG}(2n+m-1, q)$ in one of the $\mathrm{PG}^{(i)}(n+m-1, q)$;

(iii) the symbol (∞).

Lines are of two types :

(a) the n-dimensional subspaces of $\mathrm{PG}(2n+m, q)$ which intersect $\mathrm{PG}(2n+m-1, q)$ in a $\mathrm{PG}^{(i)}(n-1, q)$;

(b) the elements of $O(n, m, q)$.

Incidence in $T(n, m, q)$ is defined as follows. A point of type (i) is incident only with lines of type (a); here the incidence is that of $\mathrm{PG}(2n+m, q)$. A point of type (ii) is incident with all lines of type (a) contained in it and with the unique element of $O(n, m, q)$ contained in it. The point (∞) is incident with no line of type (a) and with all lines of type (b).

Theorem 5.1 (8.7.1 of Payne and Thas [11]) $T(n, m, q)$ *is a TGQ of order* (q^n, q^m) *with base point* (∞). *Conversely, every TGQ is isomorphic to a* $T(n, m, q)$. *It follows that the theory of the TGQ is equivalent to the theory of the sets* $O(n, m, q)$.

Corollary 5.2 *The following hold for any* $O(n, m, q)$:

(i) $n = m$ *or* $n(c + 1) = mc$ *with* c *odd*;

(ii) *if* q *is even, then* $n = m$ *or* $m = 2n$.

Let $O(n, 2n, q)$ be an egg of $\mathrm{PG}(4n-1, q)$. We say that $O(n, 2n, q)$ is *good* at the element $\mathrm{PG}^{(i)}(n-1, q)$ of $O(n, 2n, q)$ if any $\mathrm{PG}(3n-1, q)$ containing $\mathrm{PG}^{(i)}(n-1, q)$ and at least two other elements of $O(n, 2n, q)$, contains exactly $q^n + 1$ elements of $O(n, 2n, q)$.

Theorem 5.3 *Let* $\mathcal{S}^{(p)}$ *be a TGQ of order* $(s, s^2), s \neq 1$, *with base point* p. *Then the dual net* $\mathcal{N}_L^*$ *defined by the regular line* L, *with* $p \, I \, L$, *satisfies the Axiom of Veblen if and only if the egg* $O(n, 2n, q)$ *which corresponds to* $\mathcal{S}^{(p)}$ *is good at its element* $\mathrm{PG}^{(i)}(n-1, q)$ *which corresponds to* L.

Proof Assume that the dual net $\mathcal{N}_L^*$ satisfies the Axiom of Veblen. Let the egg $O(n, 2n, q)$ correspond to $\mathcal{S}^{(p)}$ and let $\mathrm{PG}^{(i)}(n-1, q)$ correspond to L. We have $s = q^n$. The dual net has $q^n + 1$ points on a line and q^{2n} lines through a point. By Theorem 3.1 the dual net $\mathcal{N}_L^*$ is isomorphic to $H_{q^n}^3$. Consider the TGQ $T(n, 2n, q) \cong \mathcal{S}^{(p)}$ and let $\mathrm{PG}(3n, q)$ be a subspace skew to $\mathrm{PG}^{(i)}(n-1, q)$ in the projective space $\mathrm{PG}(4n, q)$ in which $T(n, 2n, q)$ is defined. Let $O(n, 2n, q) = \{\mathrm{PG}^{(0)}(n-1, q), \mathrm{PG}^{(1)}(n-1, q), \ldots, \mathrm{PG}^{(q^{2n})}(n-1, q)\}$, let $\langle \mathrm{PG}^{(i)}(n-1, q), \mathrm{PG}^{(j)}(n-1, q) \rangle \cap \mathrm{PG}(3n, q) = \pi_j$ for all $j \neq i$ (π_j is $(n-1)$-dimensional), let $\mathrm{PG}(4n-1, q) \cap \mathrm{PG}(3n, q) = \mathrm{PG}(3n-1, q)$ with $\mathrm{PG}(4n-1, q)$ the space of $O(n, 2n, q)$, and let $\mathrm{PG}^{(i)}(3n-1, q) \cap \mathrm{PG}(3n, q) = \mathrm{PG}(2n-1, q)$ with $\mathrm{PG}^{(i)}(3n-1, q)$ the tangent space of $O(n, 2n, q)$ at $\mathrm{PG}^{(i)}(n-1, q)$. Then the dual net $\mathcal{N}_L^*$ is isomorphic to the following dual net $\mathcal{N}^*$: points of $\mathcal{N}^*$ are the q^{2n} spaces $\pi_j, j \neq i$, and the q^{3n} points of $\mathrm{PG}(3n, q) - \mathrm{PG}(3n-1, q)$, lines of $\mathcal{N}^*$ are the q^{4n} n-dimensional subspaces of $\mathrm{PG}(3n, q)$ which are not contained in $\mathrm{PG}(3n-1, q)$ and contain an element $\pi_j, j \neq i$, and incidence is the natural one. Clearly the points $\pi_j, j \neq i$, of $\mathcal{N}^*$ form a parallel class of points. Let M be a line of $\mathcal{N}^*$ incident with π_j and let $\pi_k \neq \pi_j, k \neq i \neq j$. As $\mathcal{N}^* \cong H_{q^n}^3$ the elements π_k and M of $\mathcal{N}^*$ generate a dual affine plane $\mathcal{A}^*$ in $\mathcal{N}^*$, and the plane $\mathcal{A}^*$ contains q^n points $\pi_l, l \neq i$. Clearly the points of $\mathcal{A}^*$ not of type π_l are the q^{2n} points of the subspace $\langle \pi_k, M \rangle$ of $\mathrm{PG}(3n, q)$ which are not contained in $\mathrm{PG}(3n-1, q)$. Hence the q^n points of $\mathcal{A}^*$ of type π_l are contained in $\langle \pi_k, M \rangle \cap \mathrm{PG}(3n-1, q)$. It follows that these q^n elements π_l are contained in a $(2n-1)$-dimensional space $\mathrm{PG}'(2n-1,q)$; also, they form a partition of $\mathrm{PG}'(2n-1,q)-\mathrm{PG}(2n-1,q)$. Consequently for any two elements $\pi_l, \pi_{l'}, l \neq i \neq l'$, the space $\langle \pi_l, \pi_{l'} \rangle$ contains exactly q^n elements $\pi_r, r \neq i$. Hence for any two spaces $\mathrm{PG}^{(l)}(n-1,q)$ and $\mathrm{PG}^{(l')}(n-1,q)$ of $O(n,2n,q)-\{\mathrm{PG}^{(i)}(n-1,q)\}$, the $(3n-1)$-dimensional space $\langle \mathrm{PG}^{(i)}(n-1,q), \mathrm{PG}^{(l)}(n-1,q), \mathrm{PG}^{(l')}(n-1,q) \rangle$ contains exactly q^n+1 elements of $O(n,2n,q)$. We conclude that $O(n,2n,q)$ is good at $\mathrm{PG}^{(i)}(n-1,q)$.

Conversely, assume that $O(n,2n,q)$ is good at the element $\mathrm{PG}^{(i)}(n-1,q)$ which corresponds to L. As in the first part of the proof we project onto a $\mathrm{PG}(3n,q)$ and we use the same notations. Since $O(n,2n,q)$ is good at $\mathrm{PG}^{(i)}(n-1,q)$, for any two elements $\pi_l, \pi_{l'}, l \neq i \neq l'$, the space $\langle \pi_l, \pi_{l'} \rangle$ contains exactly q^n elements $\pi_r, r \neq i$; these q^n elements form a partition of the points of $\langle \pi_l, \pi_{l'} \rangle$ which are not contained in $\mathrm{PG}(2n-1,q)$. If M, M' are distinct concurrent lines of $\mathcal{N}^*$, then it is easily checked that M and M' generate a dual affine plane $\mathcal{A}^*$ of order q^n in $\mathcal{N}^*$. As $\mathcal{A}^*$ satisfies the Axiom of Veblen, also $\mathcal{N}^*$ satisfies the Axiom of Veblen. $\square$

Let $O = O(n,2n,q)$ be an egg in $\mathrm{PG}(4n-1,q)$. By 8.7.2 of FGQ the $q^{2n}+1$ tangent spaces of O form an $O^* = O^*(n,2n,q)$ in the dual space of $\mathrm{PG}(4n-1,q)$. So in addition to $T(n,2n,q) = T(O)$ there arises a TGQ $T(O^*)$ with the same parameters. The TGQ $T(O^*)$ is called the *translation dual* of the TGQ $T(O)$. Examples are known for which $T(O) \cong T(O^*)$, and examples are known for which $T(O) \not\cong T(O^*)$; see Thas [13].

6   Property (G) and the Axiom of Veblen

Let $\mathcal{S} = (P, B, \mathrm{I})$ be a GQ of order $(s, s^2), s \neq 1$. Let x_1, y_1 be distinct collinear points. We say that the pair $\{x_1, y_1\}$ has *Property* (G), or that $\mathcal{S}$ has *Property* (G) at $\{x_1, y_1\}$, if every triple $\{x_1, x_2, x_3\}$ of points, with x_1, x_2, x_3 pairwise noncollinear and $y_1 \in \{x_1, x_2, x_3\}^{\perp}$, is 3-regular; for the definition of 3-regularity see 1.3 of FGQ. The GQ $\mathcal{S}$ has *Property* (G) at the *line* L, or the line L has *Property* (G), if each pair of points $\{x, y\}, x \neq y$ and $x \mathrm{~I~} L \mathrm{~I~} y$, has Property (G). If (x, L) is a flag, that is, if $x \mathrm{~I~} L$, then we say that $\mathcal{S}$ has *Property* (G) at (x, L), or that (x, L) has *Property* (G), if every pair $\{x, y\}, x \neq y$ and $y \mathrm{~I~} L$, has Property (G). Property (G) was introduced in Payne [10] in connection with generalized quadrangles of order (q^2, q) arising from flocks of quadratic cones in $\mathrm{PG}(3, q)$.

Theorem 6.1 *Let* $\mathcal{S} = (P, B, I)$ *be a GQ of order* (s^2, s), s *even, satisfying Property* (G) *at the point* x. *Then* x *is regular in* $\mathcal{S}$ *and the dual net* $\mathcal{N}_x^*$ *satisfies the Axiom of Veblen. Consequently* $\mathcal{N}_x^* \cong H_s^3$.

Proof Let $\mathcal{S} = (P, B, \mathrm{I})$ be a GQ of order (s^2, s), s even, satisfying Property (G) at the point x. By 3.2.1 of [13] the point x is regular. Let y be a point of the dual net $\mathcal{N}_x^*$, let A_1 and A_2 be distinct lines of $\mathcal{N}_x^*$ containing y, let B_1 and B_2 be distinct lines of $\mathcal{N}_x^*$ not containing y, and let $A_i \cap B_j \neq \emptyset$ for all $i, j \in \{1, 2\}$. Let $\{z\} = A_1 \cap B_1$ and let $z \, \mathrm{I} \, M$, with $x \not{\mathrm{I}} M$. Further, let $x \, \mathrm{I} \, L$, with $z \not{\mathrm{I}} L$, let u be the point of A_1 on L, and let v be the point of B_1 on L. The line of $\mathcal{S}$ incident with u resp. v and concurrent with M is denoted by C resp. D; the line incident with z and x is denoted by N. Since $\mathcal{S}$ satisfies Property (G) at x, the triple $\{C, D, N\}$ is 3-regular. By 2.6.2 of TGQ the lines of $\mathcal{S}$ concurrent with at least two lines of $\{C, D, N\}^{\perp} \cup \{C, D, N\}^{\perp\perp}$ are the lineset of a subquadrangle $\mathcal{S}'$ of order (s, s) of $\mathcal{S}$. As x is regular for $\mathcal{S}$ it is also regular for $\mathcal{S}'$. By Theorem 2.1 the point x defines a projective plane π_x of order s. Clearly A_1, A_2, B_1, B_2 are lines of the projective plane π_x. Hence B_1 and B_2 intersect in π_x. Consequently $\mathcal{N}_x^*$ satisfies the Axiom of Veblen, and so $\mathcal{N}_x^* \cong H_s^3$. $\qquad\square$

Theorem 6.2 (Thas [13]) *A TGQ $T(n, 2n, q)$ satisfies Property (G) at the pair $\{(\infty), \bar{\zeta}\}$, with $\bar{\zeta}$ a point of type (ii) incident with the line ζ of type (b) (or, equivalently, at the flag $((\infty), \zeta)$) if and only if, for any two elements $\zeta_i, \zeta_j \, (i \neq j)$ of $O(n, 2n, q) - \{\zeta\}$, the $(n-1)$-dimensional space $PG(n-1, q) = \tau \cap \tau_i \cap \tau_j$, with τ, τ_i, τ_j the respective tangent spaces of $O(n, 2n, q)$ at ζ, ζ_i, ζ_j, is contained in exactly $q^n + 1$ tangent spaces of $O(n, 2n, q)$.*

Theorem 6.3 *Let $\mathcal{S}^{(p)}$ be a TGQ of order $(s, s^2), s \neq 1$, with base point p. Then the dual net $\mathcal{N}_L^*$ defined by the regular line L, with $p \, \mathrm{I} \, L$, satisfies the Axiom of Veblen if and only if the translation dual $\mathcal{S}'^{(p')}$ of $\mathcal{S}^{(p)}$ satisfies Property (G) at the flag (p', L'), where L' corresponds to L; in the even case, $\mathcal{N}_L^*$ satisfies the Axiom of Veblen if and only if $\mathcal{S}^{(p)}$ satisfies Property (G) at the flag (p, L).*

Proof By Theorem 5.3 the dual net $\mathcal{N}_L^*$ satisfies the Axiom of Veblen if and only if $O(n, 2n, q)$ is good at the element $PG^{(i)}(n-1, q)$ which corresponds to L. By Theorem 6.1 the egg $O(n, 2n, q) = O$ is good at $PG^{(i)}(n-1, q)$ if and only if $T(O^*)$ satisfies Property (G) at the flag $((\infty), PG^{(i)}(3n-1, q))$, with $PG^{(i)}(3n-1, q)$ the tangent space of O at $PG^{(i)}(n-1, q)$; by Theorem 4.3.2 of [13], for q even, $T(O^*)$ satisfies Property (G) at the flag $((\infty), PG^{(i)}(3n-1, q))$ if and only if $T(O)$ satisfies Property (G) at the flag $((\infty), PG^{(i)}(n-1, q))$. $\qquad\square$

Theorem 6.4 *Let $\mathcal{S}^{(p)}$ be a TGQ of order $(s, s^2), s$ odd and $s \neq 1$, with base point p. If the dual net $\mathcal{N}_L^*$ defined by some regular line L, with $p \, \mathrm{I} \, L$, satisfies the Axiom of Veblen, then $\mathcal{S}^{(p)}$ contains at least $s^3 + s^2$ classical subquadrangles $Q(4, s)$.*

Proof This follows immediately from the preceding theorem and Theorem 4.3.4 of Thas [13].   □

Theorem 6.5 *Let $\mathcal{S}^{(p)}$ be a TGQ of order (s, s^2), s odd and $s \neq 1$, with base point p. If $p\,I\,L$ and if the dual net $\mathcal{N}_L^*$ satisfies the Axiom of Veblen, then all lines concurrent with L are regular.*

Proof Let N be concurrent with $L, p \nmid N$, and let the line M of $\mathcal{S}^{(p)}$ be nonconcurrent with N. By Theorem 4.3.4 of Thas [13] the lines N, M are lines of a subquadrangle of $\mathcal{S}^{(p)}$ isomorphic to $Q(4, q^n)$. Hence $\{N, M\}$ is a regular pair of lines. We conclude that the line N is regular in $\mathcal{S}^{(p)}$.   □

7   Flock generalized quadrangles and the Axiom of Veblen

Let F be a flock of the quadratic cone K with vertex x of PG$(3, q)$, that is, a partition of $K - \{x\}$ into q disjoint irreducible conics. Then, by Thas [12], with F there corresponds a GQ $\mathcal{S}(F)$ of order (q^2, q). In Payne [10] it was shown that $\mathcal{S}(F)$ satisfies Property (G) at its point (∞).

Let $F = \{C_1, C_2, \ldots, C_q\}$ be a flock of the quadratic cone K with vertex x_0 of PG$(3, q)$, with q odd. The plane of C_i is denoted by $\pi_i, i = 1, 2, \ldots, q$. Let K be embedded in the nonsingular quadric Q of PG$(4, q)$. The polar line of π_i with respect to Q is denoted by L_i; let $L_i \cap Q = \{x_0, x_i\}, i = 1, 2, \ldots, q$. Then no point of Q is collinear with all three of $x_0, x_i, x_j, 1 \leq i < j \leq q$. In [1] it is proved that it is also true that no point of Q is collinear with all three of $x_i, x_j, x_k, 0 \leq i < j < k \leq q$. Such a set U of $q + 1$ points of Q will be called a *BLT-set* in Q, following a suggestion of Kantor [7]. Since the GQ $Q(4, q)$ arising from Q is isomorphic to the dual of the GQ $W(q)$ arising from a symplectic polarity in PG$(3, q)$, to a BLT-set in Q corresponds a set V of $q + 1$ lines of $W(q)$ with the property that no line of $W(q)$ is concurrent with three distinct lines of V; such a set V will also be called a *BLT-set*.

To F corresponds a GQ $\mathcal{S}(F)$ of order (q^2, q). Knarr [8] proves that $\mathcal{S}(F)$ is isomorphic to the following incidence structure.

Start with a symplectic polarity θ of $PG(5, q)$. Let $(\infty) \in$PG$(5, q)$ and let PG$(3, q)$ be a 3-dimensional subspace of PG$(5, q)$ for which $(\infty) \notin$PG$(3, q) \subset (\infty)^\theta$. In PG$(3, q)$ θ induces a symplectic polarity θ', and hence a GQ $W(q)$. Let V be the BLT-set defined by F of the GQ $W(q)$ and construct a geometry $\mathcal{S} = (P, B, I)$ as follows.

Points : (i) (∞); (ii) lines of PG$(5, q)$ not containing (∞) but contained in one of the planes $\pi_t = (\infty)L_t$, with L_t a line of the BLT-set V; (iii) points of PG$(5, q)$ not in $(\infty)^\theta$.

Lines : (a) planes $\pi_t = (\infty)L_t$, with $L_t \in V$; (b) totally isotropic planes of θ not contained in $(\infty)^\theta$ and meeting some π_t in a line (not through (∞)).

The incidence relation I is the natural incidence inherited from $PG(5, q)$.

Then Knarr [8] proves that $\mathcal{S}$ is a GQ of order (q^2, q) isomorphic to the GQ $\mathcal{S}(F)$ arising from the flock F defining V.

Theorem 7.1 *For any GQ $\mathcal{S}(F)$ of order (q^2, q) arising from a flock F, the point (∞) is regular.*

Proof The GQ $\mathcal{S}(F)$ satisfies Property (G) at its point (∞). Then for q even, by 3.2.1 of Thas [13], the point (∞) is regular. Now let q be odd, and consider the construction of Knarr. If the point y is not collinear with (∞), that is, if y is a point of $PG(5, q)$ not in $(\infty)^\theta$, then $\{(\infty), y\}^{\perp\perp}$ consists of the $q + 1$ points of the line $(\infty)y$ of $PG(5, q)$. As $|\{(\infty), y\}^{\perp\perp}| = q + 1$ the point (∞) is regular. $\quad\square$

Let K be the quadratic cone with equation $X_0 X_1 = X_2^2$ of $PG(3, q), q$ odd. Then the q planes π_t with equation $tX_0 - mt^\sigma X_1 + X_3 = 0, t \in GF(q), m$ a given nonsquare of $GF(q)$, and σ a given automorphism of $GF(q)$, define a flock F of K; see Thas [12]. The corresponding GQ $\mathcal{S}(F)$ were first discovered by Kantor [6], and so these flocks F will be called *Kantor flocks*. Any such GQ $\mathcal{S}(F)$ is a TGQ for some base line, and so the point-line dual of $\mathcal{S}(F)$ is isomorphic to some $T(O)$, with O an $[n - 1]$-ovoid. Also, in Payne [10] it is proved that $T(O)$ is isomorphic to its translation dual $T(O^*)$; there is an isomorphism of $T(O)$ onto $T(O^*)$ conserving types of points and lines and mapping the line ζ of type (b) of $T(O)$ onto the line τ of type (b) of $T(O^*)$, where τ is the tangent space of O at ζ.

Theorem 7.2 *Consider the GQ $\mathcal{S}(F)$ of order (q^2, q) arising from the flock F. If q is even, then the dual net $\mathcal{N}^*_{(\infty)}$ always satisfies the Axiom of Veblen and so $\mathcal{N}^*_{(\infty)} \cong H^3_q$. If q is odd, then the dual net $\mathcal{N}^*_{(\infty)}$ satisfies the Axiom of Veblen if and only if F is a Kantor flock.*

Proof Consider the GQ $\mathcal{S}(F)$ of order (q^2, q) arising from the flock F. Then $\mathcal{S}(F)$ satisfies Property (G) at the point (∞).

First, let q be even. Then by Theorem 6.1 the dual net $\mathcal{N}^*_{(\infty)}$ satisfies the Axiom of Veblen, and so $\mathcal{N}^*_{(\infty)} \cong H^3_q$.

Next, let q be odd. Suppose that F is a Kantor flock. Then the point-line dual of $\mathcal{S}(F)$ is isomorphic to some $T(O)$, and by [10] $T(O) \cong T(O^*)$. The point (∞) of $\mathcal{S}(F)$ corresponds to some line ζ of type (b) of $T(O)$. Hence $T(O)$ satisfies Property (G) at ζ. By Theorem 6.3 the dual net $\mathcal{N}^*_\tau$ which corresponds with the regular line τ of $T(O^*)$, where τ is the tangent space of O at ζ, satisfies the Axiom of Veblen. Hence also the dual net $\mathcal{N}^*_\zeta$ which corresponds with the regular line ζ of $T(O)$ satisfies the Axiom of Veblen. It follows that the dual net $\mathcal{N}^*_{(\infty)}$ satisfies the Axiom of Veblen. Conversely, suppose that the dual net $\mathcal{N}^*_{(\infty)}$ satisfies the Axiom of Veblen. Hence $\mathcal{N}^*_{(\infty)} \cong H^3_q$. In the representation of Knarr, this dual net looks as follows : points of $\mathcal{N}^*_{(\infty)}$ are the lines of $PG(5, q)$ not containing (∞) but contained in one of the planes π_t, lines of $\mathcal{N}^*_{(\infty)}$ can be identified with the threedimensional subspaces of $(\infty)^\theta$ not containing (∞), and incidence is inclusion. By point-hyperplane duality in $(\infty)^\theta$, the net $\mathcal{N}_{(\infty)}$, which is the point-line dual of $\mathcal{N}^*_{(\infty)}$, is isomorphic to the following incidence structure : points of $\mathcal{N}_{(\infty)}$ are the points of $(\infty)^\theta - PG(3, q)$, lines of $\mathcal{N}_{(\infty)}$ are the planes of $(\infty)^\theta$ not contained in $PG(3, q)$ but containing one of the lines of the BLT-set V in $PG(3, q)$, and incidence is the natural one. As the net $\mathcal{N}_\infty$ is isomorphic to the dual of H^3_q, it is easily seen to be derivable; see e.g. De Clerck and Johnson [4]. In $W(q)$ the lineset $S = \{L_0, L_1\}^{\perp\perp} \cup \{L_0, L_2\}^{\perp\perp} \cup \ldots \cup \{L_0, L_q\}^{\perp\perp}$ is a linespread containing V; see e.g. [12]. As $\mathcal{N}_{(\infty)}$ is derivable, by [3] there are two distinct lines in $PG(3, q)$, but not in $\{L_0, L_1\}^\perp \cup \{L_0, L_2\}^\perp \cup \ldots \cup \{L_0, L_q\}^\perp$, intersecting the same $q + 1$ lines of S. Then by Johnson and Lunardon [5], the flock F is a Kantor flock.   □

Corollary 7.3 *Suppose that the TGQ $T(O)$, with $O = O(n, 2n, q)$ and q odd, is the point-line dual of a flock GQ $\mathcal{S}(F)$ where the point (∞) of $\mathcal{S}(F)$ corresponds to the line ζ of type (b) of $T(O)$. Then $T(O)$ is good at the element ζ if and only if F is a Kantor flock.*

Proof This follows immediately from Theorems 5.3 and 7.2.   □

8 Subquadrangles and the Axiom of Veblen

Theorem 8.1 *Let $\mathcal{S} = (P, B, I)$ be a GQ of order $(s, t), s \neq 1 \neq t$, having a regular point x. If x together with any two points y, z, with $y \not\sim x$ and $x \sim z \not\sim y$, is contained in a proper subquadrangle $\mathcal{S}'$ of $\mathcal{S}$ of order (s', t), with $s' \neq 1$, then $s' = t = \sqrt{s}$ and the dual net $\mathcal{N}^*_x$ satisfies the Axiom of Veblen. It follows that s and t are prime powers, and that for each subquadrangle $\mathcal{S}'$ the projective plane π_x of order t defined by the regular point x of $\mathcal{S}'$ is desarguesian. Conversely, if the dual net $\mathcal{N}^*_x$ satisfies the Axiom of Veblen, then either (a) $s = t$, or (b) $s = t^2$, s and t are prime powers, x and any two points y, z with $y \not\sim x$ and $x \sim z \not\sim y$ are contained in a subquadrangle $\mathcal{S}'$ of $\mathcal{S}$ of order (t, t), and the projective plane π_x of order t defined by the regular point x of $\mathcal{S}'$ is desarguesian.*

Proof Let $\mathcal{S} = (P, B, I)$ be a GQ of order $(s, t), s \neq 1 \neq t$, having a regular point x.

First, assume that x together with any two points y, z with $y \not\sim x$ and $x \sim z \not\sim y$ is contained in a proper subquadrangle S' of S of order (s', t), with $s' \neq 1$. As x is also regular for S', the GQ S' contains subquadrangles of order $(1, t)$. Then, by 2.2.2 of FGQ, we have $s' = t = \sqrt{s}$. By Theorem 2.1 the dual net $\mathcal{N}_x'^*$ arising from the regular point x of S', is a dual affine plane of order s. Hence $\mathcal{N}_x'^*$ satisfies the Axiom of Veblen. Now consider distinct lines A_1, A_2, B_1, B_2 of the dual net $\mathcal{N}_x^*$, where $A_1 \cap A_2 = \{z\}, z \notin B_1, z \notin B_2$, and $A_i \cap B_j \neq \emptyset$ for all $i, j \in \{1, 2\}$. Let $A_1 \cap B_1 = \{u\}, A_2 \cap B_2 = \{w\}$, and let $y \in \{u, w\}^\perp - \{x\}$. Let S' be a subquadrangle of order t containing the points x, y, z of S. Then A_1, A_2, B_1, B_2 are lines of the dual net $\mathcal{N}_x'^*$. As $\mathcal{N}_x'^*$ satisfies the Axiom of Veblen, we have $B_1 \cap B_2 \neq \emptyset$. It follows that the dual net $\mathcal{N}_x'^*$ satisfies the Axiom of Veblen. Consequently $\mathcal{N}_x'^* \cong H_t^3$, and so s and t are prime powers. For any subquadrangle S' the dual net $\mathcal{N}_x'^*$ is a dual affine plane of order t, which is isomorphic to a dual affine plane of order t in H_t^3. Hence the dual net $\mathcal{N}_x'^*$, and consequently also the corresponding projective plane π_x, are desarguesian.

Conversely, assume that the dual net $\mathcal{N}_x^*$ satisfies the Axiom of Veblen. Also, suppose that $s \neq t$, that is, $s > t$ by 1.3.6 of FGQ. Then, by Theorem 3.1, we have $\mathcal{N}_x^* \cong H_q^n$ with q a prime power and $n > 2$. As $s = q^{n-1}, t = q$ and $s \leq t^2$ (by the inequality of Higman, see 1.2.3 of FGQ), we necessarily have $n = 3$. Hence $s = t^2, t = q$, and $\mathcal{N}_x^* \cong H_q^3$. Now consider any two points y, z, with $y \not\sim x, x \sim z \not\sim y$. As $\mathcal{N}_x^* \cong H_q^3$ it is easily seen that z and $\{x, y\}^\perp$ generate a dual affine plane $\mathcal{A}$ of order q in $\mathcal{N}_x^*$. Let $A_1, A_2, \ldots, A_{q^2}$ be the lines of $\mathcal{A}$. Further, let P' be the pointset of S consisting of the points of $A_1^\perp \cup A_2^\perp \cup \ldots \cup A_{q^2}^\perp$ and the points of $\mathcal{A}$. Clearly P' contains z and y, and $|P'| = q^3 + q^2 + q + 1$. Further, any line of S incident with at least one point of P' either contains x or a point of $\mathcal{A}$; the set of all these lines is denoted by B'. Also, any point incident with two distinct lines of B' belongs to P'. Then, by 2.3.1 of FGQ, $S' = (P', B', I')$ with I' the restriction of I to $(P' \times B') \cup (B' \times P')$ is a subquadrangle of S of order q. As in the first part of the proof one now shows that for any such subquadrangle S' the projective plane π_x defined by x is desarguesian. $\square$

Acknowledgement The second author is a Senior Research Associate of the Belgian National Fund for Scientific Research.

References

[1] L. Bader, G. Lunardon and J.A. Thas, Derivation of flocks of quadratic cones, *Forum Math.* **2** (1990), 163-174.

[2] A. Barlotti, Un' estensione del teorema di Segre-Kustaanheimo, *Boll. Un. Mat. Ital.* **10** (1955), 96-98.

[3] M. Biliotti and G. Lunardon, Insiemi di derivazione e sottopiani di Baer
     in un piano di traslazione, *Atti Accad. Naz. Lincei Rend.* **69** (1980),
     135-141.

[4] F. De Clerck and N.L. Johnson, Subplane covered nets and semipartial
     geometries, *Discrete Math.* **106/107**, (1992), 127-134.

[5] N.L. Johnson and G. Lunardon, Maximal partial spreads and flocks, *Des.
     Codes Cryptogr.*, to appear.

[6] W.M. Kantor, Some generalized quadrangles with parameters (q^2, q),
     Math. Z. **192** (1986), 45-50.

[7] W.M. Kantor, Note on generalized quadrangles, flocks and BLT-sets, *J.
     Combin. Theory Ser. A* **58** (1991), 153-157.

[8] N. Knarr, A geometric construction of generalized quadrangles from po-
     lar spaces of rank three, *Resultate Math.* **21** (1992), 332-334.

[9] G. Panella, Caratterizzazione delle quadriche di uno spazio (tridimen-
     sionale) lineare sopra un corpo finito, *Boll. Un. Mat. Ital.* **10** (1955),
     507-513.

[10] S.E. Payne, An essay on skew translation generalized quadrangles, *Geom.
     Dedicata* **32** (1989), 93-118.

[11] S.E. Payne and J.A. Thas, *Finite Generalized Quadrangles*, Pitman,
     Boston, 1984.

[12] J.A. Thas, Generalized quadrangles and flocks of cones, *European J.
     Combin.* **8** (1987), 441-452.

[13] J.A. Thas, Generalized quadrangles of order $(s, s^2), I$, *J. Combin. Theory
     Ser. A* **67** (1994), 140-160.

[14] J.A. Thas and F. De Clerck, Partial geometries satisfying the axiom of
     Pasch, *Simon Stevin* **51** (1977), 123-137.

J.A. Thas, Department of Pure Mathematics and Computer Algebra,
University of Ghent, Krijgslaan 281, B-9000 Gent, Belgium.
e-mail : jat@cage.rug.ac.be

H. Van Maldeghem, Department of Pure Mathematics and Computer
Algebra, University of Ghent, Krijgslaan 281, B-9000 Gent, Belgium.
e-mail : hvm@cage.rug.ac.be

Talks

S. Ball	Maximal arcs
L.M. Batten	Equitable colorings
A. Bonisoli	Two-transitive ovals with a fixed external line in projective planes of even order: do they really exist?
P.J. Cameron	Finite geometry after Aschbacher's Theorem
W.E. Cherowitzo	Flocks of non-quadratic cones
J.H. Conway	Projective planes
F. De Clerck	New partial geometries derived from old ones
S. Dent	An incidence structure of partitions
L. Di Martino	Carter subgroups in classical groups
J. Doyen	Pythagoras
J. Doyen	Problems about linear spaces
G.L. Ebert	Singer line orbits
C.-A. Faure	The fundamental theorem of projective geometry
K. Fleming	An infinite family of non-embeddable quasi-residual designs with $k < v/2$
P. Gibbons	Uniform orthogonal group divisible deigns with block size three
D. Gray	Incidence maps and the representation of the infinite symmetric group
O. Grošek	The Walsh-Hadamard transform and finite groups
W.H. Haemers	Disconnected vertex sets and equidistant code pairs
Y. Ionin	Embeddability and construction of affine α−resolvable pairwise balanced designs
V. Jha	Baer central collineations
D. Jungnickel	Recent results in difference sets
J.D. Key	Computational results for the known biplanes of order 9
G.B. Khosrovshahi	Maximal anti-Pasch 3−hypergraphs
G. Korchmáros	The embedding of a k−arc into a conic in a finite plane
R. Laue	Simple 7−designs with small parameters
C.C. Lindner	Very small embeddings for partial cycle systems
R. Mathon	Searching for spreads and packings

E. Mendelsohn	Intercalates (and other partial Latin squares on 4 cells) everywhere
A.C. Niemeyer and C.E. Praeger	Are block-transitive point-imprimitive linear spaces really "counter-examples"?
E.A. O'Brien	Projective geometries and matrix groups
C.M. O'Keefe	A survey of results on generalized quadrangles of order (q, q^2)
T. Penttila	Flocks of quadratic cones in $PG(3, q)$
A. Pott	Problems about negacirculant conference matrices
A.R. Prince	Uniform parallelisms of $PG(3, 3)$
A. Rosa	Large sets of MAD Steiner triple systems not from Steiner quadruple systems
L. Satko	Traces in semigroups
R. Shaw	Configurations of planes in $PG(5, 2)$
M.S. Shrikhande	Resolvable pairwise balanced designs
E.E. Shult	A survey of (c^*, c) geometries
S.D. Stoichev	Coding of graph vertices partitions and an algorithm for code determination
L. Storme	Cyclic caps on quadrics and Hermitian varieties
A.P. Street	Designs in agroforestry
J.A. Thas	Generalized quadrangles and the Axiom of Veblen
V.D. Tonchev	Steiner triple systems and their codes
Tran Van Trung	An inequality for t–designs
S. Tsaranov	Signed graphs and related reflection groups
J. Ueberberg	Frobenius collineations in finite projective planes
B. Webb	Steiner triple systems with exceptional automorphisms

Participants

L. Bader, Rome II

S. Ball, Eindhoven

L.M. Batten, Manitoba

G. Bini, Rome I

A. Blokhuis, Eindhoven

A. Bonisoli, Basilicata

A.E. Brouwer, Eindhoven

J.M. Brown, York (Toronto)

P.J. Cameron, QMWC (London)

P.V. Ceccherini, Rome I

W.E. Cherowitzo, Colorado

C.J. Colbourn, Waterloo

J.H. Conway, Princeton

M.J. de Resmini, Rome I

F. De Clerck, Ghent

S. Dent, East Anglia

D. Deriziotis, Athens

L. Di Martino, Milan

J. Dinitz, Vermont

J. Doyen, Brussels

G.L. Ebert, Delaware

C.-L. Faure, Brussels

K. Fleming, Central Michigan

P. Gibbons, Auckland

D. Gray, East Anglia

H.-D. Gronau, Rostock

O. Grošek, Bratislava

W.H. Haemers, Tilburg

J.W.P. Hirschfeld, Sussex

S. Hobart, Wyoming

P. Houlis, Western Australia

Y. Ionin, Central Michigan

V. Jha, Glasgow Caledonian

V.R. Job, Marymount (Arlington)

D. Jungnickel, Augsburg

J.D. Key, Clemson

G.B. Khosrovshahi, Tehran

D. Klinke, Heidelberg

G. Korchmáros, Basilicata

R. Laue, Bayreuth

R.A. Liebler, Colorado State

C.C. Lindner, Auburn

E. Loukakis, Thassaloniki

S.S. Magliveras, Nebraska

S. Magliveras, Thassaloniki

R. Mathon, Toronto

V. Mavron, Aberystwyth

E. Mendelsohn, Toronto

K. Metsch, Giessen

R. Mullin, Waterloo

E.A. O'Brien, Aachen

C.M. O'Keefe, Adelaide

K. Nemoga, Bratislava

D. Nicolacopoulos, Athens

A.C. Niemeyer, Western Australia

S.E. Payne, Colorado

T. Penttila, Western Australia

A. Pott, Augsburg

C.E. Praeger, Western Australia

A.R. Prince, Heriot-Watt (Edinburgh)

A. Rosa, McMaster (Hamilton)

L. Satko, Bratislava

R. Shaw, Hull

M.S. Shrikhande, Central Michigan

257

E.E. Shult, Kansas State
J. Siemons, East Anglia
S. Simopoulos, Athens
R. Stanton, Manitoba
S.D. Stoichev, Sofia
L. Storme, Ghent
A.P. Street, Queensland
J.A. Thas, Ghent

V.D. Tonchev, Michigan Technical
Tran van Trung, Essen
S. Tsaranov, Moscow
K. Tselekis, Athens
J. Ueberberg, Giessen
S.K.J. Vereecke, Sussex
B. Webb, Open (Milton Keynes)
B. Williams, Western Australia

Printed in the United States
By Bookmasters